POLYMER-SURFACTANT SYSTEMS

SURFACTANT SCIENCE SERIES

POLYMER-SURFACTANT SYSTEMS

edited by

Jan C. T. Kwak

Dalhousie University
Halifax, Nova Scotia, Canada

CRC Press
Taylor & Francis Group
Boca Raton   London   New York

CRC Press is an imprint of the
Taylor & Francis Group, an informa business

Library of Congress Cataloging-in-Publication Data

Polymer-surfactant systems / edited by Jan C.T. Kwak.
  p.   cm. — (Surfactant science series; v. 77)
  Includes bibliographical references and index.
  ISBN 0-8247-0232-8 (acid-free paper)
  1. Surface active agents. 2. Polymers. I. Kwak, Jan C.T. (Jan Cornelis Theodorus).
  II. Series.
  TP994.P66   1998
  668'.1—dc21
                                                                              98-39464
                                                                              CIP

Headquarters
Marcel Dekker
270 Madison Avenue, New York, NY 10016
tel: 212-696-9000; fax: 212-685-4540

The publisher offers discounts on this book when ordered in bulk quantities. For more information, write to Special Sales/Professional Marketing at the headquarters address above.

Preface

Aqueous systems containing both polymers and surfactants display a wide variety of properties useful to the formulator or industrial researcher, and fascinating to the academic investigator. Processes such as hydrophobic aggregation, viscosity enhancement, gel formation, solubilization, phase separation, etc. can all be controlled and even tailored to particular applications by making use of the wide variety of surfactants available in combination with the infinite possibilities of polymer synthesis and modification. Applications of solutions containing both surfactants and polymers include personal care products, pharmaceutical and biomedical formulations, food products, household and industrial cleaning and specialty products, paints and coatings, oil drilling and recovery fluids, and many more. Precisely because of the seemingly endless number of variations possible, the informed application of polymer-surfactant systems relies on an understanding of the fundamental, molecular parameters which govern the desired system properties.

Investigations of polymer-surfactant solutions date back more than four decades. In recent years, such research has accelerated with the introduction of new methodology and instrumentation including in particular various spectroscopic and structural tools. At the same time, many classical techniques from the arsenals of physical chemistry and colloid science remain relevant, while the use of newly synthesized or modified polymers has opened up many novel applications. This volume chronicles recent advances in our knowledge of polymer-surfactant systems, combining up-to-date descriptions of the use of techniques such as NMR, time-resolved fluorescence, and calorimetry with discussions of more classical methodologies and detailed surveys of specific system properties. Other chapters present new insights on phase equilibria and theoretical models. Current applications of polymer-surfactant systems are addressed early, thereby emphasizing the traditional close relationship between fundamental studies and applications research in this area.

The editor wishes to express his gratitude to the international group of authors who have contributed to this volume. Their conscientious efforts and timely submissions are highly appreciated. The early editing of this book was done while I was attached to the Eastern Indonesia Universities Development Project (EIUDP) at Universitas Pattimura, Ambon, Maluku, Indonesia. I thank both EIUDP and Dalhousie University for giving me the opportunity to complete this task, and my colleagues at UNPATTI for their hospitality and scientific cooperation. I thank the editing and production staff at Marcel Dekker for their valued assistance. Finally, I thank Dalhousie University and the Natural Sciences and Engineering Research Council of Canada (NSERC) for support throughout the years.

Jan C.T. Kwak

Contents

Contents

Contributors

K.P. Ananthapadmanabhan   Unilever Research, U.S., Edgewater, New Jersey

Dan F. Anghel   Institute of Physical Chemistry, Bucharest, Romania

E. Desmond Goddard   Consultant, Haworth, New Jersey

Per Hansson   Physical Chemistry 1, Center for Chemistry and Chemical Engineering, Lund University, Lund, Sweden

Katumitu Hayakawa   Department of Chemistry and BioScience, Kagoshima University, Kagoshima, Japan

Gunnar Karlström   Theoretical Chemistry, Center for Chemistry and Chemical Engineering, Lund University, Lund, Sweden

Jan C.T. Kwak   Department of Chemistry, Dalhousie University, Halifax, Nova Scotia, Canada

Björn Lindman   Physical Chemistry 1, Center for Chemistry and Chemical Engineering, Lund University, Lund, Sweden

Per Linse   Physical Chemistry 1, Center for Chemistry and Chemical Engineering, Lund University, Lund, Sweden

Gerd Olofsson   Division of Thermochemistry, Center for Chemistry and Chemical Engineering, Lund University, Lund, Sweden

Lennart Piculell   Physical Chemistry 1, Center for Chemistry and Chemical Engineering, Lund University, Lund, Sweden

Sudarshi T.A. Regismond   Department of Chemistry, McMaster University, Hamilton, Ontario, Canada

Andrew P. Rodenhiser   Department of Chemistry, Dalhousie University, Halifax, Nova Scotia, Canada

Shuji Saito   Consultant, Takarazuka, Japan

Keishiro Shirahama   Department of Chemistry and Applied Chemistry, Faculty of Science, Saga University, Saga, Japan

Peter Stilbs   Physical Chemistry, Royal Institute of Technology, Stockholm, Sweden

Geng Wang   Division of Thermochemistry, Center for Chemistry and Chemical Engineering, Lund University, Lund, Sweden

Françoise M. Winnik   Department of Chemistry, McMaster University, Hamilton, Ontario, Canada

Raoul Zana   Institut C. Sadron (CNRS-ULP), Strasbourg, France

1
Polymer-Surfactant Systems: Introduction and Overview

ANDREW P. RODENHISER and **JAN C.T. KWAK**   Department of Chemistry, Dalhousie University, Halifax, Nova Scotia, Canada B3H 4J3

I. INTRODUCTION

Compared to the ever increasing variety of synthetic polymer structures reported, or even in relation to the number of polymers commercially available the group of water-soluble polymers is relatively small. If we include the common water-soluble biopolymers and their synthetic derivatives, this combined group, although still not numerous in comparison to all synthetic polymers available, has a large number of industrial, environmental, household, and medical applications, and is of significant commercial importance [1,2].

The solubility of a polymer in water is determined by the balance between the interactions of the hydrophilic and hydrophobic polymer segments with themselves and with the solvent. Similarly, the aggregation of surfactants in aqueous solution is governed by the subtle balance of hydrophilic, hydrophobic, and ionic interactions. As a result, aqueous solutions containing both polymers and surfactants display an apparently infinitely varied and indeed sometimes bewildering pattern of properties, due to the many variations in molecular structures available to the investigator or formulator. Research in the area of polymer-surfactant interactions has accelerated rapidly over the last few

1

decades, driven in part by the many current or foreseen applications for instance in pharmaceutical formulations, personal care products, food products, household and industrial detergents, paints and coatings, oil drilling and enhanced oil recovery fluids, etc., but inspired also by fundamental interest in intermolecular interactions and hydrophobic aggregation phenomena. Quantitative aspects of such studies may include the direct measurement of properties such as viscosity, conductance, volumetric and thermochemical parameters, the determination of surfactant binding isotherms by various methods, and the elucidation of the often highly complex phase diagrams. Qualitatively, from the early studies on investigators have relied on model descriptions and the systematic understanding of polymer-surfactant interactions to guide them in designing new systems and new applications. Modern spectroscopic tools such as NMR and fluorescence methods have been highly instrumental in increasing our understanding of the molecular basis for such models.

The number of reviews of polymer-surfactant research reflects the growing number of published studies. Early reviews by Breuer and Robb [3], Robb [4], Goddard [5,6], and Saito [7] were instrumental in pointing out the many research opportunities. More recent reviews highlight both applications and fundamental studies [8-14]. Many early polymer-surfactant studies involved proteins, including what is possibly the first report, by Bull and Neurath [15]. Much of this early work on protein solutions may be found in the reviews of Steinhardt and Reynolds [16] and Lapanje [17]. More recent overviews are presented by Jones [18], Ananthapadmanabhan [10] and Dickinson [19]. The topic of surfactant-protein interactions, although not a primary focus, is also addressed in chapters 3, 4, and 9 of this book.

It has been customary to classify polymer-surfactant interactions according to polymer or surfactant charge, and according to concentration region. Most studies concerned with the determination of surfactant binding to polymers are carried out at low polymer concentration, with surfactant concentrations determined by the binding region. On the other hand, phase equilibria and phase diagrams are normally studied at higher concentrations. Although the findings of low concentration binding studies may be expected to be applicable to phase equilibrium studies at higher concentration, one has to be careful in making such assumptions. Indeed, the concepts of binding and interactions in general should be distinguished with some care. In addition, much current research focuses on systematic variations of polymer structure, including in particular the introduction of hydrophobic modifications in polymers which are normally considered hydrophilic. In many such systems concepts traditionally used in polymer-surfactant studies, such as the critical aggregation concentration, cac (sometimes still referred to as T_1) and the degree of binding are not applicable or need modification. Already some of the earliest studies in polymer-surfactant systems, especially those involving nonionic surfactants, point out that considerable changes in system properties can be observed even though there is no observable change in critical micelle concentration, cmc [7]. On the other hand, for polyelectrolytes and surfactants of opposite charge, surfactant binding is clearly observable and may start at concentrations two or three orders of magnitude below the cmc.

Another distinction of interest which has been widely discussed concerns the difference between anionic and cationic surfactants. In particular, the observation that anionic surfactants display a relatively strong and cooperative interaction with nonionic hydrophilic polymers such as poly(ethyleneoxide) (PEO) and polyvinylpyrrolidone (PVP), while cationic surfactants do not exhibit interactions, remains relevant. This may be another example where one has to be careful in applying apparently simple concepts such as polymer charge or induced charge or dipolar effects. A better understanding of the microscopic structure of the aggregates and the role of hydrophobic interactions may be needed, as exemplified by the difference in micellar surface structure of anionic and cationic surfactants, evident from NMR and SANS studies on alkylsulfate micelles [20,21] and on alkylammonium surfactants [22].

The precise structure of a polymer-surfactant complex will of course depend on the structures of the molecules involved, in particular the hydrophobicity and molecular weight of the polymer, and the charge and shape of the surfactant. For linear polymers, a general model has emerged, often referred to as the "necklace" or "beads-on-a-string" model, in which one or more small surfactant micelles reside within the random coil of the polymer [23]. There are proven exceptions to this structure, particularly for surfactants which form rod-shaped micelles or vesicles [24], and with hydrophobically modified polymers; however, the model has become accepted as the typical structure of a polymer-surfactant complex. In this introduction we can only present a brief history of the development of this model, and of polymer-surfactant studies in general. The reader is referred to earlier reviews [3-14] and subsequent chapters in this volume for more complete citations.

It is interesting to note that interactions of surfactants with proteins have been known since the 1930s [15,25], well before the first studies for synthetic polymers were published. Anionic surfactants such as sodium dodecylsulfate (SDS) were found to bind to positive charge sites on proteins in stoichiometric proportions, in some cases opening up the protein conformation, and at higher surfactant concentrations with much higher ratios of surfactant molecules per protein molecule. The possibility that this excess binding to proteins was micelle-like was discussed nearly twenty years before the "necklace" model was proposed for binding of surfactants to synthetic polymers [26,23]. Isemura and co-workers observed that poly(vinyl formal), poly(vinyl butyral), poly(vinyl acetate) (PVAc) and poly(vinyl alcohol) (PVA), which have poor solubility both in organic solvents and in water, could be dissolved in an SDS solution [27,28]. Such solutions could be diluted well below the cmc of SDS without precipitation of the polymer, indicating that these were not the result of polymer being solubilized in SDS micelles. Electrophoresis experiments demonstrated the negative charge of the complex, attributed to bound surfactant. The electrophoresis measurements also indicated Langmuir-type binding, with the number of adsorbed anions increasing with increased hydrophobicity of the polymer. The authors considered several modes of binding and decided that the binding was driven by hydrophobic interactions. Studies by Saito and co-workers, summarized in chapter 9 of this book, were among the first to discuss the role of headgroup charge and nature of the counterion [29]. A few years later, White and co-workers examined the binding of cationic surfactants to anionic cellulosic polymers, and concluded that the initial binding of surfactant cations was due to ion exchange, and was followed by clustering of surfactant-counterion pairs on those binding sites [30-32]. Decreased adsorption was noted in hydroxyethyl celluloses, where some fraction of the anionic carboxyl sites had been removed.

We may speak of the start of a new era of polymer-surfactant studies with the work of Jones on PEO-SDS and PVP-SDS systems [33,34]. He defined the critical aggregation concentration (cac or T_1) as observed by conductance and surface tension measurements, and the saturation point (T_2), which was found to increase linearly with polymer concentration. Although Jones postulated a polymer-nucleated micelle, the structure suggested followed the model of anionic azo dye-PVP complexes presented by Eirich and co-workers [35], involving individually bound surfactant molecules with their tails parallel to the polymer chain and their headgroups well separated. Lewis and Robinson observed a critical concentration for binding of SDS to several polymers, which they found consistent with a hydrophobic bonding mechanism similar to micelle formation [36]. They also observed that surfactant binding broke up existing interpolymer aggregation which allowed aggregation of further surfactant to fresh polymer surface. They considered the saturated complex to be a mixed micelle.

Shinoda and co-workers noted the cooperativity of polymer-surfactant binding [37]. They made surface tension, dialysis and dye solubilization measurements of the cmc and cac for sodium alkyl sulfates in the absence and presence of PVP, finding a stoichiometric ratio of surfactant bound to the polymer at saturation, regardless of polymer concentration. From the cmc and cac data, the Gibbs energy of transfer of a CH_2 group between the aqueous phase and the aggregate was calculated and found to be the same for both the polymer-free surfactant micelle and the polymer-surfactant complex, leading to the conclusion that even at the initial stages of adsorption, bound surfactant molecules were in contact with each other rather than uniformly distributed along the PVP chain. The authors implied but did not explicitly state that the aggregates were micelle-like.

Schwuger investigated the pH dependence of the interactions of cationic and anionic surfactants with PEO and poly(propyleneglycol) (PPO), and found that cationic surfactants exhibit only weak binding [38]. In addition, binding of anionic surfactants was found to decrease with higher pH, leading to the suggestion that surfactant headgroups interact with the ether oxygens of the polymer. Most subsequent studies have not supported this hypothesis.

Shirahama *et al.* investigated the free-boundary electrophoresis of SDS-polypeptide and SDS-PVP complexes [23]. The authors determined the electrophoretic mobility to be largely independent of polymer molecular weight, as is the case with polyelectrolytes. Contrary to the earlier model of a compact, prolate ellipsoid in SDS-polypeptide complexes [39], they realized that the structure must be an open, random coil, free-draining with respect to water and counterions. Such a model would predict that the electrophoretic mobility of the complex is the same as that of the SDS clusters, and is dependent on the polymer molecular weight only in the same minor way as cluster size and charge are. Since the clusters are like small beads strung on a randomly tangled chain, they referred to this as a "necklace" model. This model has since been supported by the results of many other studies and methods, in particular the work of Cabane and Duplessix [40,41]. The polymer chain is believed to be coiled around a surfactant aggregate which is often smaller than a free micelle. Long polymer chains may bind several micelles distributed within the polymer coil and separated by lengths of free chain. The polymer is generally considered to be bound at the micellar surface, and may shield parts of the micelle hydrophobic core from the aqueous environment. The charge, hydration, hydrophobicity and stiffness of the polymer chain are important in considerations of its conformation and location within the complex. Detailed aspects of both the historical development of models, and current theories for the polymer-surfactant aggregation process are presented in chapters 4 and 5 of this book.

Most of these early binding studies concerned the interaction between nonionic polymers and anionic surfactants. As already stated, the case of polyelectrolytes and oppositely charged surfactants posed considerable experimental difficulties because the free surfactant concentrations to be measured are extremely low. Robb, in his 1981 review, wrote that "investigations of polymer/surfactant systems should have as a high priority the determination of the binding isotherm," and he noted in addition the necessity of systematic studies: "The points of attachment and the relative importance of surfactant headgroups, alkyl chainlengths, and the groups in the polymer should be determined." As will be shown in the next section, and in all chapters of this book, new methodologies have allowed answers to these questions. In particular, the use of ion-selective electrodes has solved the low surfactant concentration problem [42], and modern experimental and theoretical methods have given insight into the structural questions, as described in chapters 5-11. However, in addition to the low concentration region binding studies, phase equilibria in polymer-surfactant systems have become of increasing importance and interest. This work may be said to have started with the study of Goddard and Hannan [43], showing the boundary regions for phase separation and subsequent resolubilization at high surfactant to polymer ratios in a system of anionic surfactant and a cationically modified biopolymer. At this point, we should note the early and remarkably insightful studies of Bungenberg de Jong and co-workers on phase equilibria in aqueous systems containing biopolymers of opposite charge, as summarized many years later in a colloid science classic [44]. In recent years, phase equilibrium studies and the elucidation of the structure of the concentrated polymer-surfactant phases has become one of the most exciting areas of research, necessitating not only extensive and detailed study of the phase equilibria involved, but also the application of the full array of available modern analytical, spectroscopic and other structural tools. Chapter 3 of this book by Piculell, Lindman and Karlström summarizes the many developments in this area. Discussions of specific spectroscopic techniques may be found in chapters 6, 7, and 10.

II. EXPERIMENTAL METHODS

Research on polymer-surfactant systems has seen input from many disciplines including virtually all chemistry sub-disciplines, physics, biochemistry and engineering. Equally, many key developments have resulted from the application of new experimental methods, or from the careful collection and interpretation of data from so-called classical physical chemical methods. In the brief overview of experimental methods used in polymer-surfactant systems which follows, we have no intention to be comprehensive or historically complete. Rather, we will highlight some methodologies and typical results, recognizing that many of the major techniques are expertly covered in the subsequent chapters of this book. We will rather loosely classify techniques as either "classical" physical chemical methods, or spectroscopic methods. This classification is not meant to indicate a preference or superiority in methodology. Classical methods are by no means obsolete. For instance, as chapter 8 by Olofsson and Wang proves new developments in calorimetry have only very recently allowed their addition to our arsenal of techniques, and methods such as conductance and viscometry or rheology remain indispensable.

A. Classical Physical Chemical Methods

Binding isotherms.

The binding isotherm expresses the amount of "bound" surfactant as a function of the free surfactant concentration. In this form, it is equivalent to the adsorption isotherm at the S-L or G-L interface. In the early stages of polymer-surfactant research, binding isotherms were derived from changes in viscosity or surface tension. Such methods are often of only approximate validity, suffering from a number of questionable assumptions. Preferred methods are equilibrium dialysis or the use of surfactant selective electrodes, the latter applicable to certain ionic surfactants only. In order to determine the binding isotherm, the equilibrium free surfactant concentration needs to be determined in the solution containing both polymer and surfactant. In the equilibrium dialysis method, this is done by determining the surfactant concentration, using any suitable method, in the equilibrium, polymer-free solution. This method requires long equilibration times, a large number of samples, and is subject to considerable errors. Nevertheless, in some cases it is the only method available.

More recently, the availability of surfactant-selective electrodes [45-49] has made it possible to measure the free surfactant concentration (more correctly the surfactant activity) directly in the polymer solution. Although most reported measurements have been with cationic surfactants, the use of polymeric plasticizers in the PVC membrane has made the determination of anionic surfactants feasible. In the method, the surfactant electrode EMF relative to a standard electrode is determined for a calibration curve in the absence of polymer, followed by EMF measurements in the presence of a constant polymer concentration (Figure 1a, open and closed symbols, respectively). When there is a sharp cac, as is the case with polyelectrolyte and oppositely charged surfactant, the EMF in the presence of polymer deviates sharply from the calibration curve at the cac, reaching saturation at higher total surfactant concentration. The amount of bound surfactant and the degree of surfactant binding, β, defined as mole bound surfactant per mole polymer repeating unit or ionic group, can be determined. A few considerations should be noted. First, the method indeed determines free surfactant activities. Only when the added salt concentration substantially exceeds the surfactant ion concentration can we equate the determined values with concentrations. Nevertheless, the method has also been applied in salt-free solution, based on the observation that even in that case the calibration curve is still linear and has a Nernstian slope. The only reason for this can be that the surfactant solution does not exhibit a Debye-Hückel-type lowering of the activity coefficient. Surprisingly, this unusual observation has to the knowledge of the authors not

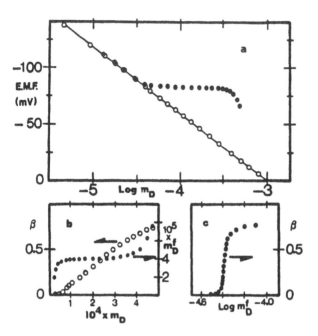

FIG. 1 Example of binding isotherm determination with cationic surfactant electrode. System: sodium dextransulfate (0.0005m)-DTABr-NaCl (0.01m). a: EMF vs. total surfactant concentration (m_D); b: free surfactant concentration (m_D^f) vs m_D; c: binding isotherm.

been fully investigated. Second, the polymer concentration needs to be kept constant. In many cases it is convenient to use a titration procedure to add surfactant, minimizing polymer use and random errors. In such a procedure, the polymer concentration can be kept constant by adding a polymer-surfactant mixture, or, when this is not possible because of phase separation, by adding a second volume of polymer solution at each titration step. Finally, care must be taken to ensure stability of the calibration curve over the course of the measurement, best achieved by "sandwiching" calibration series or individual calibration measurements between unknown determinations. Because of the rapid response of the electrodes, full binding isotherms can be determined in relatively short times, and with minimal material use. For cationic surfactants, measurements have been extended to concentrations as low as 10^{-6} M, most reports with anionic surfactants extend to below 10^{-4} M. For systems of ionic surfactants and nonionic polymers, binding curves may not be cooperative as in Figure 1, but exhibit continually increasing binding without a definite cac. The same is the case with hydrophobically modified neutral polymers. Such binding often takes place at surfactant concentrations relatively close to the cmc, and continues above the cmc [50,51].

Binding isotherms have played a key role in the development of polymer-surfactant system research, yielding information about the nature of the binding process and allowing conclusions about the structure of the aggregates. The degree of binding is also a necessary parameter, for instance, in the determination of the surfactant aggregation number from fluorescence probe measurements. Binding isotherms are normally determined at low polymer concentrations, and care should be exercised in applying such binding results to systems of high polymer concentration, where the binding process may not necessarily be the same.

Phase equilibrium.

Phase separation is commonly observed in polymer-surfactant systems. In the case of oppositely charged polymers and surfactants, true solutions may only exist at very low surfactant and polymer concentrations. The reader is referred to the chapter by Piculell, Lindman and Karlström (ch. 3) for a detailed and exhaustive description of phase equilibria. Experimental methods normally rely on visual observation of the cloud point (CP), and on the analytical determination of the composition of equilibrium phases. In many cases the CP is determined by ascending and descending temperature ramps at a given polymer and surfactant concentration. Results can be as reproducible as ± 0.5 °C, but slightly higher uncertainties are typical. A more general definition recognizes that the CP represents the same point on the phase diagram, whether traversing the phase boundary with temperature at constant composition, or with varying composition at constant temperature.

Comparison between results from different groups is often difficult due to the fact that most polymers are polydisperse and sometimes even of uncertain composition. Polymer molecular weight is an important variable determining phase boundaries, and needs to be known for such comparisons. For this reason, studies with low M_w/M_n polymer samples of known and systematically varied chemical composition are needed. In general, the study of phase equilibria, and investigations of the molecular structure of the concentrated phases and gels are among the most active and rewarding areas of current research.

Conductance and potentiometry.

Specific or molar conductivity is useful to determine the cmc of pure ionic surfactants in salt-free systems. In addition, an estimate of the degree of counterion dissociation can be obtained from the slope of the conductance curve above and below the cmc. In the case of polymer-surfactant systems, the cac can be determined by this method, and additional transition points may be observed at higher concentrations. The method can not be used to determine counterion binding to the polymer-surfactant aggregate (for nonionic polymers and ionic surfactants), because the necessary assumption concerning the micellar vs. the monomer surfactant mobility does not hold. An early example of the use of conductance to determine the cac is the study of Botré et al. [52]. An obvious disadvantage of the method is the fact that observed changes in conductance become very small in salt added systems.

Similarly, suitable ion selective electrodes can be used to monitor counterion activities, and therefore can be used to show breakpoints at the cac for anionic or cationic surfactants. In salt added systems the method again is less effective, not because of a breakdown in selectivity, but because for most common ions the interactions of the micellar surface with the counterions are only slightly dependent on the nature of the counterion.

Surface tension.

The measurement of surface tension, one of the fundamental techniques of colloid science, found very early application in the study of polymer-surfactant systems. Jones' classical study showed that a plot of surface tension vs. surfactant concentration has a change of slope at the cac which was found to be independent of polymer concentration, and coincided with the pure surfactant surface tension at saturation [33]. Surface tension can be used in added salt systems. For instance, Horin and Arai found that the cac for PVAc-SDS systems shift to lower concentration with increasing NaCl concentration. Similarly, the first binding isotherms for neutral polymer-anionic surfactant systems were determined from surface tension measurements [37]. For systems of ionic polymers, surface tension measurements show that the complex with oppositely charged surfactants is itself highly surface active.

Viscometry.

Similar to surface tension, viscosity measurements have played an important role in early studies, and continue to be useful especially in recent studies on systems with hydrophobically modified polymers. Much of the early work of Saito on the role of counterions, and on interactions with nonionic polymers, relied on viscosity measurements [7,29,54], as reviewed in chapter 9. Capillary viscometry can be used at low polymer concentrations, but in many cases shear rate dependence is informative, necessitating the use of cone and plate or rotating cylinder viscometers, preferably with low-shear capability. Non-absolute viscometers such as the Brookfield viscometer are useful for qualitative comparisons.

An interesting and simple technique for the qualitative determination of surface viscoelasticity in polymer-surfactant solutions has recently been reported by Winnik and co-workers [55,56]. The "talc test," following an early description by Lord Rayleigh, who used sulfur [57], involves observing the motion of fine talcum powder, which is applied to the surface and then disturbed by a short pulse of air flow. The results allow for the determination of a "surface phase diagram," showing differences in surface layer structure as a function of bulk solution concentration.

Dye solubilization.
Oil-soluble dyes such as Yellow OB or Orange OT are solubilized in micelles or in polymer-surfactant aggregates, and this technique can be used to determine cac values. In principle applicable to both ionic and nonionic surfactants, the method is less accurate for systems with low cac, due to the low solubilizing capacity. Care must be taken to establish equilibrium, requiring long equilibration times, and the possibility of the dye influencing the cac value must be considered.

Calorimetry.
Heat effects in micellar and polymer-surfactant solutions are relatively small, but modern calorimetric methods have been developed to allow for the application of this important technique. Isothermal calorimetric methods can be used to measure critical aggregation concentrations and to assess the relative strength of interaction between different series of polymers or surfactants. Heat capacity measurements are especially sensitive to changes in hydration. Differential Scanning Calorimetry (DSC) measures heat capacity as a function of temperature and can be used to determine phase transitions. Reports of the use of calorimetry in earlier polymer-surfactant investigations have been fairly scarce [58-61], but new instrumental developments have made the technique more amenable to overcome experimental difficulties such as sample viscosity and small heat effects. Theoretical and experimental details, and the impact of such new methodologies are are reviewed in chapter 8 by Olofsson and Wang. Volumetric data are often determined in conjunction with heat capacities [60], and give similar information if performed with ppm-level precision.

Chromatographic and other separation techniques.
Separation methods applicable to the study of polymer-surfactant systems include electrophoresis, capillary electrophoresis, gel filtration or size exclusion chromatography (SEC), ultracentrifugation, and ultrafiltration. Electrophoresis measurements were vital in the development of the "necklace" model of polymer-surfactant aggregates, as described by Shirahama in chapter 4. The bimodal distribution of electrophoretic mobilities observed in dextran sulfate-

cationic surfactant solution presents evidence for non-uniform aggregate distribution on the polymers [62]. As explained in chapter 4, such behavior would be predicted by small system thermodynamics.

SEC is of course an indispensible technique in the characterization of water-soluble polymers, but several researchers have also used the method to study polymer-surfactant interactions. Various experimental modes have been described [63], but normally the procedure involves injection of a polymer-surfactant mixture into a surfactant mobile phase for SEC, or water as eluent for low pressure gel filtration. Sasaki *et al.* used low pressure gel filtration to determine cac, saturation concentration, and binding ratios in PEO-SDS mixtures [63]. More recent results from Wyn-Jones *et al.* on the interactions of SDS with PVP and PPO could not reproduce the perfect plateaus reported by Sasaki *et al.*, but still allowed the determination of cac and saturation concentration [64,65]. Veggeland and Austad used high pressure SEC to study PEO-SDS as well as alkylarylsulfonate-xanthan mixtures used in enhanced oil recovery [66].

Ultracentrifugation has rarely been applied to polymer-surfactant systems, although the method is of course standard for proteins. François *et al.* used the technique to monitor the growth of the PEO-SDS complex relative to the pure polymer [67]. Combined with viscosity measurements, they determined the increase in molecular weight of the complex, and the SDS to PEO binding ratio.

Micellar relaxation kinetics.

Fluorescence methods to study micellar relaxation are well established (chapters 7, 10), but non-spectroscopic techniques can also be used in polymer-surfactant systems. Tondre used a temperature jump technique to study the relaxation kinetics of micelle formation and dissociation [68]. The perturbation of the micellar equilibria was monitored by the change in absorption of a solubilized acridine dye. Combination of the data with a mass action model predicted the locations of the cac and cmc quite well. In PEO-SDS solutions, with 10,000 and 1,000 molecular weight PEO, binding was observed only for the larger polymer, with only one or two surfactant aggregates on such a relatively short chain.

D'Aprano *et al.* measured the frequency dependence of ultrasonic relaxation in PVP-SDS. This technique yields relaxation times, which can be used, again in combination with a mass action model, to calculate surfactant aggregation number, aggregate size polydispersity, and the exit rate of the surfactant monomer. The aggregation number reported for the PVP-SDS complexes was 19±3, much smaller than the accepted value of 65 for free SDS micelles, but in accordance with the generally lower values in polymer-surfactant systems determined from fluorescence quenching experiments.

B. Spectroscopic Methods

Spectroscopic methods covering the full range of the electromagnetic spectrum have been applied to the study of polymer-surfactant systems. In most cases, such methods provide information about local structure and environment of a component of the system or a probe molecule; however, some methods can also provide quantitative information for instance on aggregation numbers or aggregate sizes, diffusion coefficients, solubilization parameters, or kinetic parameters. Many of these techniques are expertly described in the following chapters of this book. For this reason this brief introduction is not meant to be exhaustive, and will only highlight some specific cases of interest.   The methods are presented in order of increasing frequency of incident radiation.

Nuclear Magnetic Resonance (NMR).
   The explosive development of NMR has also resulted in important applications in the study of polymer-surfactant systems. Much of this work has been pioneered by the Swedish groups in Lund and Stockholm. Chemical shift, relaxation measurements and multidimensional NMR, employing ^{1}H, ^{2}H, ^{13}C ^{23}Na and other nuclei have been applied.   The review of Stilbs (chapter 6) provides theoretical background, discusses experimental conditions and cautions, and surveys recent work. We should note the important use of NMR, in particular the FT-PGSE technique, to obtain diffusion coefficients in multicomponent systems. This method can be applied in dilute solutions and in concentrated phases. The diffusion coefficients, representing a time-average, in turn can be used to determine for instance solubilizate distribution coefficients. Recent improvements in the FT-PGSE technique, described in chapter 6, allow its application to the polymer as well as to the surfactant component.
   Chemical shift measurements can be used to obtain information about the environment of the nucleus under study.   One example is the intermolecular aromatic ring current shift, used for the first time in micellar solutions by Eriksson [70], later extended to mixed micellar systems by Tokiwa and Tsujii [71]. Recognizing the observation by Stilbs that it is necessary for the aromatic group and the group under investigation to form a weak collision complex of preferred geometry [72], we note that indeed the absence of an intermolecular ring current shift does not exclude proximity. In micellar solution, shifts are normally referenced relative to the HOD resonance, which is not influenced by aromatic solubilizates, and the aromatic shift will be observed for nuclei which time averaged are in the proximity of, and in non-isotropic average position

relative to the aromatic ring. One of the earliest examples of the use of ^{1}H shifts in polymer-surfactant systems can be found in the work of Oakes [73]. Studies by Cabane using ^{1}H and ^{13}C shifts were particularly influential in the microscopic description of the PEO-SDS complex [74]. Particularly noteworthy is the report of Gjerde, Nerdal and Høiland who applied nuclear Overhauser enhancement (NOESY) 2-D NMR methods to show the proximity of PEO segments and SDS methylene groups in SDS-PEO solutions [75].

In principle, chemical shift measurements can also be used to obtain quantitative information, including kinetic information in polymer-surfactant systems. An intriguing case is shown in Figure 2, presenting the ^{1}H spectrum of a 0.1% PEO-2000 solution with increasing amounts of added 10-phenyldecanoate, crossing the cac of this system at about 8mM [76]. If we look at the single PEO peak at about 3.7 ppm, we note that this peak shifts from 3.74 ppm below the cac to 3.56 ppm well above the cac, the shift due to the intermolecular ring current effect of the surfactant terminal phenyl groups [77]. More importantly, at intermediate surfactant concentrations, there is appreciable peak broadening due to the the fact that the PEO segments are distributed between two sites of different chemical shift, i.e. "bound" and "free." The frequency- and temperature dependence of this effect can be used to gain information on the distribution equilibrium and on the binding kinetics.

Electron Spin Resonance (ESR).

The application of ESR requires the use of radical molecules such as TEMPO, PROXYL or their neutral and ionic derivatives. Although a wide variety of such probes is commercially available, the technique has not yet found wide application in micellar and polymer-surfactant systems. In principle, the method can give information about probe environment, but is based on line shape rather than frequency changes. Most commonly it can provide information on aggregate size and environment microviscosity [78,79].

Infra-Red and Raman spectroscopy.

These common techniques have not found wide application in the study of polymer-surfactant systems. The strong absorption bands of water often necessitate the use of liquid or solid thin films in the application of IR, and Raman studies are normally limited to high solute concentrations. Typical of the kind of application for which vibrational techniques are most useful is the study by Duffy *et al.* on the adsorption of cationic surfactant and anionic polyelectrolytes on an Au-alkanethiol surface [80]. Ponomarenko *et al.* used FTIR to monitor the amide bands and surfactant chain C-H stretches in cast films of cationic surfactant-sodium poly(α,L-glutamate), deriving information about hydrogen bonding and conformation in the complexes [81,82].

Stack Plot

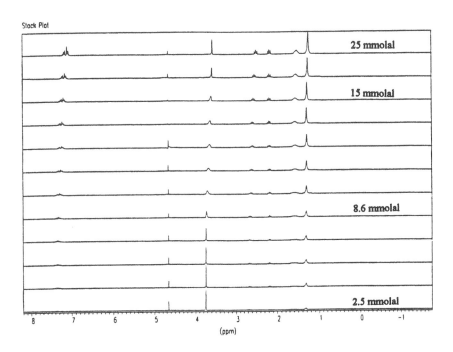

FIG. 2 Stacked ^{1}H NMR spectra of PEO-10-phenyldecanoate solutions. [PEO] 0.1 wt%, surfactant concentration from 2.5 (lowest spectrum) to 25 (top) mmolal.

Light scattering.

Static light scattering can be used in polymer-surfactant systems to obtain information on aggregate size. It can also be used in the form of simple turbidity measurements to determine phase boundaries and relative particle sizes. For instance, Dubin and co-workers have made extensive use of static and dynamic light scattering techniques to study aggregate size and the role of micellar surface charge in polymer-surfactant systems. Micellar surface charge density can be varied by using ionic-nonionic mixed surfactant systems [83,84].

UV-Visible spectroscopy.

The main use of UV-VIS spectroscopy in polymer-surfactant systems involves the interpretation of absorbance wavelength shifts of dyes solubilized in hydrophobic domains of the aggregates. Many such studies are performed in connection with fluoresence measurements, as described in chapters 7, 10, and 11. Circular dichroism (CD) can be used to provide structural information about the complexes, for example the degree of helicity in polypeptides. Some early references may be found in refs. 5 and 8.

Fluorescence.

A variety of static and dynamic fluorescence techniques have been applied to polymer-surfactant systems. Steady state techniques, including depolarization measurements, give information on such quantities as probe environment, micropolarity, microviscosity and, with certain caveats, aggregation numbers of the aggregates. Time-resolved techniques have become particularly powerful, yielding surfactant aggregation numbers and polydispersity measurements as well as kinetic information. Both small probe molecules and polymer-anchored probes may be used. Expert reviews of these methods are presented in chapters 7 and 10.

Microscopy.

Visible microscopy is generally not suitable to observe the nm-scale structures in polymer-surfactant bulk solution. Cryogenic transmission electron microscopy (Cryo-TEM) is a powerful technique allowing for the visual observation of aggregate structures. Spectacular examples include the thread-like structures sometimes observed in micellar and polymer-surfactant systems [85,86].

Neutron and X-ray scattering (SANS and SAXS).

Both neutron- and X-ray scattering are techniques providing information about long-range order. Due to its longer wavelength, neutron scattering is generally more applicable in polymer-surfactant systems, where repeating distances tend to be longer. In addition, this technique allows for contrast variation by the use of deuterated surfactants, solvents, and, at least in principle, polymers, to single out one component of the mixture. Cabane and Duplessix applied this contrast variation method to study the structure of PEO-SDS mixtures [40,41]. They found aggregate radii similar to that of a free SDS micelle. At low ionic strength the aggregates were distributed in an homogeneous manner throughout the polymer coil, but at high ionic strength the complex was slightly more compact, and the aggregate distribution was higher in the center of the polymer coil, showing the influence of electrostatic repulsion. Recently, Lee and Cabane have applied SANS to SDS-poly(N-isopropylacrylamide) systems above the polymer cloud point [87]. They found that at low surfactant concentration the polymer forms a turbid suspension of small colloidal particles stabilized by the surfactant, while at higher surfactant concentration polymer chains are solubilized by the surfactant, forming "necklaces" of bound micellar clusters.

III. CHAPTER OVERVIEW

The brief descriptions presented above are only meant to highlight the topics described in detail in the following chapters. In addition, we have tried to present a recollection of some of the early studies which in our view have been influental in the development of research in the area of polymer-surfactant systems, without the goal of being exhaustive or necessarily representative of all techniques and investigations. The chapters which follow will present the reader with expert reviews and descriptions of the wide variety of techniques, methodologies and applications of polymer-surfactant systems.

Chapter 2 by Goddard and Ananthapadmanabhan presents a wide overview of applications, with detailed sections on personal care products, detergency, foams, and interactions at the mineral surface. Chapter 3 by Piculell, Lindman and Karlström describes the recent and rapidly developing experimental and theoretical studies of phase equilibria, pioneered by the Lund groups. In chapter 4, Shirahama discusses thermodynamic theories for aggregation as well as the many proposed models. Linse, Piculell and Hansson in chapter 5 give an up-to-date review of model descriptions, including thermodynamic models, mean field lattice models, and the recent application of Monte Carlo simulations.

Chapter 6 by Stilbs presents a critical review of the use of NMR methods in polymer-surfactant systems, showing both the pitfalls and the exciting possibilities of this technique. Chapter 7 by Winnik and Regismond is the first of two chapters discussing the application of modern time-resolved fluoresence methods, with special emphasis on systems of hydrophobically modified polymers.

Calorimetry had been rarely used in polymer-surfactant systems, due to a number of experimental problems. As shown in chapter 8 by Olofsson and Wang, new instrumental developments have now brought calorimetric techniques into the mainstream of polymer-surfactant research. Chapter 9 by Saito and Anghel summarizes the extensive work on interactions between water-soluble polymers and nonionic surfactants, and also presents a discussion of the role of headgroup charge and the influence of small electrolytes. Chapter 10 by Zana is the second chapter describing fluorescence methods, including a thorough review of the determination of aggregation numbers by probe fluorescence methods, and a review of studies on weakly hydrophobic polyelectrolytes and hydrophobically modified polymers. Finally, in chapter 11 Hayakawa presents a description of dye solubilization in polymer-surfactant aggregates.

ACKNOWLEDGEMENTS

The second author (JCTK) wishes to express his gratitude to the undergraduate students, graduate students, postdoctoral fellows, and faculty colleagues at Dalhousie University and abroad who over the years have cooperated with him in polymer and surfactant research.

REFERENCES

1. R.L. Davidson, *Handbook of Water-Soluble Gums and Resins*, McGraw-Hill, New York, 1980.
2. P. Molyneux, *Water-Soluble Synthetic Polymers: Properties and Behavior*, Vol. 1, Vol. 2, CRC Press, Boca Raton, 1984.
3. M.M. Breuer and I.D. Robb, *Chem. Ind.* 530 (1972).
4. I.D. Robb, in *Anionic Surfactants: Physical Chemistry of Surfactant Action* (E.H. Lucassen Reynders, ed.), Marcel Dekker, New York, 1981, pp 109-142.
5. E.D. Goddard, *Colloids Surf. 19*: 255 (1986).
6. E.D. Goddard, *Colloids Surf. 19*: 301 (1986).
7. S. Saito, in *Nonionic Surfactants: Physical Chemistry* (M.J. Schick, ed.), Marcel Dekker, New York, 1987, Ch.15.
8. K. Hayakawa and J.C.T. Kwak, in *Cationic Surfactants: Physical Chemistry* (D.N. Rubingh and P.M. Holland, eds.), Marcel Dekker, New York, 1991, Ch. 5.
9. E.D. Goddard and K.P. Ananthapadmanabhan, (eds). *Interactions of Surfactants with Polymers and Proteins*, CRC Press, Boca Raton, 1993.
10. K.P. Ananthapadmanabhan in *Interactions of Surfactants with Polymers and Proteins* (E.D. Goddard and K.P. Ananthapabmanabhan, eds.), CRC Press, Boca Raton, 1993, Ch. 2.
11. E. D. Goddard, in *Interactions of Surfactants with Polymers and Proteins* (E.D. Goddard and K.P. Ananthapabmanabhan, eds.), CRC Press, Boca Raton, 1993, Chs. 4, 10.
12. B. Lindman and K. Thalberg, in *Interactions of Surfactants with Polymers and Proteins* (E.D. Goddard and K.P. Ananthapabmanabhan, eds.), CRC Press, Boca Raton, 1993, Ch. 5.
13. J.C. Brackman and J.B.F.N. Engberts, *Chem. Soc. Rev. 22*: 85 (1993).
14. Y.-C. Wei and S.M. Hudson, *Rev. Macromol. Chem. Phys. C(35)1*: 14 (1995).
15. H.B. Bull and H. Neurath, *J. Biol. Chem. 118*: 163 (1937).
16. J.A. Steinhardt and J.A. Reynolds, in *Multiple Equilibria in Proteins*, Academic Press, New York, 1969, p. 234.
17. S. Lapanje, in *Physiochemical Aspects of Protein Denaturation*, Wiley, New York, 1978, Chs. 3, 4, and 6.
18. M.N. Jones, *Chem. Soc. Rev. 21*: 127 (1992).
19. E. Dickinson, in *Interactions of Surfactants with Polymers and Proteins* (E.D. Goddard and K.P. Ananthapabmanabhan, eds.), CRC Press, Boca Raton, 1993, Ch. 7.

20. K.A. Dill, D.E. Koppel, R.S. Canter, J.D. Dill, D. Bendedouch, and S.-H. Chen, *Nature 309*: 42 (1984).
21. F.M. Menger and D.W. Doll, *J. Am. Chem. Soc. 106*: 1109 (1984).
22. J. Tabony, *Mol. Phys. 51*: 975 (1984).
23. K. Shirahama, K. Tsujii and T. Takagi, *J. Biochem. 75*: 309 (1974).
24. Y.-Q. Zhang, T. Tanaka and M. Shibayama, *Nature 360*: 142 (1992).
25. M.L. Anson, *Science 90*: 256 (1939).
26. R.M. Hill and D.R. Briggs, *J. Am. Chem. Soc. 78*: 1590 (1956).
27. T. Isemura and Y. Kimura, *J. Polym. Sci. 16*: 92 (1955).
28. T. Isemura and A. Imanishi, *J. Polym. Sci. 33*: 337 (1958).
29. S. Saito, *Kolloid-Z. 137*: 93, 98 (1954).
30. F.H. Sexsmith and H.J. White, Jr., *J. Colloid Sci. 14*: 598 (1959).
31. Y. Gotshal, L. Rebenfeld and H.J. White, Jr., *J. Colloid Sci. 14*: 619 (1959).
32. F.H. Sexsmith and H.J. White, Jr., *J. Colloid Sci. 14*: 630 (1959).
33. M.N. Jones, *J. Colloid Interface Sci. 23*: 36 (1967).
34. M.N. Jones, *J. Colloid Interface Sci. 26*: 532 (1968).
35. H.P. Frank, S. Barkin, and F.R. Eirich, *J. Phys. Chem. 61*: 1375 (1957).
36. K.E. Lewis and C.P. Robinson, *J. Colloid Interface Sci. 32*: 539 (1970).
37. H. Arai, M. Murata, and K. Shinoda, *J. Colloid Interface Sci. 37*: 223 (1971).
38. M.J. Schwuger, *J. Colloid Interface Sci. 43*: 491 (1973).
39. J.A. Reynolds and C. Tanford, *J. Biol. Chem. 245*: 5161 (1970).
40. B. Cabane and R. Duplessix, *J. Phys. (Paris) 43*: 1529 (1982).
41. B. Cabane and R. Duplessix, *Colloids Surf. 13*: 19 (1985).
42. K. Hayakawa and J.C.T. Kwak, *J. Phys. Chem. 86*: 3866 (1982).
43. E.D. Goddard and R.B. Hannan, *J. Am. Oil Chem. Soc. 54*: 561 (1977).
44. H.G. Bungenberg de Jong, in *Colloid Science* (H.R. Kruyt, ed.), Vol. II, Elsevier, New York 1952, ch. 8-11.
45. S.G. Cutler, P. Meares, and D.G. Hall, *J. Electroanal. Chem. 85*: 145 (1974).
46. T. Maeda, M. Ikeda, M. Shibahara, and I. Satake, *Bull. Chem. Soc. Japan 54*: 94 (1981).
47. K. Shirahama, H. Yuasa, and S. Sugimoto, *Bull. Chem. Soc. Japan 54*: 375 (1981).
48. K. Hayakawa, A.L. Ayub, and J.C.T. Kwak, *Colloids Surf. 4*: 389 (1982).
49. D.M. Painter, D.M. Bloor, N. Takisawa, D.G. Hall, and E. Wyn-Jones, *J. Chem. Soc. Faraday Trans. 1*, 84: 2087 (1988).
50. K. Shirahama, A. Himuro, and N. Takisawa, *Colloid Polym. Sci. 265*: 96 (1987).
51. C. Howley, D.G. Marangoni, and J.C.T. Kwak, *Colloid Polym. Sci. 275*: 760 (1997).
52. C. Botré, F. De Martiis, and M. Solinas, *J. Phys. Chem. 68*: 3624 (1964).
53. S. Horin and H. Arai, *J. Colloid Interface Sci. 32*: 547 (1970).
54. S. Saito, *Kolloid-Z. 158*: 120 (1958).
55. S.T.A. Regismond, F.M. Winnik, and E.D. Goddard, *Colloids Surf. A 119*: 221 (1996).
56. S.T.A. Regismond, K.D. Gracie, F.M. Winnik, and E.D. Goddard, *Langmuir 13*: 5558 (1997).

57. Lord Rayleigh, *Proc. Roy. Soc. (London) 48*: 127 (1890).
58. K. Shirahama and N. Ide, *J. Colloid Interface Sci. 54*: 450 (1976).
59. G.C. Kresheck and W.A. Hargraves, *J. Colloid Interface Sci. 83*: 1 (1981).
60. G. Perron, J. Francoeur, J.E. Desnoyers, and J.C.T. Kwak, *Can. J. Chem. 65*: 990 (1987).
61. J.C. Brackman, N.M. van Os, and J.B.F.N. Engberts, *Langmuir 4*: 1266 (1988).
62. K. Shirahama, K. Kameyama, and T. Takagi, *J. Phys. Chem. 96*: 6817 (1992).
63. T.K. Korpela and J.-P. Himanen, in *Aqueous Size Exclusion Chromatography*, (P.L. Dubin, ed.), *J. Chromatography Library 40*, Elsevier, Amsterdam, 1988, ch. 13.
64. W.A. Wan-Badhi, W.M.Z. Wan-Yunus, D.M. Bloor, D.G. Hall, and E. Wyn Jones, *J. Chem. Soc. Faraday Trans. 89*: 2737 (1993).
65. D.M. Bloor, W.M.Z. Wan-Yunus, W.A. Wan-Badhi, Y. Li, J.F. Holzwarth, and E. Wyn-Jones, *Langmuir 11*: 3395 (1995).
66. K. Veggeland and T. Austad, *Colloids Surf. A 76*: 73 (1993).
67. J. François, J. Dayantis, and J. Sabbadin, *Eur. Polym. J. 21*:165 (1985).
68. C. Tondre, *J. Phys. Chem. 89*: 5101 (1985).
69. A. D'Aprano, C. La Mesa, and L. Persi, *Langmuir 13*: 5876 (1997).
70. J.C. Eriksson, *Acta Chem. Scand. 17*: 1487 (1963).
71. F. Tokiwa and K. Tsujii, *J. Phys. Chem. 75*: 3560 (1971).
72. P. Stilbs, *J. Colloid Interface Sci. 94*: 463 (1983).
73. J. Oakes, *Eur. J. Biochem. 36*: 553 (1973).
74. B. Cabane, *J. Phys. Chem. 81*: 1639 (1977).
75. M. I. Gjerde, W. Nerdal, and H. Høiland, *J. Colloid Interface Sci. 183*: 285 (1996).
76. D. Arden and J.C.T. Kwak, unpublished results.
77. Z. Gao, R.E. Wasylishen, and J.C.T. Kwak, *J. Colloid Interface Sci.:137*: 137 (1990).
78. K. Shirahama, M. Tohdo, and M. Murahashi, *J. Colloid Interface Sci. 86*: 282 (1982).
79. F.M. Witte, P.L. Buwalda, and J.B.F.N. Engberts, *Colloid Polym. Sci. 265*: 42 (1987).
80. D.C. Duffy, P.B. Davies, and A.M. Creeth, *Langmuir 11*: 2931 (1995).
81. E.A. Ponomarenko, A.J. Waddon, D.A. Tirrell, and W.J. MacKnight, *Langmuir 12*: 2169 (1996).
82. E.A. Ponomarenko, A.J. Waddon, K.N. Bakeev, D.A. Tirrell, and W.J. MacKnight, *Macromolecules 29*: 4340 (1996).
83. P.L. Dubin and R. Oteri, *J. Colloid Interface Sci. 95*: 453 (1983).
84. Y. Li, P.L. Dubin, H.A. Havel, S.L. Edwards, and H. Dautzenberg, *Macromolecules 28*: 3098 (1995).
85. R. Zana, A. Kaplun, and Y. Talmon, *Langmuir 9*: 1948 (1993).
86. D. Süss, Y. Cohen, and Y. Talmon, *Polymer 36*: 1809 (1995).
87. L.-T. Lee and B. Cabane, *Macromolecules 30*: 6559 (1997).

2

Applications of Polymer-Surfactant Systems

E. DESMOND GODDARD Consultant, 349 Pleasant Lane, Haworth, N.J. 07941

K.P. ANANTHAPADMANABHAN   Unilever Research, U.S., 45 River Road, Edgewater, N.J. 07020

I. INTRODUCTION

Polymers and surfactants are used extensively in a wide variety of industrial, cosmetic, and pharmaceutical applications. In many, if not most, such cases their encounter is unintentional in the sense that they are added for their independent (individual) function rather than for those arising out of their mutual interactions. For example, polymers are often used in formulations for controlling the rheology of solutions and suspensions and for altering the interfacial properties of solids. Surfactants are used for altering the wettability, solubilization and emulsification properties by changing the properties of the interfaces involved. When present together, polymers and surfactants can interact with each other leading to significant alterations in properties which are often considered undesirable. On the other hand, they may well be beneficial. Though limited, there are examples of systems in which the interaction between the polymer and the surfactant is exploited or circumvented to provide the beneficial effects sought from a formulation. Some of the potential benefits arising from the positive effects of polymer-surfactant interactions have been reviewed elsewhere [1]. In this chapter further examples of application systems in which polymer-surfactant interactions play a definite role in governing the system behavior are presented together with instances involved in actual industrial, cosmetic and pharmaceutical applications in which polymer-surfactant interactions influence the system behavior significantly and where approaches to modulating the interactions and enhancing the desired benefits become possible. Non-interacting systems are not considered. Polymeric surfactants, because of their unique properties, are also finding their way into application systems and some examples of such systems are again included. Because of the nature of the subject and its rapid growth, some overlap with the previous article [1] has been inevitable. An attempt has been made to minimize repetition and to emphasize areas where activity and/or opportunities seem to be substantial. This applies especially to the "Personal Care" section, and for this reason this topic is given special emphasis.

II. RHEOLOGY AND GELATION

A.   Viscosity Enhancement

It is well established that the interaction of a nonionic polymer such as polyethyleneoxide (PEO) or polyvinylpyrrolidone (PVP) with an anionic surfactant such as sodium dodecyl sulfate (SDS) can result in "charging up" the polymer and in effect imparting polyelectrolyte properties to the nonionic polymer [2,3,4]. This effect manifests itself as a significant increase in viscosity at a certain concentration (T_1 or "c.a.c.") of the surfactant, independent of polymer molecular weight. This viscosity increase can be as high as fivefold. In addition to an increase in viscosity, these systems also exhibit considerable viscoelastic effects. Viscosity increases resulting from the addition of cationic surfactants to nonionic polymer are less common because of their relatively weaker interaction. If the hydrophobicity of the polymer is relatively high, the interactions can be enhanced to modify their rheological properties. For example, polymers such as ethylhydroxyethylcellulose (EHEC) exhibiting a lower consolute temperature (LCST) will phase separate at higher temperatures, indicating a higher hydrophobicity at elevated temperatures. The increase in viscosity and gel formation reported by Carlsson *et al.* [5] for the polymeric ether EHEC-cetyltrimethylammoniumbromide (CTAB) system in a certain temperature-surfactant concentration window may be a manifestation of the increased hydrophobicity of the polymer leading to increased interaction with the cationic surfactant (see section IV below). The gelation itself may be due to network formation and entanglement brought about by the surfactant micelles.

We have reported earlier that even larger increases in viscosity can be observed in oppositely charged polyelectrolyte-surfactant systems [6,7,8]. In one case, involving the cationic cellulosic Polymer JR 400 (Union Carbide), relatively low levels of SDS led to substantial viscosity increases (as large as 200 fold, see Figure 1) or even weak gel formation in the immediate pre-precipitation zone. Combination of SDS with the highest available molecular weight grade (Polymer JR 30 M) resulted in strong gels at polymer levels as low as 1% and surfactant levels around 0.1% [7, 8]. Several other anionic surfactants such as alkyl benzene sulfonates, alkyl ether sulfates and sulfosuccinates were also found to form such gels. Gel formation in these systems is likely to be due to the increased chain entanglement resulting from the crosslinking of bound surfactant clusters and the naturally high entanglement of the higher molecular weight polymer.

Viscosity increases and gel formation in polyanion-cationic surfactant systems are also reported by Thalberg and Lindman [9] for hyaluronan and polyacrylate polymers in combination with alkyltrimethylammonium bromide surfactants. Similarly, we have observed viscosity increases and gel formation with polymers such as CMC(carboxymethyl cellulose) and CTAB at relatively low polymer-surfactant levels [10].

Interactions of nonionic surfactants with most polymers is relatively weak and therefore significant increases in viscosity resulting from polymer-surfactant interactions do not occur. However, interaction of the EO groups of nonionic surfactants with polyacrylate type polymers is known and this can lead to alterations in viscosity [11, 12]. Hydrophobic modification of the polymer, on the other hand, can dramatically increase the level of interaction and can lead to a substantial influence on the rheology of the system (see Figure 1). Sarrazin-Cartalas *et al.* [13] report a marked increase in viscosity of hydrophobe modified polyacrylate polymers in the presence of nonionic surfactants. An important consequence, since the size

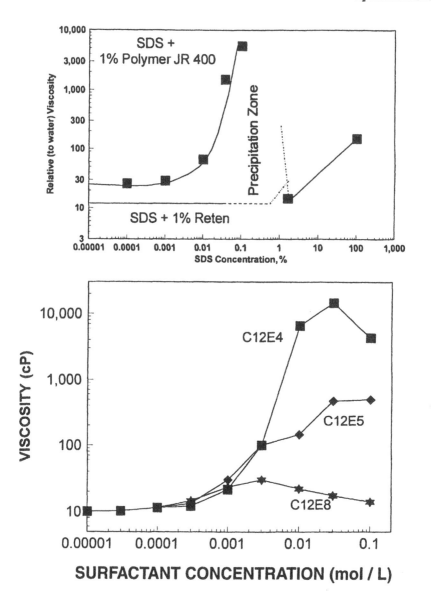

FIG. 1 Top: Relative viscosity of 1% Polymer JR and of 1% Reten 220 as a function of SDS concentration. From ref. 6, *Colloids Surf. 13, 47 (1985)*, with permission from Elsevier Science-NL, Sara Burgerhartstraat 25, 1055 KV Amsterdam, the Netherlands.

Bottom: Variation in viscosity of a hydrophobically modified polyacrylate polymer(3-C12) with the surfactant concentration. Polymer concentration 1%, Temperature 25°C, shear rate, 1.3 s^{-1}. From ref. 13 with permission from the American Chemical Society.

and shape of nonionic micelles can be appreciably influenced by temperature, is that these types of nonionic system can be used to bring about thermally induced viscosity enhancement and gelation. Sarrazin-Cartalas *et al.* [13] have attributed the observed increase in the gelation at higher temperatures, in the case of a hydrophobe modified polyacrylate-$C_{12}E_4$ system, to network formation mediated by giant surfactant vesicles and the hydrophobic groups of the polymer. We will return to the subject of thickening in such systems later.

Increases in viscosity and gel formation may not be desirable in all cases and may have to be modulated to obtain an optimum balance between stability and pourability. There are several approaches to thinning polymer-surfactant systems. The most obvious approach is to alter the polymer-surfactant ratio to move away from the optimum gelling ratio [7]. Addition of competing polymers, surfactants, or salts can also lead to viscosity reduction. Brackman and Engberts [14] have reported an interesting case of viscosity reduction on adding a water soluble polymer polypropyleneoxide (PPO) or polyvinylmethylether to the solution of a cationic surfactant (CTAB) in which the surfactant micelles are rod-like due to the presence of strongly bound salicylate anion. Competition between the polymer and the salicylate is evidently responsible for the pronounced structural reorganization of the system which is de-gelled in the thinning process.

Other instances, where addition of free polymers and anchored (hydrophobic) polymers to concentrated lamellar liquid dispersions of concentrated surfactants (lamellar phase volume 80%) reduces the viscosity of the system, have been reported [15, 16]. While the former may involve polymer-surfactant interactions, the latter in a way involves hydrophobic interactions between the hydrophobic polymer and the lamellar dispersions. In these cases, any increase in viscosity would involve the swelling of the dispersions and a reduction in viscosity would be associated with deswelling (phase volume reduction to 60%) of the dispersions by the polymer. Here a competition between the surfactant and the polymer for the solvent is responsible for reduction in the viscosity, as opposed to any direct interaction of the polymer with the surfactant. Even greater reductions (as much as 80%) can be achieved by using electrolytes. This route, however, results in salting out of the nonionic EO groups and flocculation of the lamellar droplets. Van de Pas [15] has shown that this can be overcome by using hydrophobe modified polymers with ionic functional groups which can anchor into the liquid lamellar dispersions and remain hydrated, even at high salt concentrations by virtue of the ionic group (see Figure 2). This approach results in concentrated lamellar dispersions that are substantially stable in concentrated electrolyte solutions and still possess low viscosity.

A common method to create the network structure which is characteristic of gels is to use a crosslinked version of a polymer that, unaltered, is soluble in water. Of course, if the polymer in its uncrosslinked state is reactive to added surfactant then a substantial influence on the polymer in the cross linked state can be expected on addition of a reactive surfactant. Such a polymer is poly(N-isopropylacrylamide),

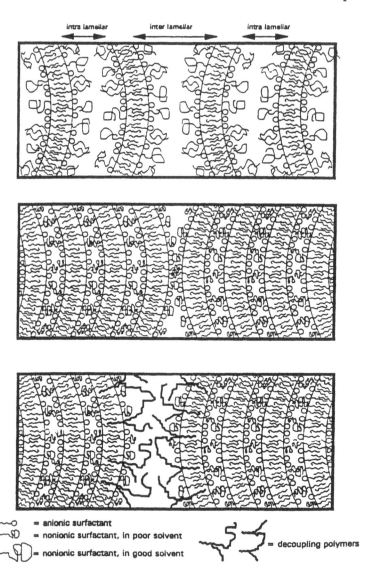

intra lamellar   inter lamellar   intra lamellar

~~~O = anionic surfactant
~~~SD = nonionic surfactant, in poor solvent
~~~SD = nonionic surfactant, in good solvent

= decoupling polymers

FIG. 2 A schematic representation of anchoring polymer (hydrophobe modified polymer)-induced decoupling of interaction forces between lamellar layers and lamellar droplets. Top: low salt concentration, repulsion between layers and droplets. Center: high salt concentration, attraction between layers and droplets due to poor solvency of nonionic head groups, thin lamellar layers and flocculated lamellar droplets.   Bottom: high salt concentration and hydrophobe modified decoupling polymer: attraction between lamellar layers and repulsion between the droplets. Reproduced from ref. 15 with permission from J.C. van de Pas.

NIPA. Some aspects of its reactivity towards surfactants will be discussed in sections VII and X of this chapter.

B. Influence of Surfactants on Swelling-Deswelling Characteristics of Polymeric Gels

The de-swelling effect of water soluble polymers on lamellar lipid dispersions of concentrated surfactant solution was discussed earlier under the section on rheology. Polymer-surfactant interaction can be expected to influence the swelling-deswelling characteristics of polymeric gels. It is well known that polyelectrolyte gels that are used as super absorbents for water under physiological conditions (e.g. saline, urine, blood) can lose their absorbing capacity because of electrolyte effects. In a recent study, Zhang et al. [17, 18] have shown that these effects can be overcome by using a nonionic polymer gel (NIPA) in combination with a surfactant such as SDS. Results reproduced in Figure 3 from reference 17 show that the swelling characteristics in the presence of high levels of salt do not change for the NIPA-SDS gel unlike the case of the SDS-free gel. The authors, on the basis of SANS results, suggest that the initial swelling power is due to the "polyelectrolyte" character of the SDS micelle bound gel and the lack of swelling sensitivity to salt is associated with the formation of cylindrical micelles around the polymer at higher salt levels. It appears that the reduced electrostatic effects are to some extent compensated by the increased hydrophobic interactions at higher salt levels and the consequent increase in size of the polymer bound micelles. The result is increased tolerance to electrolytes.

III. SOLUBILIZATION

A well known feature of surfactant solutions is their ability to dissolve a variety of oil soluble materials, e.g. hydrocarbons, esters, perfumes, dyes and so on. This property is utilized in the compounding of many formulations, the process involving the dissolution of the solubilizate in the surfactant micelles. Formation of polymer/surfactant complexes is currently regarded as a depression of the critical aggregation concentration of the surfactant. According to this picture, superior solubilization in such systems can be anticipated over that provided by the surfactant alone. There is, in fact, abundant evidence of this enhancement, e.g., in the solubilization of dyes [19, 20, 21], hydrocarbons [22] and sparingly soluble fluorescers [23] as has been observed with combinations of uncharged polymers and charged surfactants. In combinations of polyelectrolyte and oppositely charged surfactants the effect can be much larger, as shown for the oil soluble dye Orange OT in the cationic cellulosic Polymer JR/SDS system [24,25]. An important discovery of this work was that solubilization can occur at extremely low

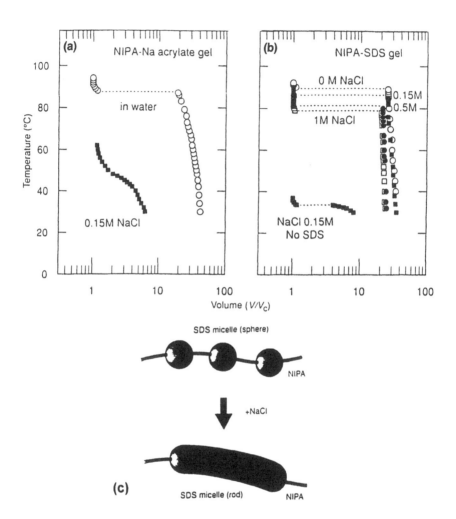

FIG. 3 (a) Temperature dependence of the volume of a copolymer gel of poly(N-isopropylacrylamide) (636 mM) and acrylic acid (64 mM ionizable groups) in 0.15 mM NaCl. V_c indicates gel volume in collapsed state at high temperatures. Swelling is completely lost in salt solutions. (b) Temperature dependence of swelling in the presence of 2% SDS. No marked change in swelling curves, indicating that gel retains high swelling power in salt solution even at 1M. (c) Illustration of the possible micellar formation of anionic surfactant in a hydrophobic gel. Ionic surfactants such as SDS form spherical micelles in water and rod-like micelles at high salt concentrations. This helps prevent collapse of the gel in concentrated salt solutions. From ref. 17 with permission.

concentrations, indicating a c.a.c. over two orders of magnitude lower than the c.m.c. of SDS [24,26]. While the degree of solubilization falls as the concentration of SDS approaches the precipitation zone, it is reestablished at higher concentration and the principal solubilization zone is wider (shifted to lower concentration) than what is observed for the micellar region of SDS by itself. A practical way to reduce or avoid the above mentioned "precipitation" zone of ionic surfactant /oppositely charged polyion pairs is to incorporate a suitable nonionic surfactant into the system [27, 28].

Because of their tendency to associate in solution, it is to be expected that surfactants and polymers can influence each other's solubility. There is abundant evidence of the ability of a surfactant to increase the range of solubility of polymers in water and this subject is treated in the next section. Less well known is the ability of a water soluble polymer to increase the solubility range of a surfactant. Schwuger and Lange [29] for example, reported that polyvinylpyrrolidone can reduce the Krafft point of sodium hexadecyl sulfate by close to 10°C, an effect evidently linked to a lowering of the monomer concentration required for aggregation in the presence of this polymer.

In some cases, polymers that are normally considered to be insoluble can be solubilized in surfactant solutions, as was demonstrated by Isemura and Imanishi [30] for the polymer polyvinylacetate and the surfactant SDS. This confirmed work carried out earlier by Sata and Saito [31]. Similar, but less strong solubilization of certain insoluble polymers by cationic surfactants has also been demonstrated by Saito [32]. These types of solubilization make it possible to regenerate the original polymer in different state, *e.g.* as a fiber or a film. Work related to this approach was reported by Lundgren [33], in which concentrated equi-weight mixtures of protein and anionic surfactant in water were extruded as continuous fiber into a magnesium sulfate coagulating bath prior to final drawing, washing, and drying.

A result of potential environmental significance [34] comes from the observed efficiency of combinations of hexadecylpyridinium chloride and maleicanhydride/ vinylmethylether (Gantrez S-95 polymer, GAF Corp.) in solubilizing chlorophenols. This combination has potential as an adjuvant in a micelle-enhanced ultrafiltration process for clean-up of polluted streams [34]. Another illustration of such a procedure is the solubilization of trichloroethylene by poly-electrolyte-surfactant (polystyrenesulfonate-cetylpyridinium chloride) complexes [35].

IV. CLOUD POINT ELEVATION

It is well known that many soluble polymers have an inverse solubility/temperature relationship, and their solutions exhibit a "cloud point." Examples are polymers based on polyalkylene oxides and those with multiple amide groups, or multiple ether/hydroxyl groups such as found in several water-soluble cellulose based materials. While the phenomenon, in general terms, is explained as being due to "dehydration" of polar groups such as ether, amide, hydroxyl, etc., it is well known that it can often be offset by the addition of ionic surfactants which can result in increases in cloud point of several degrees. This phenomenon seems to be a clear case of polymer/surfactant interaction and the formation of complexes of increased intrinsic solubility. Practically speaking, prevention of clouding in formulations can be important for maintaining viscosity and the stability of matter in suspension.

A simple representation of the phenomenon is given in Figure 4. In the "ideal" case the cloud point of the polymer is unaffected (AB) by the addition of surfactant until a critical concentration is reached; thereafter there is a linear increase (BC) of cloud point with increase of surfactant concentration. Whereas many polymer/surfactant systems follow this pattern, it is known now that the response is complicated more often following, for example, the pattern AB'C', in which a depression of the cloud point precedes the increase observed at higher surfactant concentrations. This behavior is encountered in certain cellulose based polymers]. In particular, the behavior of the polymer ethylhydroxyethylcellulose has been extensively studied by Karlström et al. [38,39,40] and modelled on the basis of Flory-Huggins concepts. At present it can be said that the response of the cloud point of a polymer to added surfactant can be quite complicated and is influenced by (i) the structure of the polymer, (ii) its molecular weight and molecular weight distribution, (iii) its tendency to self-aggregate (iv) the presence of salt, (v) the type of salt, and (vi) the nature of the added surfactant, including the sign of its charge. More work in this practically interesting area would be desirable.

The raising of the cloud point of a nonionic surfactant by an anionic surfactant can be considered to be a special case of polymer/surfactant interaction. Here the "polymer" is a hydrophobically substituted species in which the hydrophilic moiety (most often with ethylene oxide repeating units) is an oligomer rather than a true polymer. This phenomenon has been of much importance, and has long been known to formulation chemists and involves the close association of the two species in mixed micelles, i.e. the "complexes" in this case.

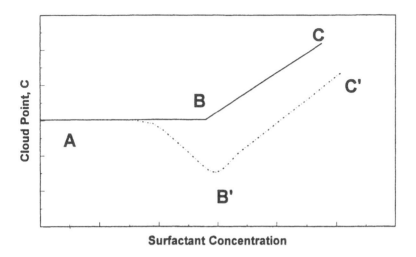

FIG. 4 Schematic representation of possible effects of ionic surfactants on the cloud point of polymers.

V. CATALYSIS AND ENZYMATIC REACTIONS

There is a limited amount of information in the literature on this subject. Of course, enzymes themselves are naturally occurring polypeptide derivatives and, as such, are expected to interact with surfactants. Interestingly, several enzymes [41], unlike the "standard" protein bovine serum albumin, are not denatured by SDS, and these enzymes have been shown not to bind significant amounts of this surfactant.

It is known that the stability of enzymes in surfactant solutions can be enhanced by polyethylene glycols (PEGs). This may be another example in which a reduction in monomer activity could play a role in the observed effect. Note, however, that PEGs, in general, do enhance enzyme stability even in the absence of surfactants and this may be related to the protective action of the PEG itself. In a recent patent, Hessel *et al.* [42] have claimed that incorporation of nonionic ethylene glycol containing hydrophobic copolymer enhances the stability of the enzymes in liquid detergent systems containing linear alkyl benzene sulfonate surfactant. In a complex liquid detergent system it is rather difficult to predict the role of the polymer. In any case, a possible role of the hydrophobic polymer in reducing the monomer activity and, in turn, reducing the surfactant induced denaturation of the enzyme cannot be ruled out.

Turning from bio-systems to chemical systems, one also finds that study of the influence of polymer surfactant interaction on reaction kinetics has been very limited. Of course, in the same way that micelles can provide an environment to catalyze various reactions, a subject which has been widely investigated [43], it can be expected that the aggregates in various polymer/surfactant systems will behave similarly. However, the involvement of the polymer itself in such combinations in influencing the kinetics of reactions has not received much attention. One such study has, however, been recently reported [44]. It concerns photocleavage of the chromophoric group in the copolymer N-acryloylpyrrolidine (98.6%)/4-vinylpyridinedicyanomethylide (1.4%). From measurements of e.m.f. with an "SDS electrode," and of changes in λ_{max} in the absorption spectrum of the polymer's chromophore, this copolymer was shown to bind SDS. Evidence of a polymer/surfactant interaction effect came from the observation that the rate of cleavage of the yluric bonds of the chromophore (on irradiation with a laser at 406.7 nm), as a function of SDS concentration, followed almost exactly the binding curve of SDS by the polymer. Evidently the chromophoric groups become more "accessible" as a result of this binding.

VI. THE PERSONAL CARE INDUSTRY

A.   Reduction of Surfactant Monomer Concentration: Mildness of Surfactants

Polymers which reduce the effective c.m.c. of the surfactant, by promoting aggregation at lower concentrations than the formal surfactant c.m.c., in effect reduce the surfactant monomer concentration to lower levels than that in polymer free solutions. As a consequence, interactions which depend solely on monomer activity, and which are not influenced by the polymer-surfactant complex, can be modulated by the presence of the polymer. As can be expected, the stronger the interaction, the lower the monomer activity. Thus, the monomer activity of a charged surfactant can be more effectively lowered by an oppositely charged polymer than by an uncharged polymer. Introduction of hydrophobic groups on the polymer will lead to a further lowering in activity compared to what can be achieved using the unmodified polymer (45). For example, the results given in Figure 5 show that introduction of hydrophobic groups on a cationic HEC (Quartisoft LM 200, Union Carbide) can lower the c.a.c. to significantly lower values than that achieved by the cationic polymer Polymer JR 400.

The practical implications of the modulation of surfactant monomer activity by polymers can be significant. For example, even though the mechanisms of skin

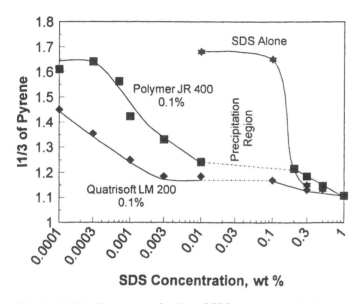

SDS Concentration, wt %

FIG. 5 (a) I_1/I_3 of pyrene as a function of SDS concentration in the presence and absence of Quatrisoft LM 200 (hydrophobe modified cationic HEC), or Polymer JR 400 (cationic HEC). Lower values of I_1/I_3, indicate lower polarity of the complex. The results show that the presence of hydrophobic and cationic groups lower the onset of interaction much more than cationic groups alone.

irritation by surfactants are not fully understood, the irritation potential of surfactants has been related to their monomer activity in solution by several investigators [46]. Reduction in surfactant monomer activity by adding polymers may result in reduced surfactant binding and possibly in reduced irritation. In support of this argument, there exists evidence in the literature which clearly shows that the addition of polymer reduces the irritation caused by a surfactant [47,48]. Incorporation of hydrated cationic polymers in toilet bars [49,50] and cationic proteins in surfactant solutions [51] has also been reported to enhance the mildness of surfactant towards skin. Other mechanisms could also be contributing to the observed reduction in irritation [52]. The effect of the polymer also can be viewed as the polymer/protein providing an alternate substrate for surfactant binding and thus competing with the substrate for the surfactant molecules. Note, however, that for the polymer to have a positive effect, the c.a.c. in its presence should be less than the critical concentration required for the binding of the surfactant to the protein substrate.

From a practical point of view, obtaining information on surfactant monomer activity may not be a simple task. One approach is to use predictive models such as the one developed by Nikas and Blankschtein *et al.* [53] using a phenomenological-*cum*-thermodynamic model for calculating the activity of surfactant (SDS) monomers in the presence of different levels of a polymer like PEG. Typical results given in Figure 6 show the relative levels of polymer needed to maintain a certain level of monomer activity at different levels of SDS. Thus, in principle, it is possible to estimate the amount of polymer needed to maintain a particular surfactant monomer concentration. Interestingly, the activity vs. concentration plots reported by Nikas *et al.* [53] are, in fact, mirror images of the surface tension vs. log concentration plots, suggesting that from a practical point of view the surface tension plots may be a simple way to predict the activity profiles [54]. This approach may not, however, provide information in cases in which the polymer-surfactant complex itself is significantly surface active, as in the case of oppositely charged polymer/surfactant systems [55]. In such cases, surfactant binding data would allow the construction of such diagrams.

B. Surface Conditioning: Hair

The term "conditioning" is of great significance when applied to a variety of domestic cleaning procedures, *e.g.*, for bathroom and bedroom linen, wearing apparel, hair shampooing, body cleansing, and so on. In most of these, conditioning is virtually synonymous with the use of cationic adsorption agents. The reason for this is that many surfaces, including the natural fibers, hair and cotton, are predominantly negatively charged. Cleaning (washing) formulations normally employ anionic surfactants (less often nonionic surfactants) and, without a conditioning component, can leave the fiber containing mass (be it a bath towel, a mass of human hair, and so on) in a less than optimum state as a result of high electrostatic charge, high interfiber friction, or other less well specified properties of the cleaned substrate. Experience has taught that the "right" conditioning agent can alleviate this situation. Elaboration of mechanisms is not appropriate here: suffice it to say that cationic surfactants are predominantly used in textile conditioning and that cationic polyelectrolytes (and to a much lesser extent cationic surfactants) are used in hair conditioning, undoubtedly as a result of their high adsorption efficiency and the intrinsic properties of their adsorbed layers.

We return now to the main subject of the text, *i.e.*, polymer/surfactant interaction. Subjecting a substrate such as a hair fiber conditioned with an adsorbent layer of polycation to a relatively large amount of anionic surfactant (as in a normal shampooing operation) carries the consumer's expectation of removal of the conditioner and regeneration of the virgin surface. Technically, this presents a situation in which a polyelectrolyte strongly preadsorbed through electrostatic forces is exposed to a solution of a surfactant with which it also interacts strongly. Model

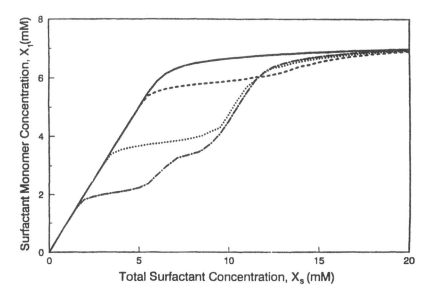

FIG. 6 Predicted surfactant concentration X_1, as a function of total surfactant concentration, X_s at 25°C, for aqueous solutions containing a model anionic surfactant such as SDS and a model nonionic polymer such as PEO. The solid lines corresponds to the pure surfactant solution in the absence of the polymer. From ref. 53 with permission.

experiments carried out on hair fibers with preadsorbed conditioning polymer (cationic cellulosic, radiolabelled, Polymer JR*) confirmed that there was substantial (~70%) removal of this polymer on exposure to 0.1 M SDS solution [56], presumably the result of the overwhelming extent of the micelle/water interface[57]. Similar results on stripping by a polymer/surfactant interaction mechanism were obtained using ESCA as the analytical tool [58].

An interesting, if more complicated, case comes in the current and wide use of "2 in 1" or conditioning shampoos. In these, the conditioning polymer, at a relatively low level (~1%), is incorporated directly into the shampoo formulation which contains about 10% (to as high as about 20%) of a surfactant, or blend of surfactants, as a rule predominantly of the anionic type. The last two decades have seen an explosive growth of this type of product and the patent literature is replete with product claims in this area. The commercial success of this category, whose scientific merit may not be immediately obvious, has abundantly demonstrated that practically speaking "2 in 1" shampoos are effective.

Much work has been done on model systems to investigate mechanisms. The most widely investigated systems have contained Polymer JR, the cationic cellulosic polymer referred to above. Model studies were carried out on the adsorption of

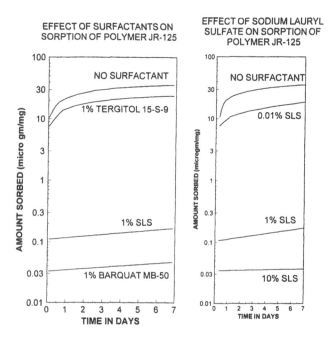

FIG. 7 Uptake of Polymer JR by hair in the presence of various surfactants. From ref. 59 with permission.

radio-tagged polymer in the presence of excess SDS on hair fibers. Under these conditions none of the polymers should exist in uncomplexed form in solution. Results in Figure 7 show that, whereas the uptake of polymer is considerably reduced in the presence of 1% SDS, it is still substantial [59]. Two "non-interacting" surfactants, on the other hand, affect the adsorption in very different ways. The nonionic surfactant Tergitol 15-S-9 (12-15C secondary alcohol with 9E0 units, Union Carbide), even at the high concentration of 10%, scarcely influences the adsorption. However, at 1% level the cationic surfactant Barquat MB50 (myristyldimethylbenzylammonium chloride, Lonza) reduces the adsorption of Polymer JR by a factor of close to one hundred. The interpretation in the latter case is that the surfactant itself is strongly adsorbed by the negatively charged fibers and, being of much lower molecular weight and of higher diffusion rate, preempts the negative adsorption sites on the fibers.

The mechanism of adsorption of a polycation from a solution containing an excess of interacting (anionic) surfactant has invited considerable speculation: with this excess of surfactant the overall charge of the "complex" will be negative [26] and hence there would be a net repulsion from the negatively charged fiber surface.

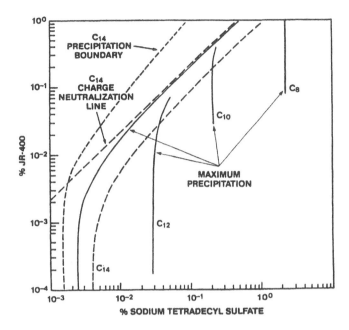

FIG. 8 Solubility diagram of the Polymer JR 400/sodium tetradecylsulfate system. Maximum precipitation lines for the corresponding C_{12}, C_{10}, and C_8 sulfates included. From ref. 60 with permission.

However, the polymer/surfactant complex will have a dynamic structure and, statistically, there is a finite chance of the disintegration of a polymer bound micelle in the vicinity of the fiber substrate so exposing a segment of the polycation chain to the latter. If adsorption of the segment takes place then, statistically, there is a finite chance of the next segment's adsorbing, and so on.

The above hypothesized mechanism of hair conditioning by cationic polymers in "2 in 1" formulations concerns the breakdown kinetics of the polymer/surfactant complex and the kinetics of adsorption of the "freed" polymer. There is, however, a very different and quite widely held hypothesis concerning the utility of "2 in 1" shampoos as (wet) conditioning agents. The mechanism concerns the dilution characteristics of the shampoo concentrate and can be understood by reference to the solubility diagram of the model system, Polymer JR400/SDS, shown in Figure 8 [60]. As reported and analyzed earlier, the line corresponding to maximum precipitation (stoichiometric charge equivalence) departs from a 45° slope of the log-log plot on sufficient dilution of the system. This means that a composition that was initially clear, containing say ten times as much surfactant as polymer, could on sufficient dilution yield a turbid system (solid line in Figure 8). The hypothesis of conditioning by this mechanism asserts that fine precipitate particles, which are intrinsically slippery, are formed and entrained when the washed hair is rinsed, so facilitating the subsequent wet combing process. Of course, the presence of polymer itself on the hair fibers after drying can contribute to conditioning by other postulated mechanisms, such as the reduction in static electrification during combing.

Finally, there is one other general way in which keratin surfaces exposed to polymer/surfactant combinations can be influenced. The deposited polymer or polymer/surfactant combination can favorably influence the receptivity of the surface to a second conditioning species. This effect is claimed, for example, in a patent issued to Colgate-Palmolive Co. [61] in which the presence of a cationic polyelectrolyte in an anionic surfactant shampoo apparently leads to more efficient deposition of a silicone oil conditioner. The mechanism (unspecified) probably concerns heteroflocculation of the silicone droplets onto the hair fibers modified by the adsorption of the first polymer (or complex). A patent on a similar topic has recently been issued to Procter and Gamble [62]. Use of cationic polymers for deposition of particulate materials dispersed in anionic surfactant systems has been the subject of seveal other patents [63-66]. In spite of this patent activity, mechanistic understanding of deposition from such surfactant systems is limited. Some fundamental studies on this general subject have, however, been reported by Berthiaume and Jachowicz [67]. Employing hair fibers pretreated with a cationic polymer (poly (methylacrylamidopropyltrimethyl) ammonium chloride) they examined the deposition of silicone oil from surfactant stabilized emulsions. In this case, the results were unpredictable inasmuch as deposition was higher from the emulsion stabilized with a cationic surfactant than with an anionic surfactant. A tentative explanation was offered in terms of competitive adsorption of the surfactant onto ionic sites on the keratin.

Precipitation of the polymer-surfactant complex is apparently not a requirement for polymers to enhance the deposition of other actives onto substrate such as hair. For example, Sime [65], using polymer-surfactant combinations in which the systems do not exhibit any precipitation tendency during dilution, has shown that the deposition of zinc pyridinethione can be enhanced markedly onto hair by using Jaguar C 13 S (Rhone Poulenc) as the cationic polymer in a lauryl ether sulfate containing shampoo systems. According to Sime, the low charge density of the polymers, in this case less than 0.0017 (charge density = number of charges per monomer unit divided by the molecular weight of the monomer unit), reduces the precipitation region and therefore does not influence the deposition. Specifically, Sime has shown that the extent of deposition depends on the type of surfactant as well as the relative concentration of the surfactant and the polymer. Thus, for a given polymer concentration, the deposition shows a maximum with increase in the surfactant concentration. Similarly, for a given surfactant concentration, the deposition shows a maximum as a function of the polymer concentration. The reduction in the deposition at high polymer concentration may be similar to the stabilization effect in classical bridging systems at high flocculant levels due to adsorption of the polymer at both surfaces, resulting in a repulsion between the surfaces, rather than bridging. The effect of surfactant concentration may be due to the following. At very low surfactant level, there is excess polymer to adsorb on the particles as well as on the surface, resulting in repulsion between the two. With increasing surfactant levels, the polymer is likely to be partitioned among the surface, the particle and the micelle, leading to a reduction in binding to both the particle and the surface. The consequence is a decrease in the repulsion between the particle and the surface. At very high surfactant levels, however, the polymer is predominantly adsorbed on the micelle surface and is not available for bridging.

C. Surface Conditioning: Skin

While the discussion above has been restricted to shampoo systems and the keratin substrate, *i.e.* hair, the reasoning can also be applied to the cleaning and conditioning of the other major keratin substrate, skin, by invoking similar mechanisms. Indeed, there are many parallels between the two and it is not uncommon these days to incorporate "conditioning" cationic polymers into liquid skin cleaning surfactant systems, such as "body shampoos." The patent literature in the personal care area shows significant activity in the use of polymers, especially cationic polymers, in skin cleansing formulations. It has been claimed that cationic polymers enhance skin feel and mildness [49,50], and provide enhanced moisturization. It has also been claimed that the polymers enhance the tactile properties of lather during rinsing [68]. The mechanisms of such action have not yet been established. It is, however, instructive to examine the role of polymers in cleansing systems and to speculate on some of the possible mechanisms of action.

First of all we should state that the practice of incorporating polymers in cleansing bars has been rather limited, one possible reason being the processing issues involved. However, with the rapid growth in liquid skin cleansers, use of polymers in such formulations has become a common practice, the incorporation of cationic polymers in shampoos having in many respects paved the way for this usage. Since cleansing formulations, in general, contain anionic surfactants interactions of a cationic polymer with the anionic surfactant can be expected, as stated above. In addition to anionic surfactants, these formulations may contain nonionic or amphoteric surfactants. We have referred to the fact that cationic polymer-anionic surfactant systems can exhibit rather complex phase behavior involving precipitation zones, gel formation zones, etc. [55]. We note again that the precipitation behavior can be modulated by choosing appropriate ratios of polymer to surfactant, and adding co-surfactants, electrolytes and even other polymers [1,27,28,55]. An understanding of the polymer-surfactant phase diagram, especially in the presence of other relevant additives, is essential for manipulating the processibility and enhancing the delivery of functional benefits from such formulations.

As mentioned earlier, cationic polymers can enhance the mildness of anionic surfactants towards skin. One of the factors in enhancing the mildness is related to the ability of the polymer to reduce the activity of the surfactant monomer and, in turn, to lower its binding to the corneum. It is not, however, possible to isolate this effect from other possible mechanisms involving the ability of the polymer to bind to the skin surface itself and the consequent effects on the skin surface.

The presence of the cationic polymer in a typical cleansing system can be expected to influence the lather characteristics of the system. While the polymer can stabilize and enhance the lather under non-precipitation conditions by increasing the surface viscosity in the liquid film lamellae, it can reduce/destroy the lather under precipitation conditions by lowering the levels of surfactant available in the solution phase. This will be discussed in section IX.

Yet another function of the polymer in a cleanser system, as noted above, is to enhance the deposition of agents such as emollient oils, feel enhancers such as silicone oils and other skin benefit agents [61-66]. One mechanism is by classical polymer bridging type, even though the dynamics needed to promote such deposition may be rather involved.

Direct evidence of the adsorption of cationic polymers on skin from an anionic surfactant system has been reported by Goddard and Leung [69]. The authors determined the uptake of Polymer JR 125, on stratum corneum in the presence of three anionic surfactants (SDS, alpha olefin sulfonate (AOS), and alkyl 2EO sulfate) and a coco-betaine surfactant (see Figure 9). Even though this experiment exposed the substrate for 30 minutes, the trends observed are indicative of those expected under normal use conditions. At low salt concentration, uptake of the polymer decreases markedly with increase in surfactant concentration. Interestingly, at surfactant concentrations above about 10%, the polymer uptake increases and at 20% the extent of uptake is higher than that obtained in the absence of the surfactant. Note that the surfactant concentrations are similar to the levels present in liquid cleansers and shampoos. Essentially, these results show that despite the presence of a high concentration of anionic surfactant, cationic polymers can still bind to skin. Reduction in polymer uptake with increasing anionic surfactant concentration can be attributed to competition for the polymer between the micelles and the substrate. The observation that alkyl EO sulfates depress uptake more than SDS is also interesting. The reasons for the increase in polymer uptake at extremely high levels of the anionic surfactant, on the other hand, are not clear at present. A possible explanation is that at such high surfactant levels, because of the corresponding high ionic strength of the contacting solution, the interaction of the polymer with the surfactant micelle is likely to be inhibited to some extent. Also, the high ionic strength may cause some salting out of the polymer. Both of these effects

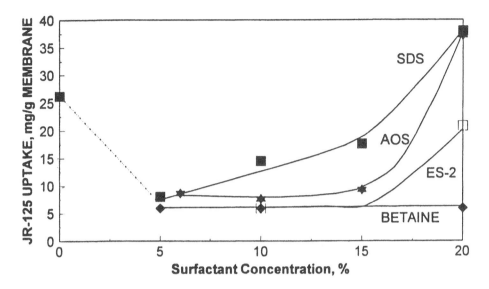

FIG. 9 Uptake of Polymer JR 125 (1%) by stratum corneum from surfactant solutions, Exposure 30 minutes, room temperature. Significant polymer binding evident even at 20% anionic surfactant.

should enhance the binding of the polymer to the substrate. From a practical point of view, the fact that it is possible for a cationic polymer to adsorb from an anionic surfactant solution onto a negatively charged substrate is important for surface conditioning as well as for enhancing the deposition of other materials on the surface.

VII. POLYMER-SURFACTANT INTERACTION AT THE MINERAL-SOLUTION INTERFACE

In contrast to personal care systems, much information concerning detailed adsorption and competitive adsorption effects exists for mineral/solid surfaces. Since both polymer and surfactant can adsorb, there has been much interest and considerable work done to determine what effect each has on the adsorption of the other. Most of the work done has involved mineral or latex solids. The effects are both cooperative and competitive, depending upon the substrates, pH, concentration of the polymer and the surfactant, and so on. The situation is similar to that encountered in the personal care and detergent systems in that polymer conditioning results from deposition to a solid surface. Reported cases consist mainly of the polymer activating or enhancing the adsorption of the surfactant or the surfactant promoting or enhancing the adsorption of the polymer.

Interaction of a nonionic polymer, such as PEO or PVP, with an anionic surfactant, such as SDS, has been used to modify the properties of surfaces such as alumina [70], silica [71], iron oxide [72, 73] and coal [74]. For example, in the case of coal, SDS adsorbs with its tail towards the hydrophobic surface and the hydrophilic head groups towards the solution. In this case, PEO interacts with the headgroup region of adsorbed SDS molecules and, in adsorbing on such a coal surface, provides a steric hydrophilic layer that enhances the wettability of the coal [74]. An opposite situation exists in the case of silica where the SDS molecules do not normally adsorb because of the negative surface charge under normal pH conditions. However, PEO has a tendency to adsorb on silica. The work of Maltesh et al. [71] shows that in the presence of PEO, SDS adsorbs on silica. The extent of SDS adsorption was higher when the polymer was preadsorbed on the surface rather than when it was adsorbed from a mixture of surfactant and the polymer. In another study, Cosgrove et al.[75] have examined the conformation of the adsorbed PEO on a silica surface in the presence of SDS. Results of their photon correlation study reproduced in Figure 10 show that the hydrodynamic thickness of adsorbed polymer exhibits a minimum with respect to surfactant concentration. The authors have attributed the decrease in binding with increase in SDS level to complexation of the polymer with SDS and its consequent charging which leads to increased repulsion between the polymer and the substrate. In fact, around the surfactant c.m.c., almost all the polymer appears to have desorbed from the surface. Interestingly, with further increase in the SDS levels, the hydrodynamic thickness appears to increase suggesting that polymer with some bound micelles adsorbs at the interface in an extended form. To support the latter conclusion, simultaneous measurements of adsorption of surfactant and polymer are necessary. Whereas SDS by itself does not adsorb on silica, it adsorbs strongly in the presence of PEO. In this case, the polymer has promoted the binding of the surfactant to the solid.

In general, the extent of interaction of the polymer is a function of the type of surfactant. For example, Esumi et al. [70] have investigated the adsorption of PVP on alumina in the presence of single or double chain anionic surfactants. The authors found that the adsorption of PVP in the presence of Aerosol OT (sodium bis (2-ethylhexyl) sulfosuccinate) was significantly greater than that in the presence of Li-dodecyl sulfate. These results are consistent with the extent of interaction of the polymer with the individual surfactants.

Surfactants can modify the surface properties, especially the wettability, of minerals to separate them from complex ores containing several minerals. In such cases, polymers are often used to modify the interaction of surfactants with the solid surface. For example, Somasundaran and co-workers [76,77] have shown that cationic polymers can be used to enhance the adsorption of anionic surfactants on silica/quartz surfaces and, in the process, enhance the hydrophobicity of the silica-

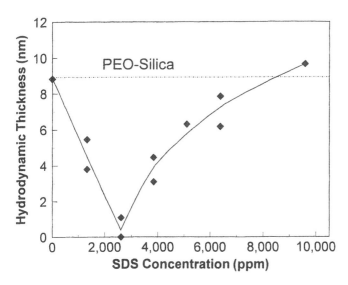

FIG. 10 Hydrodynamic thickness of PEO adsorbed onto silica *vs.* SDS concentration in the plateau region of the PEO adsorption isotherm. From ref. 75 with permission.

water interface. Cationic polymers at low levels, on the other hand, do not reduce the adsorption of cationic surfactant on the surface, but do alter the surface hydrophobicity [77]. In the absence of the polymer, cationic surfactant adsorption at low levels makes the surface hydrophobic, but in the presence of the polymer co-adsorption takes place, with the adsorbed polymer layer determining the wettability of the exposed solid-liquid interface (see Figure 11 for a schematic of the possible mechanism). At high polymer concentration, however, the adsorption of cationic surfactant adsorption can be inhibited.

In another study, Shubin [78] has examined the adsorption of hydrophobically modified cationic HEC (Quatrisoft LM 200) on silica in the presence of SDS. The results obtained show that at a given polymer level, low levels of SDS addition increases the binding of the polymer (see Figure 12). At higher levels, however, the binding of the polymer is reduced markedly. The amount of polymer in the absence of the surfactant is sufficient to impart a positive charge of 18 mV to the silica surface. According to the author this positive charge, in effect, limits continued binding of the polymer. At low added levels of SDS, the net charge on the surface is reduced by the binding of the SDS to the polymer's cationic charges and this results in a much higher level of polymer adsorption. On the other hand, at higher surfactant levels competition between micelles and the surface for the polymer determines the net adsorption and this explains the observed, significant reduction in

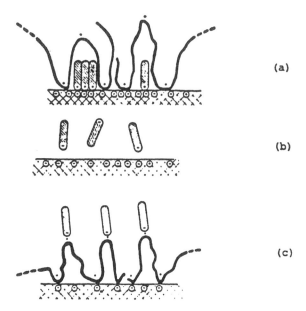

FIG. 11 (a) Schematic depiction of the cationic polymer PAMA and dodecylamine co-adsorption on quartz particles resulting in their flotation (hydrophobicity) depression. (b) Schematic depiction of quartz dodecylsulfonate system. PAMA and dodecylsulfonate co-adsorption on quartz particles resulting in their flotation (hydrophobicity) activation. From ref. 77, *Colloids Surf. 13*, 73 (1985) with permission from Elsevier Science-NL, Sara Burgerhartstraat 25, 1055 KV Amsterdam, the Netherlands.

the binding of the cationic polymer on the silica surface. Clearly the amount of cationic polymer adsorbed from a concentrated surfactant onto a negatively charged surface is limited because of the interaction in solution of the polymer with the surfactant, as was evident in the simulated shampoo systems discussed earlier.

In a recent study, Magdassi *et al.* [79] have shown that polymer-surfactant interactions at the solid-liquid interface can be effectively used to flocculate montmorillonite suspensions. In this case, montmorillonite was treated initially with a cationic polymer and subsequently exposed to SDS solutions. Optimal flocculation was obtained when the molar ratio of the cationic charge to SDS was 1:1. The mechanism of flocculation proposed by the authors involves hydrophobic interactions among SDS molecules bound to the cationic sites of the polymer that is already attached to the clay surface.

In addition to the absolute amount of each solute adsorbed, a potentially important aspect of mixed polymer/surfactant adsorption at the solid/water interface is the configuration of the adsorbed polymer. Working with the hydrophilic and

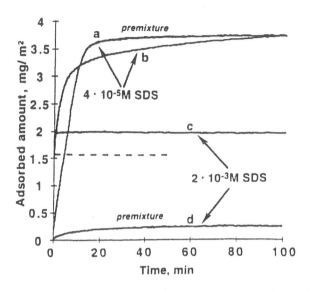

FIG. 12 Total amount of material adsorbed onto the SiO₂ surface from 34 ppm Quatrisoft (hydrophobe modified cationic HEC, Union Carbide Corp.) solutions at two different SDS concentrations: 4×10^{-5} M (curves a and b) and 2×10^{-3} M (curves c and d). The surfactant was introduced into the system at t = 0 in two different ways: after completion of polymer adsorption (b,c) and simultaneously with the polymer (a,d). The level of Quatrisoft adsorption from surfactant-free solution is shown by a dashed line. From ref. 78 with permission.

hydrophobic suspensions of silica, Otsuka *et al.* [80] found that the adsorption of PVP is enhanced at lower levels of added surfactant (lithium dodecyl sulfate and lithium perfluorooctanesulfonate). A notable finding, from ESR and photon correlation spectroscopy measurements on the silica suspensions, is that this trend is accompanied by a decrease in thickness of the adsorbed polymer layer which is interpreted as a "remarkable" change to a conformation of "trains" from one of "loops and tails," in particular on the hydrophobic silica surface.

Within the context of this section dealing with mineral/water interfaces it is appropriate to conclude with citations of recent work on some related interfaces. The first concerns suspensions of graphon in water: Gabrielli *et al.* [81] studying the adsorption of polyvinylpyrrolidone and of sodium dodecyl sulfate in these systems found improved stability with combinations of these species over that observed when either was used alone. The second system - a novel one - concerns a polystyrene latex with grafts of PNIPAM on the particles [82]. The influence of temperature on the dimensions of the grafted polymer was determined by dynamic light scattering -- all as a function of added SDS concentration. Without surfactant, and at temperatures providing better than theta conditions, the polymer grafts exist as coils but these start to collapse as the theta temperature is approached. An interesting finding is that, starting around the c.a.c. of SDS [83] and above, added SDS reduced and eventually eliminated, this collapse; also, transition to the globular form of the polymer which occurs at the theta point became much sharper in the presence of the surfactant.

Because of the biological importance of proteins at interfaces many model studies on their interaction (and removal) with surfactants have been reported [84]. A rececnt example concerns the removal of T4 lysozyme from adsorbed layers on the surface of silica by SDS [85]. Adsorption (and removal) was monitored by ellipsometry. It was found that removal of the protein, a process clearly involving a polymer-surfactant interaction, could be effected at concentrations well below the c.m.c. of the surfactant.

VIII. DETERGENCY

Removal of proteins from interfaces by surfactants represents one of the fundamental aspects of the detergency process. For example, the cleaning of cooking and tableware generally involves the removal of food and milk proteins - a very practical case of polymer/surfactant interaction. A similar situation exists in the washing of clothes where the removal of a range of proteins, e.g. food and blood stains, is often involved. Consequently, it has long been a standard practice to include proteins in model soils evaluating the efficiency of detergent systems [86].

There is, however, another aspect of fabric cleaning which involves the addition of polymers to aid the overall detergency process: selected water soluble polymers are used in laundry detergent compositions, generally at a level of 1%. They are employed as "antiredeposition agents", i.e., their function is to prevent particles of dirt from redepositing onto the cleaned fabric during the ensuing detergency cycle. Their action involves adsorption onto the fabric and also the particles, and the polymer operates by minimizing hetero- and homo-flocculation in the particle/fabric system. Interesting evidence of the practical importance of polymer/surfactant interaction came with the introduction of polyester wearing apparel in the 1950s. The traditional anti-redeposition agent for cotton, viz., sodium carboxymethyl cellulose (CMC), was found to be ineffective for polyester fabrics. In this case, one of the most effective agents was methyl cellulose, but its activity was far greater in nonionic surfactant based detergent formulations: in anionic surfactant systems it had limited activity. An obvious explanation is that, whereas methylcellulose is not reactive towards nonionic surfactants, it reacts strongly with anionic surfactants, and one can conclude that the adsorption characteristics of the polymer are seriously altered as a result of complex formulation. Being negatively charged, CMC is relatively unaffected by anionic surfactants (and also by nonionic surfactants). On the other hand, some interaction with the former may be expected in the presence of Ca^{++} or Mg^{++} to give a bridged polymer/surfactant complex.

In a recent study, Jaeger [87] investigated the possible interactions of surfactants and polymers in laundry detergents. The systems investigated include surfactants such as alkylbenzenesulfonates and C13 ethoxylated alcohols, and polymers such as polycarboxylates, acrylate-maleate copolymers, polyvinylpyrrolidone and modified PEGs. Specifically, the effect of polymer on soil removal by the surfactant and the effect of surfactant on antiredeposition action of the polymers were tested. The essential conclusion was that there was no synergy between polymers and surfactants with respect to soil removal or antiredeposition action. However, unlike the soil removal, the antiredeposition was influenced by the specific combination of surfactant and polymer.

A recent trend in the liquid detergents area has been the move towards highly concentrated formulations. Viscosity control of such formulations can be a difficult task. As discussed in the earlier section on rheology, hydrophobically modified polymers can act as "decoupling" agents to reduce the viscosity of concentrated suspensions by preventing the flocculation of lamellar dispersions and this represents a significant progress in this area (see also reference 15).

The development of polymers beneficial to the detergency process continues to be an active area of research on the part of both detergent supply and marketing companies. For example, a recent article [88] refers to the development by Rhone-Poulenc, Inc. of an "anti-encrustation" terpolymer for use in zeolite/calcium carbonate based detergents which effectively prevents the entrapment of calcium carbonate crystals in the fabric during the washing process. For this and other types of soil release polymers, it is clear that possible interaction of the polymer and the surfactant(s) present in the formulation must be carefully assessed. As stated above, a mitigating factor would exist if the added polymer were a polyelectrolyte and there were charge/charge repulsion between it and the surfactant(s) present.

IX. FOAMS

Although scientific interest in foams is quite old, it can be said that the "modern" era of research on foams was ushered in about a half-century ago by the advent of highly purified, single component surfactants, specifically sodium dodecyl sulfate (SDS). Both foam stability and single lamella drainage studies of SDS solutions showed unusual sensitivity to a "third component," subsequently identified as parent dodecanol: as little as 1% dodecanol in the SDS specimen sufficed to completely change (improve) the stability of foam and to alter the drainage kinetics of single lamellae, often changing the latter from "fast draining" to "slow draining." Accompanying studies showed that this small amount of dodecanol could, by mixed film formation, dramatically decrease the surface tension and increase the surface viscosity of SDS solutions [89,90,91,92,93]. A current assessment of this field is that, while work of great scientific elegance on single films and model foams has been carried out, it has not yet led to a general theory of foaming but has emphatically established the potential importance of "third components" in the single film and foaming behavior of surfactant solutions. In the case of fatty acid soaps the third component is recognized to be free fatty acid.

History may now be repeating itself in the sense that a "new" type of third component is becoming recognized which can influence the filming and foaming behavior of surfactant solutions. In this case, instead of a monomeric species of high surface activity such as dodecanol, we are speaking of water soluble polymers, usually of moderate or low intrinsic surface activity. The role of water soluble polymers in improving the stability of foams has traditionally been viewed as a bulk phase thickening phenomenon to reduce the rate of film drainage unless, of course, the polymer by itself is appreciably surface active [94]. In the case of mixed polymer and surfactant systems, while interactions in bulk will have to be considered, it is clear that a full understanding of foaming behavior will also have to take into account the properties of mixed films which may be present at the air/water interface. Here, we will be concerned with selected studies of surfactant/polymer pairs in which the polymer is either charged, *i.e.*, a polyion, or carries no charge. A short section is included on polymers which are themselves surface active. In general, although restricted to model systems, it is hoped that the treatment will help to encourage wider investigation of actual foams.

Polyion/surfactant pairs.

Our starting point concerns the polycation, Polymer JR 400 (Amerchol Corp.), whose surface and bulk properties in mixed aqueous solution with various surfactants were first reported about twenty years ago [95, 26]. A notable point is that this polymer itself is very weakly surface active at the air/water interface. On the other hand, in the presence of an anionic surfactant like SDS, marked synergistic lowering of surface tension was observed. Obviously, strong electrostatic interaction is involved since this polycation showed no interaction with either a nonionic surfactant or a betaine surfactant of net zero charge and no interaction with cationic surfactants can be confidently assumed. Nor did the uncharged parent polymer, hydroxyethylcellulose, show any interaction with SDS. These results are illustrated in Figure 13. The strong synergistic lowering of surface tension observed in the Polymer JR/SDS system led the authors of this work to propose the existence of a surface complex (or a series of surface complexes) comparable to those which form in bulk phase (see phase map, Figure 14). It was also proposed that beyond the region of precipitation in bulk, *i.e.*, in the resolubilization zone, the growing micellar interface competed with the air/water surface for polymer and eventually denuded it of the latter.

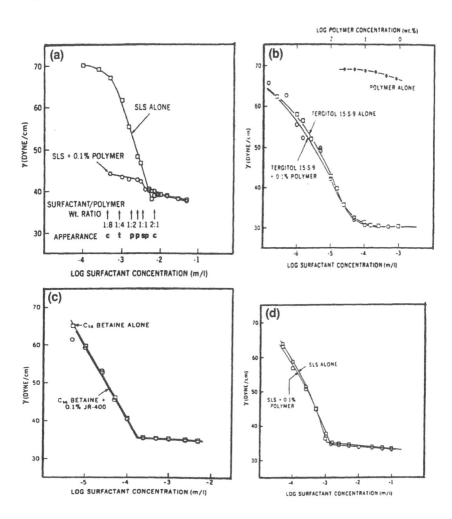

FIG. 13 Surface tension, concentration curves for various aqueous solutions at 25°C. (a) Na Lauryl sulfate with and without 0.1% Polymer JR; (b) Tergitol 15-S-9, an ethoxylated secondary alcohol with and without 0.1% Polymer JR; (c) tetradecylbetaine with and without 0.1% Polymer JR; and (d) Na Lauryl sulfate with and without 0.1% Cellosize (hydroxyethylcellulose).

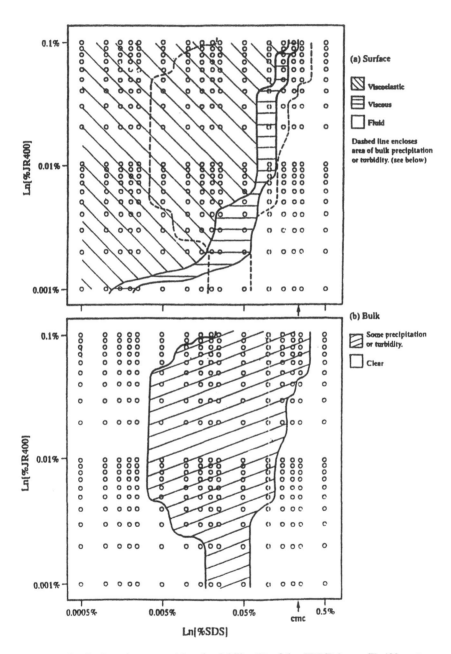

FIG. 14 a,b Surface-phase map (a) and solubility (b) of the SDS/Polymer JR 400 system. From ref. 57 with permission.

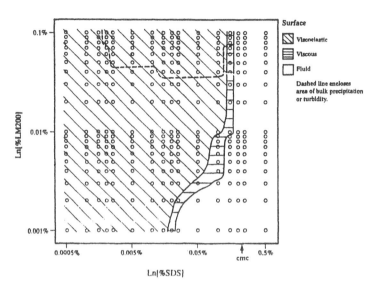

FIG. 14 c Surface-phase map and solubility of the SDS/Quatrisoft LM 200 system. Open circles represent the observation points. From ref. 57 with permission.

A schematic representation of various bulk and interfacial interactions in cationic polymer/hydrophobically modified cationic polymer-anionic surfactant systems is given in Figure 15. Further evidence of formation of surface complexes came from insoluble monolayer studies on sodium docosyl sulfate where the introduction of a minute amount (10 ppm) of Polymer JR into the substrate led to a substantial increase in surface pressure and the development of viscoelasticity in the monolayer.

Both bulk phase viscosity [96] and, especially, surface phase viscosity effects would be expected to influence the foaming properties of these mixed systems. Surface "phase mapping" of the Polymer JR 400/SDS system, using a qualitative test of surface viscoelasticity has recently been reported [56]. A considerable portion of the diagram (Figure 14) reveals the existence of surface viscoelasticity, suggesting that even the smallest level of polymer could affect the foaming properties of SDS. Only cursory testing of the foaming properties of these systems was reported [26], but was sufficient to indicate definite enhancement effects. Very recent work [97] has shown that surface viscoelasticity and enhanced foaming effects are also obtained in Polymer JR combinations with the lower chain length homologs (C_8 and C_{10}) of SDS, leading to foaming performance superior to the latter system.

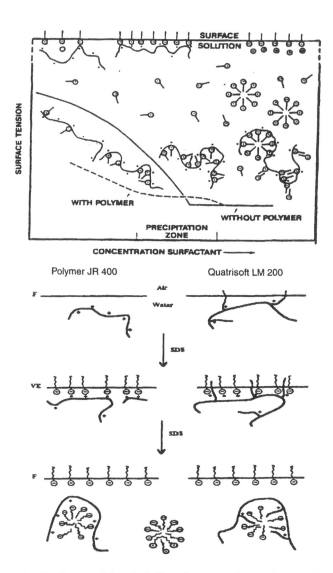

FIG. 15 Top: conditions in bulk and surface of a solution containing a polycation (fixed concentration) and anionic surfactant. Full line: hypothetical surface tension/concentration curve of the surfactant alone; dotted line: mixture with polycation. Simple counter cations are depicted only in the surface zone. Bottom: schematic representation of the polycation/anionic surfactant interaction as a function of surfactant concentration in the bulk and at the air/water interface, in the cases where the polycation is Polymer JR 400 (left) and Quatrisoft LM 200 (right), F, fluid; VE, viscoelastic.

Data showing synergistic lowering of surface tension (g) similar to that in Figure 13a were reported by Buckingham et al. [98] for the polycation, poly-L-lysine, SDS system. These authors derived the following form of the Gibbs adsorption equation for their system

$$\delta\gamma = -RT\Gamma\delta\ln[SDS]$$

an implication being that the surface excess, Γ, of the DS anion, though highly influenced by the presence of polycation, is rather insensitive to its actual concentration. This equation (or, more correctly, its equivalent form) has recently been re-derived by Asnacios et al. [99] for the system copolymer acrylamide-acrylamidesulfonate/dodecyltrimethylammoniumbromide (AM-AMPS/DTA$^+$Br$^-$). Points of interest from this, and an accompanying, paper [100] are as follows. The high area per molecule ($\sim$80Å^2/mol.) for DTA$^+$ in the presence of the polymer at surfactant levels just prior to, or at, the c.a.c. and c.m.c. compared to the value of about 45Å^2/mol. for DTA$^+$Br$^-$ alone, suggests complex formation between the polymer and the surfactant at the interface. Evidence for heterogeneities in the films was also deduced from thin film disjoining pressure curves, when the surfactant was at or above the c.a.c. Considerable stability of the film under these conditions has clear implications for foaming. Clear evidence of stratification in the film when the DTAB concentration was well below the c.a.c was also noted.

Mixed polyion/surfactant films appear to be ripe for further investigation. Parameters such as polyion molecular weight, structure and charge density are obvious candidates for investigation, as well as variation in the surfactants themselves. It is to be hoped that such studies will include properties of actual foams.

Uncharged polymer/surfactant pairs.

The early work (mid 1950s to early 1970s) on polymer/surfactant interaction involved a variety of uncharged water soluble polymers, e.g., polyethyleneoxide (PEO), polyvinylpyrrolidone (PVP), methylcelullose (MeC), polyvinylalcohol/acetate (PVOH/Ac) and polypropyleneoxide (PPO). In their review of 1973, Breuer and Robb [101] pointed out that a reactivity index could be assigned to the polymer: with anionic surfactants the "reactivity" followed the sequence PVOH<PEO<MeC<PVAc<PPO~PVP; and with cationic surfactants the sequence was PVP<PEO<PVOH<MeC<PVAc<PPO. Reactivity seemed to increase with increasing hydrophobic nature of the polymer. The inversion of the position of PVP in the two sequences was explained in terms of the slight effective positive charge carried by the pyrrolidone group. Subsequently, Goddard and Ananthapadmanabhan [102] noted that the sequence also seemed to be in accord with the surface activity of the polymers, i.e., their ability to lower the surface

tension of water. Thus, polymers like hydroxyethylcellulose (HEC) and polyacrylamide (PAAm), which lead to negligible or very small lowering of water surface tension, have been found to be unreactive to surfactants in solution while ethylhydroxy-ethylcellulose and polyisopropylacrylamide, which both have appreciably surface activity, are reactive. An important implication is that those polymers which are reactive form mixed films at the surface of water when mixed with surfactants. Note that even though Jones [103], the pioneer of the surface tension method to assess interaction between surfactant and uncharged polymer in aqueous solution, utilized such measurements as an indicator of interaction in solution, this author does not seem to have drawn any inferences of mixed film formation at the air/water interface or its implications at least in his early work.

The case of adsorption in a mixed surfactant and uncharged polymer system has been analyzed thermodynamically by de Gennes [104]. If the polymer adsorbs one can expect a change in adsorption of the surfactant. This effect has been demonstrated experimentally for the system SDS/PVP by use of specular neutron reflection [105] and by a radiotracer method [106]. De Gennes draws attention to the consequences of co-adsorption of surfactant and polymer on colloid stability in general and on the drainage of "soap" films, in particular: drainage rate should be decreased a) when the layers of adsorbed polymer overlap and b) also by increased viscosity of the surface. The consequences anticipated in both cases would be more stable films and more stable foams generated from these mixed systems.

To date, testing of these ideas has been rather limited. For example, Cohen-Addad *et al.* [107,108] carried out film drainage tests on a series of polymer/surfactant systems and found rather small specific effects of the added polymer, even in the case of the "strongly interacting" pair, PEO/SDS. Black films were ultimately formed in all cases. It should be noted that all concentrations of SDS were well above the cmc. Other work, using X-ray reflectivity, led this group to conclude that there is a layer of PEO adsorbed under the SDS polar heads in the black films [109], at least when the PEO was of high molecular weight.

As with the polyelectrolyte category discussed above we can expect much more work on film formation on these mixed systems which, hopefully, will include work on actual foams.

Surface active (hydrophobically substituted) polymer/surfactant pairs.

This is the final category considered: when the polymer is itself markedly surface active the situation becomes very different; in many cases the polymer alone will be able to sustain a foam generated from its aqueous solution. An example of this is the hydrophobically substituted cationic cellulosic polymer, Quatrisoft LM200 (Amerchol, Inc.) [110]. There are many other examples in the literature, including proteins themselves and their derivatives. On addition of a surfactant, mixed adsorbed films will form and the film and foaming characteristics will depend very much on the specifics of the components themselves and the nature of their interaction.

Some implications regarding real systems

Developers of domestic and industrial foaming products are well aware that there are a number of properties by which foams and foaming are characterized by consumers. These include:

a. Ease of generation (foamability")
b. Stability (rate of draining, integrity of bubbles)
c. Richness (bubble size, liquid content, foam viscosity)
d. Openness (bubble size, foam viscosity)
e. Quality (combination of b. to d. above)
f. Resistance to foam breakers (fats, oils, solid particles)
g. Rinsability (destruction on dilution)
h. Pumpability

To achieve these characteristics, as desired, developers call upon a body of practical information, which includes lists of "lathering additives" or "foam boosters" -- the "third component" referred to above. Little use is made of the scientific research which has been carried out on thin films. This situation still represents a challenge and an opportunity to basic researchers in this field [111].

X.   HYDROPHOBE MODIFIED POLYMERS (POLYMERIC SURFACTANTS)

Hydrophobe modified polymers (HB-P) constitute a relatively new and versatile group of materials which are receiving much attention since they combine polymer and surfactant properties. It is known also that they are generally reactive towards surfactants in solution and at interfaces. One can find many recent scientific publications on this subject (see also chapters 3, 7, 9 and 10 of this book). For example, Danino *et al.* [112] report an extensive study, using several techniques, of the polyamphile poly(disodium maleate-co-alkylvinylether) and its mixtures with anionic or nonionic surfactants which have a disruptive effect on the association on the association of this polymer. Many properties of hydrophobe modified polymers, *e.g.* thickening, gelling, foaming, emulsifying, etc., can be changed by the addition of surfactants. In particular, the first two properties (thickening and gelling) have received much attention recently [1,9,112]. A depiction of association structures of hydrophobically modified polymers is given in Figure 16.

There are many examples of the use of hydrophobe modified polymers in commercial products, *e.g.* paints, emollients, cleansers, and so on. In view of what has been said above, it is clear that several of their properties such as thickening,

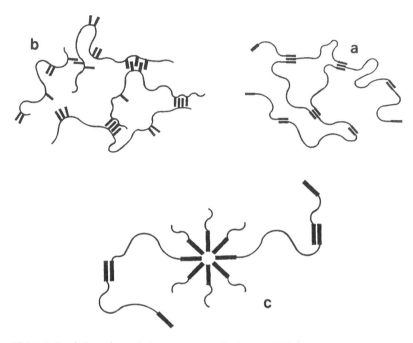

FIG. 16 Depiction of association structures of hydrophobically substituted polymers: (a) end-substituted, (b) "comb" (c) end-substituted polymer associating via a surfactant micellar bridge. Reproduced from E.D.Goddard, ref. 1, with permission.

emulsifying, foaming, wetting etc., can be fine tuned by added surfactants. One can predict that this area will continue to be profitable for basic research.

Goddard (ref. 1, Chapter 10) has presented a brief review (up to 1992) of the subject of association in solution of hydrophobe modified polymers and surfactants. A polymer type currently receiving much attention in this respect is referred to as HEUR - hydrophobically ethoxylated urethane, the reaction product of a PEG and a diisocyanate, end-capped with a long chain alcohol or amine. HEURs have much potential as thickeners in coating formulations. Their structure suggests they will show pronounced interaction with surfactants in solution and this is indeed the case. Here we refer to two recent studies. The first, by Hulden [113] included surface tension studies of the polymer with both anionic (SDS) and nonionic (NPE$_{10}$) added surfactants. Clear indications of critical association concentrations ("T_1" and "T_2") in the former case, and "T_2" in the latter case, were found. Furthermore, characteristic viscosity increases (actually maxima) when surfactants were added to solutions of HEUR polymer (*i.e.* boosting of thickening) were generally observed, a clear indication of association in solution. The second study by Annable *et al.* [114]

essentially confirms the interaction patterns and presents an in-depth mechanical analysis of the flow properties of these systems. One additional finding was that, at higher polymer concentrations, additions of surfactant no longer lead to rheological maxima.

The high level of current interest in mixed hydrophobe modified polymer/surfactant systems is illustrated by the fact that a score of papers from well known domestic and international schools were presented at a recent symposium (ACS, Spring Meeting 1997, Surface & Colloid Division) on polymer/surfactant interaction. The symposium, organized by P.L. Dubin, while including many studies on phase and rheological behavior, also described the use of such varied techniques as microcalorimetry, fluorescence and light scattering, and included investigations of the interaction of hydrophobe modified polymers with liposomes.

A. Polymeric Surfactants as Emulsifiers

As stated above, polysoaps have great potential utility as emulsion and suspension stabilizers and also as surface conditioners. This arises from their combined functionality as surfactant and polymer. Some of the recently introduced polymeric emulsifiers are multifunctional in the sense that they have more than one type of pendant functionality allowing the backbone to remain at the interface with the pendant groups in the two bulk phases to be compatibilized. As evident from the following examples, these structures provide significantly enhanced stability to the emulsions compared to the conventional bifunctional emulsifiers.

Hameyer and Jenni [115], in an elegant study have recently shown that multifunctional polymeric surfactants, cetyldimethicone-co-polyol and polymethyl acrylate polyalkyl polyether copolymer, can effectively stabilize W/O emulsions. These comb-like polymer have molecular weights around 7000 and HLB values lower than 6. According to the authors, primary emulsions based on multifunctional polymeric emulsifiers can be easily transformed to ternary systems by simple stirring in the presence of monofunctional ionogenic or nonionic surfactants with high HLB values (>30). The cetyldimethicone-c-opolyol has three functional groups, the cetyl group which can partition into the oil phase, polyol groups which are highly hydrophilic, and the silicone group that is relatively incompatible with both the aqueous phase and hydrocarbon oil phase. Thus, it is possible that the silicone group essentially remains at the interface with the hydrocarbon group in the oil phase and the polyol group in the aqueous phase. Similarly, the polymethyl acrylate polyalkyl polyether also has three functional groups. Again, the backbone of the polymer is likely to stay at the interface with the alkyl group in the oil phase and the hydrophilic polyether group in the aqueous phase. Such a multifunctional polymer, from an energetics point of view, is likely to prefer the oil-water interface rather than either of the bulk phases and therefore should enhance the stability of O/W/O emulsions.

Multiple emulsions of the inverse type (W/O/W) emulsions are finding application as carriers of sensitive ingredients or as controlled delivery aids. For example, isolation of two water soluble, but reactive ingredients, such as an enzyme and a denaturant, could be achieved by entrapping the former in the inner water phase and the latter in the outer aqueous phase. They can also be used for controlling the rate of release of actives from the inner phase. Yet another method of achieving such controlled delivery systems is by using highly architectured polymers often referred to as "dendrimers," described below.

XI. MISCELLANEOUS SYSTEMS AND APPLICATIONS

Several novel polymer/surfactant systems have been reported in the recent literature and these may find interesting applications in the coming years. Some of the interesting examples are briefly reviewed below.

A. Dendrimer-Surfactant Complexes

Dendrimers are a relatively novel class of macromolecules synthesized from various initiator cores to which radially branched layers, referred to as "generations," are covalently attached [116]. The exterior layer of the dendrimer can be modified to provide terminal functionalities. Tomalia and co-workers have pioneered the work on the polyamidoamine dendrimers, often referred to as "star burst dendrimers" [117]. Because of the nature of the synthesis of these compounds, their composition and constitution, including the molecular weight and number of terminal groups, are well defined. The synthesis of starburst dendrimers leads to either carboxylate groups (half-generation) or amine groups (full generation) as terminal functionalities. The molecular weight of dendrimers varies from 924 (half-generation, diameter about 24 A) to as high as 639,247 for the 9.5 generation (diameter about 140 A). The number of terminal groups for the half generation is only 9 whereas that for the 9.5 generation is 3072. Interestingly, the lowest molecular weight dendrimer is smaller than a surfactant micelle, but the higher generation dendrimers can be significantly larger than a conventional spherical micelle.

Using molecular probes and photophysical measurements, Turro et al. [118,119] have shown that the interior of the starburst dendrimer is hydrophilic. They have also shown that dendrimers and surfactant micelles exhibit similar abilities to mediate electron-transfer processes conducted on their surface. Interaction of dendrimers with oppositely charged surfactants has also been examined by Turro et al. [119]. As can be expected, surfactants interact strongly with dendrimers, forming inverse micelle-like structures with a hydrophilic interior and a hydrophobic exterior. Ottaviani et al. [120] have proposed various types of

1) **Micelles (pure or mixed)**
   in equilibrium
   with surfactant monomers

Spin Probe

2) **Low concentration SBDs interacting**
   with surfactant aggregates:

(a) Later generation SBDs

(b) Earlier generation SBDs

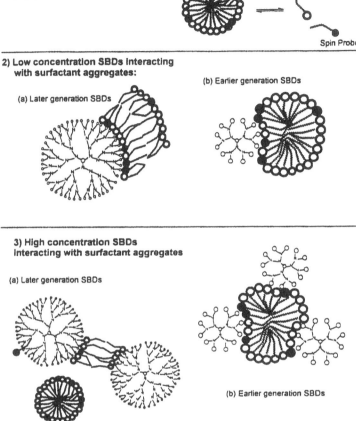

3) **High concentration SBDs**
   interacting with surfactant aggregates

(a) Later generation SBDs

(b) Earlier generation SBDs

FIG. 17 Scheme, in a bidimensional projection, of the suggested supramolecular structures which formed by adding increasing concentrations of SBD-COO$^-$ (Starburst dendrimers - early generations, G<4, left side) to mixed micellar solutions of CAT 16 (cationic ESR probe) and CTAC (cetyltrimethylammonium bromide). From ref. 120, *Colloids. Surf. A 115*, 9 (1996), with permission from Elsevier Science-NL, Sara Burgerhartstraat 25, 1055 KV Amsterdam, the Netherlands.

supramolecular structures as a function of dendrimer size (generation) and the concentration of both the dendrimer and the surfactant (see Figure 17). The potential utility of these complexes has not been fully exploited and is clearly an area for further research.

B.   Amphiphilic Block Copolymers

Amphiphilic block copolymers with relatively large size hydrophobic groups compared to the hydrophilic blocks, and thus somewhat similar to the conventional surfactants, have been reported recenlty [121,122]. These are somewhat similar to the ethoxylated nonionic surfactants or the EO-PO copolymer surfactants, but often differ in terms of the relative size of their hydrophobic and hydrophilic moieties. Their di-block structure makes them "polymeric surfactants" as opposed to "hydrophobe modified polymers" and therefore they can be expected to exhibit phase behavior similar to that of conventional surfactants.   For example, Zhang and Eisenberg [123,124] have reported that polystyrene-polyacrylic acid copolymers with 410 polystyrene units and about 25 acrylic acid units form spherical micelles, rod like micelles and uni- and multi- lamellar vesicles in aqueous solutions. They differ from the earlier reported [125] micelle forming block copolymers in that the size of hydrophobic core is significantly larger than the corona. The core diameter of such a spherical micelle is around 300 nm and it increases markedly with increase in salt concentration.   These "giant spherical micelles" and other structures may offer interesting opportunities in the area of solubilization, delivery, and solution structuring.

XII. CONCLUDING REMARKS

A scan of the patent literature for the last 5 years turned up nearly 200 examples of compositions that employ a combination of polymer and surfactant. This attests to the growing industrial importance of these materials when used in concert. The most common category by far is the hair treatment segment, followed by skin treatment, including soap bars. Detergent systems (fabric and hard surface) follow. Among the residual categories are antibacterial (dandruff), textile, foaming, dyeing, flocculation, pigment, antistatic, oral and dye treatment products.

At present, and as outlined in the introduction, it is frequently not possible to ascertain whether the combined presence is the result of a choice based on individual performance or one based on the properties of the two components used in concert. What is clear is that in a great number of systems it will be the properties of the combined polymer, surfactant system which will contribute to the performance of the system. It is this which represents a continuing challenge and opportunity for the colloid scientist.

REFERENCES

1. E.D. Goddard, in *Interactions of Surfactants with Polymers and Proteins*, (E.D.Goddard and K.P.Ananthapadmanabhan, eds.) CRC Press, Boca Raton, 1993, p. 395.
2. M.N. Jones, *J.Colloid Interface Sci.* 23: 36 (1967).
3. J. François, J.Dayantis, and J.Sabbadin, *Eur. Polym. J.* 21: 165 (1985).
4. J.C. Brackman, *Langmuir* 7: 469 (1991).
5. A. Carlsson, G. Karlström, and B. Lindman, *Colloids Surf.* 47: 147 (1990).
6. P.S. Leung, E.D. Goddard, C. Han and C. Glinka, *Colloids Surf.* 13: 47 (1985).
7. E.D. Goddard and P.S. Leung, *Langmuir* 7: 608 (1991).
8. E.D. Goddard, K.P.A. Padmanabhan, and P.S. Leung, *J.Soc.Cosmet.Chem.* 42: 19 (1991).
9. K. Thalberg and B. Lindman, *Colloids Surf.* 47: 147 (1990).
10. K.P. Ananthapadmanabhan and E.D. Goddard, unpublished results, 1988
11. S. Saito and T.J. Tanaguchi, *J. Colloid Interface Sci.* 44: 114 (1973)
12. S. Saito in *Nonionic Surfactants: Physical Chemistry*, (M.J. Schick, ed.) Marcel Dekker, New York, 1987, p.881.
13. A. Sarrazin-Cartalas, I. Iliopoulos, R. Audebert, and U. Olsson, *Langmuir* 10: 1421 (1994).
14. J.C. Brackman, and J.B.F.N. Engberts, *J.Am.Chem.Soc.* 112: 872 (1990).
15. J. C. van de Pas, Th.M. Olsthoorn, F.J. Schepers, C.H.E. de Vries, and C.J. Buytenhek, *Colloids Surf.* 85: 221 (1994).
16. J.G. Doolan and C.A.Gody, *US Patent 5,425,806* (1995).
17. Y.Q. Zhang, T.Tanaka and M. Shibayama, *Nature* 360: 142 (1992).
18. E. Kokufuta, Y.Q. Zhang, T. Tanaka and M. Mamada, *Macromolecules* 26: 1053 (1993).
19. H. Lange, *Kolloid Z. Z. Polym.* 243: 102 (1971).
20. S. Saito, *Kolloid Z.,* 154: 19 (1957).
21. M.N. Jones, *J. Colloid Interface Sci.* 26: 532 (1968).
22. S. Saito, *J. Colloid Interface Sci.* 24: 227 (1967).
23. N.J.Turro, B.H.Baretz, and P.L.Kuo, *Macromolecules* 17: 321 (1984).
24. E.D. Goddard, R.B. Hannan, and G.H. Matteson, *J. Colloid Interface Sci.* 60: 214 (1977)
25. K. Hayakawa, *this volume*, ch. 11
26. E.D. Goddard and R.B. Hannan, *J. Colloid Interface Sci.* 55: 73 (1976).
27. P.L. Dubin and D.Davis, *Colloids Surf.* 13: 113 (1985).
28. P.L.Dubin, M.E.Y. Vea, M.A. Fallon, S.S. The, D.R. Rigsbea, and L.M. Gan, *Langmuir* 6: 1422 (1990).
29. M.J. Schwuger and H. Lange, *Proc. 5th Int. Congr. Deterg. Barcelona (1968)*, Vol. 2, Ediciones Unidas S.A., Barcelona, 1969, p. 955.
30. T. Isemura and A. Imanishi, *J. Polym. Sci.* 33: 337 (1958).
31. N. Sata and S. Saito, *Kolloid. Z.* 128: 154 (1952).
32. S. Saito and Y. Mizuta, *J. Colloid Interface Sci.* 23: 604 (1967).

33. H.P. Lundgren, *US Patent* 2,459,708 (1949).
34. B.H. Lee, S.D. Christan, E.E. Tucker, and J.F. Scamehorn, *Langmuir 1*: 1332 (1991).
35. J.F. Scamehorn, *Chemical Engineering. News (ACS), 3/10/1997*.
36. E.D. Goddard, unpublished results, 1975.
37 C.J. Drummond, S. Albers, and D.N. Furlong, *Colloids Surf. 62*: 95 (1991).
38. G. Karlström, A, Carlsson, and B. Lindman, *J. Phys. Chem. 94*: 5005 (1990).
39. A. Carlsson, G. Karlström, and B. Lindman, *Langmuir 2*: 536 (1986).
40. A. Carlsson, G. Karlström, B. Lindman, and O. Stenberg, *Colloids Polym. Sci. 226*: 1031 (1988)
41. K.P. Ananthapadmanabhan, *in Interactions of Surfactants with Polymers and Proteins*, (E.D.Goddard and K.P. Ananthapadmanabhan, eds.), CRC Press, Boca Raton, 1993, p. 347.
42. J.F. Hessel and M.P. Aronson, *US Patent* 4,908,150 (1990).
43. E.H. Cordes (ed.), *Reaction Kinetics in Micelles*, Plenum Press, New York, 1973.
44. D.M. Bloor, W.A.Wan-Bahdi, J.F. Holzwarth, and E.Wyn-Jones, *J. Phys. Chem. 97*: 5793 (1993).
45. K.P. Ananthapadmanabhan, P.S. Leung, and E.D. Goddard, in *Polymer Association Structures: Microemulsions and Liquid Crystals*, ACS Symp. Ser. 384, American Chemical Society, Washington, D.C., 1988, p. 297.
46. L.D. Rhein, C.R. Robbins, K. Fernee, and R. Cantore, *J. Soc. Cosmet. Chem. 37*: 125 (1986).
47. F.J. Prescott, E. Hahnel, and D. Day, *Drug and Cosmet. Ind. 93*: 443 (1963).
48. J.B. Ward and G.J. Sperandio, *J. Soc. Cosmet. Chem. 15*: 32 (1964).
49. R.F. Medacliff Jr., M.O. Visscher, J.R. Knochel, and R.M. Dahgren, *US Patent* 4,820,447 (1989).
50. N.W. Jordan, W.M. Winkler, M.O. Visscher, S.A. Seaman, and O. McGuffey, *US Patent* 3,076,953 (1991).
51. E.A. Tavss, E. Eigen, V. Temnikow, and A.M. Kligman, *J. Am. Oil Chem. Soc., 630*: 574, 1986.
52. J.A. Faucher, E.D. Goddard, R.B. Hannan, and A.M. Kligman, *Cosmet. Toiletries 92*: 39 (1977).
53. Y.J. Nikas and D. Blankschtein, *Langmuir 10*: 3512 (1994).
54. E.D. Goddard, *Colloids Surf. 19*: 255 (1986).
55. E.D. Goddard, *Colloids Surf. 19*: 301 (1986).
56. R.B. Hannan, E.D. Goddard, and J.A. Faucher, *Textile Res. J. 48*: 57 (1978).
57. S.T.A. Regismond, F.M. Winnik, and E.D. Goddard, *Colloids Surf. A 119*: 221 (1996)
58. E.D. Goddard and W.C. Harris, *J. Soc. Cosmet. Chem. 38*: 233 (1987).
59. J.A. Faucher and E.D. Goddard, *J. Colloid Interface Sci. 55*: 313 (1976).
60. E.D. Goddard and R.B. Hannan, *J. Am. Oil Chem. Soc. 54*: 561 (1977).
61. S. Sukhvinder, C.R. Robins, and W.M. Cheng, *WO 94/06409 A1* (1994).
62. R.L. Wells, *US Patent* 5,573,709 (1996).

63. J.J. Parran, *US Patent* 3,761,418 (1973).
64. J.J. Parran, *US Patent* 3,723,325 (1973).
65. J. Sime, *EP* 0 093 601 (1983).
66. P. Somasundaran, K.P. Ananthapadmanabhan, M. Fujuwara, and L. Tsaur, US Patent 5,476,660 (1996).
67. M.D. Berthiaume and J. Jachowicz, *J. Colloid Interface Sci. 141*: 299 (1991).
68. A. Saud, *U.S. Patent* 4,704,224 (1987).
69. E.D. Goddard and P.S. Leung, *Cosmet. and Toiletries 97*: 55 (1982).
70. K. Esumi, Y. Yakaku, and H.Otsuka, *J. Jap. Soc. of Colour Material 67*: 431 (1994).
71. C. Maltesh and P. Somasundaran, *J. Colloid and Interface Sci. 153*: 298 (1992).
72. K. Esumi and K. Meguro, *J. Colloid Interface Sci. 129*: 217 (1989).
73. C. Ma and C.L. Li, *Colloids Surf. 47*: 17 (1991).
74. H.W. Kilau and J.I. Voltz, *Colloids Surf. 57*: 17 (1991).
75. T. Cosgrove, S.J. Mears, L. Thompson, and I. Howell, in *Surfactant Adsorption and Surface Solubilization*, (R. Sharma, ed.) ACS Symp. Ser. 615, 1995, p.196.
76. P. Somasundaran and L.T. Lee, *Sep. Sci. & Technol. 16*: 1475 (1981).
77. P. Somasundaran and J. Cleverdon, *Colloids Surf, 13*: 73 (1985).
78. V. Shubin, *Langmuir 10*: 1093 (1994).
79. S. Magdassi and B. Rodel, *Colloids Surf. 119*: 51 (1996).
80. H. Otsuka, K. Esumi, T.A. Ring, J.T. Li, and K.D. Caldwell, *Colloids Surf. 116*: 161 (1996).
81. G. Gabrielli, F. Cantale, and G. Guarini, *Colloids Surf. 119*: 163 (1996).
82. P.W. Zhu and D.H. Napper, *Langmuir 12*: 5992 (1996).
83. H.G. Schild and D.A. Tirrell, *Polym. Prepr. 30*: 350 (1989).
84. J. Brash and T. Horbett, (eds), *ACS Symp. Ser. 602*: (1995).
85. M. Wahlgren and T. Arnebrant, *Langmuir 13*: 8 (1997).
86. K. Durham, (ed.) *Surface Activity and Detergency*, McMillan, London, 1961, p.235.
87. H. Jaeger, *Commun. Jorn. Com. Esp. Deterg. 24*: 165 (1993).
88. E.M. Kirschner, Chemical and Engineering News (ACS) 1/27/1997, p. 30.
89. K.J. Mysels, K. Shinoda, and S. Frankel, *Soap Films-Studies of their Thinning*, Pergamon Press, New York, 1959.
90. K.J. Mysels, *J. Phys. Chem. 68*: 3441 (1964).
91. K.J. Mysels, *J. Gen. Physiol. 52*: 113s (1968).
92. T.D. Miles, *J. Phys. Chem. 49*: 71 (1944).
93. L. Shedlovsky, *Ann. N.Y. Acad. Sci. 49*: 279 (1948).
94. *Kirk Othmer Encyclopedia of Chemical Technology*, 3rd Ed; Wiley, New York (1980).
95. E.D. Goddard, T.S. Phillips, and R.B. Hannan, *J. Soc. Cosmet. Chem. 26*: 461 (1975).
96. P.S. Leung and E.D. Goddard, *Colloids Surf. 13*: 47 (1985).
97. S.T.A.Regismond, F.M.Winnik, and E.D.Goddard, *Colloids Surf.* in press (1998).

98. J.H. Buckingham, J. Lucassen, and F. Hollway, *J. Colloid Interface Sci. 67*: 423 (1978).

99. A. Asnacios, D. Langevin, and J.F. Argillies, *Macromolecules 29*: 7412 (1996).

100. V. Bergeron, D. Langevin, and A. Asnacios, *Langmuir 12*: 1550 (1996).

101. P.M. Breuer and I.D. Robb, *Chem. Ind.*: 530 (1972).

102. E.D. Goddard and K.P. Ananthapadmanabhan, *Interactions of Surfactants with Polymers and Proteins*, CRC Press, Boca Raton, 1993, p.112.

103. M.N. Jones, *J. Colloid Interface Sci. 23*: 36 (1967).

104. P.G. de Gennes, *J. Phys. Chem. 94*: 8407 (1990).

105. I.P. Purcell, R.K. Thomas, J. Penfold, and A.M. Howe, *Colloids Surf. 94*: 125 (1995).

106. K. Chari and T.Z. Hussain, *J. Phys. Chem. 95*: 3302 (1991).

107. S. Lionti-Addad and J.M. deMeglio, *Langmuir 8*: 324 (1992).

108. S. Cohen Addad and J.M. diMeglio, *Langmuir 10*: 773 (1994)

109. S. Cohen-Addad, J.M. diMeglio, and R. Ober, *C.R. Acad. Sci. Set. 2 315*: 39 (1992).

110. E.D. Goddard and D.B. Braun, *Cosmet. and Toiletries 100*: 41 (1985)

111. R.K.Prud'homme and S.A.Khan, (eds.), *Foams: Theory, Measurements, and Applications*, Marcel Dekker, New York, 1995.

112. D.Danino, A.Kaplan, Y.Talmon, and R.A.Zana, in *Structure and Flow in Surfactant Solutions*, ACS Symp. Ser. 578: 105 (1994).

113. M. Hulden, *Colloids Surf. 82*: 263 (1994).

114. T. Annable, R. Buscall, and R. Ettelaie, *Colloids Surf. 112*: 97 (1996).

115. P. Hameyer and K.R. Jenni, *Paper presented at the 18th international IFSCC congress*, October, 1994, Venezia, Italy.

116. D.A. Tomalia, A.M. Naylor, and W.A. Goddard III, *Angew. Chem. Int. Ed. 29*: 138 (1990).

117. D.A. Tomalia, H. Baker, J. Dewald, M. Hall, G. Kallos, S. Martin, J. Rock, J. Ryder, and P. Smith, *Macromolecules 19*: 2466, (1986).

118. G. Caminati, N.J. Turro, and D.A. Tomalia, *J. Am. Chem. Soc. 112*: 8515 (1990).

119. M.F. Ottaviani, E. Cossu, N.J. Turro, and D.A. Tomalia, *J. Am. Chem. Soc. 117*: 4387 (1995).

120. M.F. Ottaviani, N.J. Turro, S. Jockusch, and D.A. Tomalia, *Colloids Surf. A: 115*: 9 (1996).

121. A. Halperin, M. Tirrell, and T.P. Lodge, *Adv. Polym. Sci. 100*: 31 (1992).

122. L. Zhang and A. Eisenberg, *Science 268*: 1728 (1995).

123. L. Zhang and A. Eisenberg, *Macromolecules 29*: 8805 (1996).

124. L. Zhang and A. Eisenberg, *J. Am. Chem. Soc. 118*: 3168 (1996).

125. R. Xu and M.A. Winnik, *Macromolecules 24*: 87 (1991).

3
Phase Behavior of Polymer-Surfactant Systems

LENNART PICULELL and BJÖRN LINDMAN Physical Chemistry 1, Center for Chemistry and Chemical Engineering, Lund University, Box 124, S-221 00 Lund, Sweden

GUNNAR KARLSTRÖM Theoretical Chemistry, Center for Chemistry and Chemical Engineering, Lund University, Box 124, S-221 00 Lund, Sweden

I.   INTRODUCTION

The tendency to demix is generally much stronger for mixtures involving polymeric species than for mixtures of only small molecules. This is because the entropy of mixing, at a given volume fraction, is a much weaker thermodynamic driving force for large molecules. The same holds for self-assembled aggregates of surfactant molecules. Phase separation phenomena are therefore abundant in both polymer and surfactant solutions, as well as in their mixtures, which will concern us here.

In practice, mixtures of polymers and surfactants are employed for a variety of reasons. One purpose may be to introduce a feature typical of one of the systems to the other: to impart a high viscosity (a polymer feature) to a surfactant solution, or a capacity to solubilize hydrophobic molecules (a surfactant feature) to a polymer solution. In such cases miscibility is typically desired, and demixing, if it occurs, is an unwanted side effect. On the other hand, the reason for adding the other component may well be to deliberately affect the phase behavior. Added surfactant may enhance the solubility of a polymer in certain cases, whereas in other cases a concentrated mixed phase will form, which is in equilibrium with almost pure water. A third possibility is the formation of two liquid phases of similar solvent content, each one enriched in one of the colloidal components. Such systems may be used, for instance, in separation applications.

To control the phase behavior of mixed polymer/surfactant solutions is thus of considerable importance, and a list of practically relevant questions may include the following.

   - Are there any general recipes for achieving monophasic systems?
   - What happens on removing solvent from a mixture
   (drying/concentration)?
   - Are there ways to design mixtures that do not take up much solvent
   (water)?
   - Is it possible to design mixtures that swell considerably, but not
   indefinitely, in excess solvent?
   - What are the effects of added polymer on the morphology and stability
   of surfactant structures?

- What are the consequences of adding low-molecular compounds, simple salts in particular?

In this chapter it is our ambition to provide at least some guiding principles that may help to answer questions of this type. This requires a good understanding of the molecular mechanisms that underlie polymer/surfactant phase behavior. Such mechanisms, as we understand them at present, will therefore be the focus of this chapter.

All mixtures dealt with here contain at least one low-molecular solvent (water). The simplest cases considered are thus ternary mixtures of polymer, surfactant and water, and the simplest types of phase separation in such mixtures involve only two phases. Two main classes of biphasic demixing may be distinguished, on the basis of the distribution of the two macromolecular solutes: they may either segregate into different phases of similar total concentration, or associate into one concentrated phase in equilibrium with a very dilute solution. Accordingly, we recently introduced the terms *segregative* and *associative*, respectively, for these two types of phase separation [1]. Figure 1 illustrates segregative and associative phase diagrams for the particularly simple cases where both of the separating phases are isotropic liquids, and both solutes are completely miscible with the solvent. These two types of phase separation were first described, many years ago, for mixed polymer solutions [2], and we have previously drawn attention to the useful analogies between such mixtures and polymer/surfactant mixtures as regards their phase behavior [1]. These analogies, which are particularly useful in cases where the surfactant forms finite, identifiable "association polymers" (micelles), will be explored further here.

Much recent research has dealt with more complex systems, however, where the simple picture associated with the diagrams in Figure 1 is insufficient. Obvious cases concern mixtures where the surfactant forms infinite or ordered assemblies, as in liquid crystalline phases or microemulsions. A different type of complication occurs for mixtures of oppositely charged polymer and surfactant--the most common type of associatively phase separating mixtures. These are actually four-component systems, and a quasi-ternary representation, as in Figure 1b, is therefore incomplete, at best. We note that the complications just mentioned are not specific to polymer/surfactant mixtures; polymer solutions may also form liquid crystalline structures, and the multicomponent nature of oppositely charged mixtures holds for polyelectrolytes as well.

There are, however, cases when the simple polymer/polymer analogy breaks down altogether. One is when the polymer contains strongly hydrophobic groups which can self-associate into micellar structures. Such

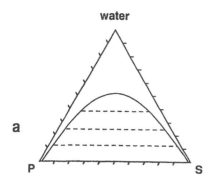

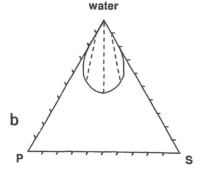

FIG. 1 Schematic ternary phase diagrams showing segregative (a) and associative (b) phase separation in mixed solutions of polymer and micelle-forming surfactant.

structures can solubilize individual surfactant molecules; hence a description in terms of interactions between polymer molecules and aggregates of self-associated surfactant molecules is not always appropriate. The latter picture should apply only in the limit when the surfactant molecules grossly outnumber the surfactant hydrophobes. Instead, analogies with mixed surfactant systems are more useful, over an important part of the mixing range.

Often, the binary polymer/water or surfactant/water systems already display multiphase behavior. In such cases, it is natural to describe the phase behavior of the ternary mixture with reference to the binary phase equilibrium, and to see how this is perturbed by the addition of (small amounts of) the third component. The classification according to associative-segregative behavior is still useful, though. As an example we may consider a case where the polymer solution phase-separates into a dilute and a concentrated phase. If the added component has the role of a "nonsolvent" (for the polymer) it will enrich in the dilute phase, and this will be a segregating system. Conversely, an associative phase separation is one where the added component is enriched in the concentrated phase. In this context, we should recall that association or segregation in a mixed solution are not always driven by the interactions between the macromolecules. Interactions with the solvent are also important [3-6].

Guided by the above considerations, we have chosen to organize this chapter as follows. The next three sections (II-IV) all focus on simple cases, where the polymers are linear and flexible, and the surfactant aggregates finite and micellar. Our main classification here is not based on charge as is usual, but on the degree of involvement of hydrophobic association between polymer and surfactant. Thus, in section II, we discuss mixtures of intrinsically water-soluble ("hydrophilic") polymers with pure surfactant micelles. Here the colloidal components are separated from each other by at least some layer of water molecules. The phase separation in these mixtures may be either segregative (nonionic mixtures) or associative (oppositely charged mixtures), and the polymer/polymer analogy is helpful. The other extreme case, where the polymer is strongly amphiphilic and self-associating, is treated in section III. In such mixtures also, both associative and segregative phase separation occur, but segregation is limited to systems where the surfactant is in large excess.

In section IV, we turn to the intermediate case of slightly hydrophobic main chain polymers. Such polymers are slightly soluble in surfactant aggregates, but they are not sufficiently hydrophobic to form hydrophobic aggregates on their own. Moreover, slightly hydrophobic polymers often phase separate from water at elevated temperatures, and many studies deal with the effects of added surfactant on this phase separation. Globular proteins, which are also hydrophobic polymers, but with some important distinguishing features, are discussed separately in section V. The specific features associated with mixing polymers with infinite surfactant structures are considered in section VI.

Past studies have shown that the Flory-Huggins theory for polymer solutions can provide important insight into those mixtures where the polymer/polymer analogy applies. Such model predictions are discussed in section VII, which includes calculations that have not been presented previously in the literature. Finally, the chapter ends with a recapitulation of some important conclusions in section VIII.

As should already be apparent, we have chosen to present here what we believe to be generic types of phase behavior for the different types of situations which we have just described, involving comparatively simple types of polymer in water-continuous phases. We have illustrated these with experimental studies on a few representative systems, rather than attempting a comprehensive coverage of published work in this very active area. Slightly more space is given to areas where important recent progress has been made that have not been covered in previous reviews [1,7].

II. HYDROPHILIC POLYMERS WITH PURE SURFACTANT MICELLES

Here we describe the phase behavior of isotropic aqueous mixtures of surfactants with hydrophilic polymers. The polymers considered are well soluble in water, and no short-range attractions (hydrophobic or otherwise) between the polymer and surfactant are present. Hence, no molecularly mixed polymer-surfactant aggregates (such as mixed micelles) are formed. The polymer molecules and the surfactant aggregates - even if oppositely charged - are always separated by at least some layer of hydration water. In most cases, the phase separation will be segregative, but association may occur if the polymers and surfactant aggregates are oppositely charged. The association vanishes, however, in the presence of sufficient amounts of added salt, and may eventually be replaced by segregation. A prerequisite for both segregative and associative phase separation is that the surfactants have self-associated into aggregates. We will focus on the simplest cases when the aggregates are finite and micellar.

A. Segregating Mixtures

A number of recent studies [1,8-15] have concerned mixtures where there is no net attraction (hydrophobic or electrostatic) between the polymer and the surfactant. The phase separation is here found to be segregative, as for the analogous mixed polymer solutions [16]. As in the latter mixtures, the extent of the segregation in the polymer/surfactant mixtures is expected to depend on the relative strengths of the short-range interactions between the various components (solute-solute and solute-solvent) and on the entropy of mixing. Differences in the short range interactions generally favor segregation; miscibility is thus an effect of the entropy of mixing. The latter is largely determined by the number of molecules or aggregates that the system contains at a given weight composition. For nonionic mixtures, this is given by the degree of polymerization of the polymer and the aggregation number of the surfactant micelle. For large molecules and aggregates the entropy of mixing becomes quite weak, and phase separation results. For charged solutes, however, and in the absence of external salt, the entropy of mixing may again become large owing to the numerous small counterions originating from the polyelectrolyte and/or the surfactant.

1. Nonionics

The behavior of segregative nonionic polymer/surfactant mixtures is illustrated in Figures 2-4, giving the results of a recent study where the effects of temperature, micellar aggregation number, and chemical similarity/dissimilarity of the micelles and the polymers were investigated [12]. Different combinations of polymer and surfactant, based on either glucose (dextran and octyl glucoside, respectively) or ethylene oxide (PEO and $C_{12}E_5$ or $C_{12}E_8$) were mixed, and the micellar aggregation numbers of the $C_{12}E_m$ surfactants were varied both by changing the length (m) of the oligo (ethylene oxide) headgroup and by changing the temperature. At sufficiently high solute concentrations, a biphasic demixing occurred in all mixtures, but the extent of the two-phase area varied with the investigated variables.

Figure 2 compares the ternary phase diagrams at 10°C for the "chemically dissimilar" mixtures of dextran with $C_{12}E_5$ or $C_{12}E_8$, respectively. Effects of varying the temperature from 10 to 25°C are shown in Figure 3. At these low temperatures, $C_{12}E_8$ forms essentially spherical micelles, whereas the $C_{12}E_5$ micelles are large and rodlike, and grow significantly in size with increasing temperature [17-20]. Clearly, increasing the micellar size, either by shortening the oligo (ethylene oxide) headgroup or by increasing the temperature, results in an increasing two-phase area, as expected. On the other hand, when there is no micellar growth (as for $C_{12}E_8$ in Figure 3), the tendency to demix decreases with increasing temperature. Both these trends were confirmed also for the other mixtures investigated, and the effect of temperature (and, presumably, micellar growth) has been documented in independent studies on similar mixtures [8]. As expected, it has also been found that the miscibility decreases, for a given surfactant, when the molecular weight of the polymer is increased [10,11,14].

"Chemical" effects are illustrated in Figure 4, where different combinations of dextran or PEO with alkyl glucoside or EO-type surfactants are compared. The results show that the couples having the greatest miscibility are those that are chemically similar, as might be expected. For the large $C_{12}E_5$ micelle, however, the demixing with the chemically similar PEO is stronger than with dextran. This deviation from the main trend could be due to the fact that the average size of the random coil of PEO 40000, despite its lower molecular weight, is actually larger than that of dextran T70 ($M_W = 70000$).

2. Nonionic/ionic and similarly charged mixtures

The segregation in a mixed polymer/surfactant solution is strongly affected by the introduction of small fractions of ionic groups on one or both of the polymer

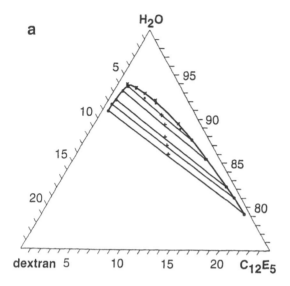

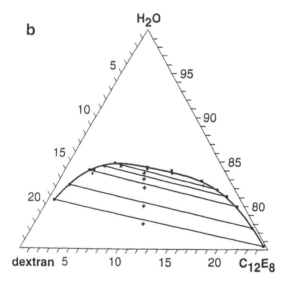

FIG. 2 Ternary phase diagrams (compositions in wt%) at 10°C for aqueous mixtures of Dextran T70 (M_w 70000) with a) $C_{12}E_5$ and b) $C_{12}E_8$. From ref. 12 with permission.

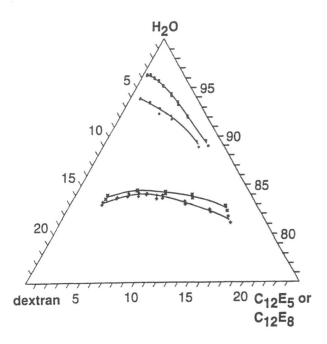

FIG. 3 Phase boundaries (compositions in wt%) for aqueous mixtures of dextran T70 with (from top to bottom) $C_{12}E_5$ at 25°C, $C_{12}E_5$ at 10°C, $C_{12}E_8$ at 10°C, and $C_{12}E_8$ at 25°C. From ref. 12 with permission.

or surfactant micelle [13,21]. Figures 5 and 6 illustrate these effects with results[13] obtained for mixtures of $C_{12}E_5$ and dextran where small fractions of SDS and covalent sulfate esters were used to modify the micelles or polymers, respectively (*cf.* Figure 2). In the thus modified systems, x_{SDS} denotes the mole fraction of SDS of the total surfactant content and DS_{Dx} the average number of sulfate groups per glucose unit in the dextran. The systems were studied with no external salt present.

The addition of charges on only one of the segregating components results in a large increase in the miscibility, as shown in Figure 5. Similar effects for segregating aqueous polymer mixtures have previously been demonstrated by Iliopoulos *et al.* [22], and they have been understood to be a consequence of the large entropy of mixing of the many small counterions to the polyion [1,21,23]: since each separating phase has to be electroneutral, a segregation leads to an unfavorable enrichment of these counterions in the polyelectrolyte-rich phase.

Interestingly, it has been shown both theoretically [21] and experimentally [13,21] that the effect of charge on one of the macromolecular components could be balanced if the correct amount of similar charge was introduced also on the other macromolecule. The experimental results, shown in Figure 6,

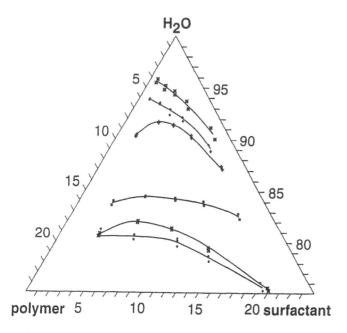

FIG. 4   Phase boundaries (compositions in wt%) at 10°C for aqueous mixtures of polymer/surfactant couples, from top to bottom: PEO 40000/ $C_{12}E_5$, dextran T70/ $C_{12}E_5$, PEO 40000/octylglucoside, dextran T70/ $C_{12}E_8$, dextran T70/octylglucoside, and PEO 40000/ $C_{12}E_8$. From ref. 12 with permission.

demonstrated that the two-phase area of the balanced mixture nearly coincided with that of the uncharged reference mixture. Analyses of the counterion contents in the separating phases suggested that the balanced mixture corresponded to one where the counterion concentration was the same in the entire system, whereas different counterion concentrations in the two separating phases were generally found in the imbalanced mixtures. It thus seems that the balanced state is one where the entropy of mixing of the counterions is maximal for those phase compositions that also correspond to the lowest free energy of the uncharged reference system.

       One of the most important practical applications of segregating aqueous polymer solutions is as non-denaturing separation systems for biological macromolecules [16]. As a corollary to the study in ref [13], Sivars *et al.* investigated the partition of globular proteins in slightly charged polymer/surfactant systems [24]. Slightly imbalanced systems of neutral polymer and mixed surfactant were found to be useful for "tuning" the partition of charged proteins; the latter could be steered to one phase or the other, depending on the charge of the added surfactant. Two of the virtues of these

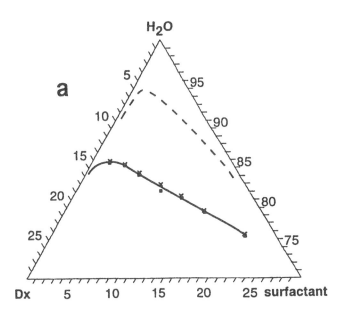

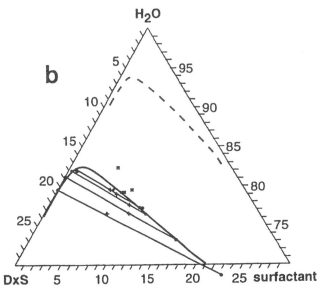

FIG. 5 Quasi-ternary phase diagrams showing the effects of charge on the segregation in aqueous dextranT70/ $C_{12}E_5$ mixtures at 10°C: $x_{SDS} = 0.01$, $DS_{Dx} = 0$ (a) and $x_{SDS} = 0$, $DS_{Dx} = 0.12$ (b). Dashed lines refer to the phase boundary of the uncharged mixture (Figure 2, top). From ref. 13 with permission.

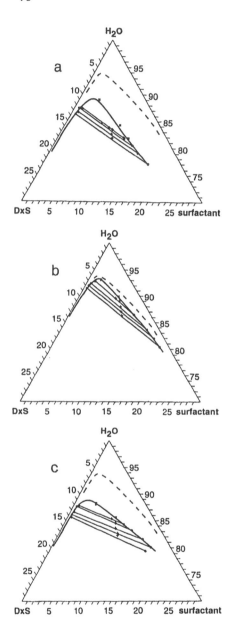

FIG. 6 Quasi-ternary phase diagrams at 10°C for mixtures of slightly sulfated dextranT70 ($DS_{Dx} = 0.025$) with mixed $C_{12}E_5$/SDS micelles where $x_{SDS} = 0$ (a), 0.021 (b) or 0.083 (c). Dashed lines refer to the phase boundary of the uncharged mixture (Figure 2, top). From ref. 13 with permission.

systems were (i) that the net charge of the micelles could be tuned easily and independently of pH, and (ii) that no precipitation occurred in mixtures of slightly charged micelles and oppositely charged proteins. The latter observation is interesting, in view of the fact that charged proteins typically are precipitated by oppositely charged surfactants at very low concentrations (*cf.* section V below). The reason for the absence of precipitation in the mixed micellar systems must be that the activity of the charged surfactant (roughly equal to the monomer concentration) is extremely low in mixed micellar solutions that are dominated by the nonionic surfactant.

Segregation also occurs in salt-free mixtures of purely ionic micelles with highly charged polyelectrolytes of the same sign. This is shown for the NaHy/SDS mixture [9] in Figure 7, where it is also seen that the segregation increases markedly on addition of large amounts of salt. Two mechanisms discussed above could contribute to this salt effect. First, added salt is expected to increase micellar growth, which results in increased segregation. Second, it is unlikely that the salt-free segregating mixtures of this system

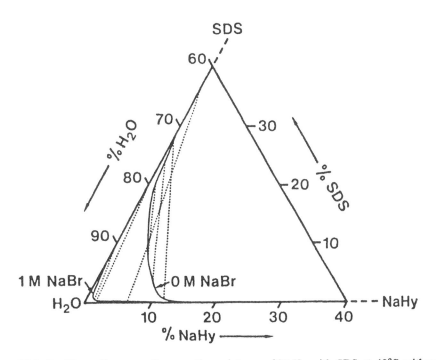

FIG. 7   Phase diagrams of segregating mixtures of NaHy with SDS at 40°C without and with added salt. From ref. 9, *Colloids Surf. A 76*, 283 (1993) with permission of Elsevier Science-NL, SaraBurgerhartstraat 25, 1055 KV Amsterdam, the Netherlands.

happen to be balanced with respect to the counterion concentrations, in the sense discussed above. Added salt should therefore reduce the imbalance, and increase the segregation.

3.   Polymer/surfactant vs. polymer/polymer mixtures

Although the general trends are the same, there are some significant quantitative differences in the phase diagrams of segregating mixtures of polymers and surfactant micelles as compared to the analogous mixed polymer solutions. A general feature of the former systems is that the surfactant-rich phase is more concentrated than the polymer-rich phase [9-13]. This effect is exemplified for nonionic mixtures in Figure 2 and for ionic surfactants and polymers in Figures 5-7. This means that, at equal concentrations, the osmotic pressure of a solution of a flexible polymer is higher than that of a solution of surfactant micelles. This finding is expected from general considerations. The dominating contribution to the osmotic pressure of a semi-dilute solution of a flexible polymer originates from the internal degrees of freedom of the polymer: the number of ways to fold each chain becomes severely restricted as the concentration increases above the overlap concentration [25]. This contribution is absent for a solution of rigid particles, which is a more appropriate reference system for a micellar solution.

Comparisons of phase diagrams also indicate that the miscibility is greater (the two-phase area is smaller) for polymer/surfactant mixtures [12] than for comparable polymer/polymer mixtures [16]. This difference has been ascribed [12] to the fact that micelles are much more compact than flexible polymer chains. Thus, the number of unfavorable short-range contacts should be fewer in a polymer/micelle than in a polymer/polymer mixture, if similar overall concentrations are compared.

B.   Oppositely Charged Mixtures

Highly charged polymers, aggregates or particles differ strongly from the corresponding nonionic species. There is a strong enrichment of counterions close to the polyion and this uneven counterion distribution lowers the entropy markedly. For ionic surfactants specifically, this unfavorable entropic effect strongly counteracts micellization, and this is the major reason behind the fact that the CMC is two orders of magnitude higher for ionic surfactants than for comparable nonionics [26,27]. In the presence of electrolyte, micelles are stabilized, the more so the higher the valency of the oppositely charged ion. Particularly efficient in reducing the CMC is an oppositely charged polyelectrolyte, which can be viewed as a multivalent counterion. The polyelectrolyte-induced lowering of the CMC of oppositely charged surfactants (sometimes referred to as a cooperative binding of the surfactant to the polyelectrolyte) is long known and has been extensively studied and reviewed [28,29]; *cf.* chapters 4 and 5 in this volume.

Another early observation in oppositely charged mixtures was an associative phase separation ([30] and refs. therein) occurring at quite low surfactant concentrations, often only slightly above the CMC. The concentrated phase could be a solid or a highly viscous gel-like solution. It was also observed that addition of larger amounts of electrolyte inhibited the phase separation and that "redissolution" could be obtained by adding larger concentrations of surfactant. However, there were in the early work no studies of the coexisting phases and thus no attempts at a systematic description in terms of phase diagrams. In hindsight this is surprising since both for polymers and for surfactants separately, the study of phase diagrams has been invaluable for the progress of our understanding of these systems.

1. General features

We recall that we are here concerned with the simplest type of mixtures where the polymer is intrinsically hydrophilic, so that no molecularly mixed polymer-surfactant aggregates are formed. A polymer that meets this criterion is the anionic polysaccharide NaHy. The description of polyion/ionic surfactant systems in terms of phase diagrams will be illustrated by investigations of mixtures of NaHy with a cationic surfactant, $C_{14}TABr$ [31-37]. A large number of other mixtures of charged homopolymers and oppositely charged surfactants has also been investigated with analogous findings [38-43].

Phase diagrams for three-component systems are generally described using the Gibbs' triangle and the same approach has been used for polyelectrolyte/oppositely charged surfactant/water systems. However, it should be observed that, as for any system composed of two electrolytes and a solvent, this is not correct, since at least a four-component representation is required. The aspects of phase diagram representation in two and three dimensions have been dealt with in some detail [35]. A convenient way of presenting the three-dimensional phase diagram is the pyramidal one of Figure 8. The base of the pyramid has four apices, each representing one of the possible neutral salts. In the following, we will denote these four salts as polyelectrolyte (polyion + simple counterion), surfactant (surfactant ion + simple counterion), complex salt (polyion + surfactant ion) and simple salt. The water concentration is given along an axis perpendicular to the base. In the simplest case, corresponding to a situation where the surfactant does not form any liquid

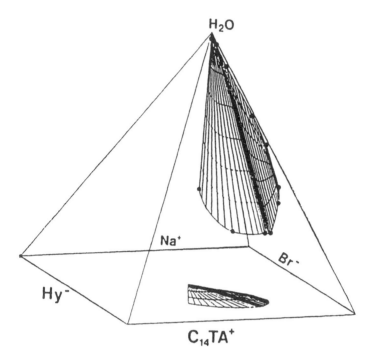

FIG. 8 Pyramid presentation of the phase diagram of a system of a polyelectrolyte, an oppositely charged ionic surfactant and water, here exemplified by sodium hyaluronate (NaHy) and tetradecyltrimethylammonium bromide (C_{14}TABr). The "sail" hanging from the water apex corresponds to the associative two-phase region; its projection on the pyramid base, corresponding to a total concentration of ionic species of 1M, is also shown. From ref. 35 with permission.

crystalline phase in the composition range considered, we obtain an extensive connected region of homogeneous isotropic phase and two two-phase regions, at low and high concentrations of salt, respectively (Figure 9). The tie-lines of the two two-phase regions run in entirely different directions. Moreover, they do not--especially not in the low-salt two-phase region--lie in the plane defined by the water, polyelectrolyte and surfactant apices. The latter plane contains all possible global compositions that can be achieved by mixing these three species, and we will therefore refer to it as their *mixing plane*. The conventional Gibbs' triangle representation refers to this mixing plane of polyelectrolyte, surfactant and water (or, occasionally, brine).

The low-salt two-phase region, suspended like a droplet from the water apex, corresponds to the aforementioned associative phase separation. The dilute phase is clearly low-viscous, while the concentrated phase can range

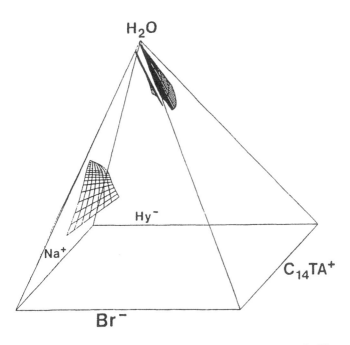

FIG. 9 Pyramid phase diagram for the same system as in Figure 8 showing phase separation surfaces for three concentrations of added NaBr, 0, 0.075 and 1.0 M (from upper right to lower left). The two upper surfaces belong to an associative two-phase body and the lower one to a segregative one. (The base of the pyramid corresponds to a total ionic concentration of 2.5 M.) From ref. 36 with permission.

from a moderately viscous solution to a stiff "gel." It is isotropic and is best characterized as a more or less concentrated solution. In accordance with the original definition [2] we will refer to this type of phase separation, where both of the separating phases are liquids, as a coacervation, to distinguish it from cases where the concentrated phase is solid-like. In the high-salt two-phase region there is a coexistence of one solution rich in surfactant and one rich in polymer; both solutions are typically of low or moderate viscosity. This is clearly a segregative phase separation. In a rather broad salt concentration range intermediate between the associative and the segregative two-phase regions there is a single homogeneous solution phase, corresponding to slightly attractive interactions between the solutes.

For several systems, partial phase diagrams of mixtures of polyelectrolyte and surfactant have been determined, both in the presence and in the absence of added salt. Data obtained include both the compositions of separate phases (tie-lines), and the stability ranges of the one-phase areas. Phase boundaries of the latter type are obtained by titrating a one-phase mixture with one component (salt, surfactant or polyelectrolyte) until the onset of phase separation is observed as a "clouding" of the solution. This type of phase boundary is therefore commonly referred to as a cloud-point curve. It has been observed that, in the conventional Gibbs' triangle representation, the cloud-point curves do not at all coincide with phase boundaries defined by the end-points of tie-lines from systems well inside the biphasic region [38,43]. This "discrepancy" is readily understood with the help of the pyramidal phase diagram (Figures 8 and 9). In a salt-free mixture of polyelectrolyte, surfactant and water, the cloud-point curve corresponds to the intersection of the mixing plane with the surface of a two-phase volume (*e.g.*, the "droplet"in Figure 8). A boundary defined by tie-lines, on the other hand, is obtained by a projection of the the endpoints of the tie-lines onto the mixing plane. Such a phase boundary is called a quasi-ternary coexistence curve, and its location in the quasi-ternary representation is obviously not unique; it must depend on the global compositions of the separating systems from which it was derived. Unfortunately, the cloud-point curve has been referred to as a spinodal in the literature [43]. This is not correct; a spinodal is a boundary in phase space that separates a metastable one-phase region from an unstable region. As explained above, the difference between the cloud-point curve and the quasi-ternary coexistence curves is not related to metastability.

Once the above limitations are appreciated, it is still instructive and convenient to use the conventional quasi-ternary representation to look at general features of the phase behavior. We exemplify by the effects of added salt on the NaHy/C_{14}TABr mixture in Figure 10. In order to study electrolyte effects, it is most illustrative to use the electrolyte composed of the two counterions. Otherwise, the number of components increases further and a meaningful representation of the phase behavior becomes virtually impossible. In Figure 10, where all the diagrams show quasi-ternary coexistence curves, we see that electrolyte addition shrinks the two-phase region, and that addition of further salt may eliminate phase separation completely. At very high electrolyte concentrations, there appears as mentioned above a segregative phase separation, which is the typical situation in mixtures of two species of high molecular weight. The system illustrated is thus intrinsically segregative, but segregation is inhibited in the absence of added electrolyte due to the strong effective attraction between the oppositely charged polyelectrolyte and surfactant.

Even though no other attempts at three-dimensional phase diagrams have appeared, available studies suggest that the phase diagram depicted in Figure 9, and the patterns shown in Figure 10, apply quite generally to mixtures of intrinsically hydrophilic polyelectrolytes with oppositely charged surfactant micelles. It should be noted, however, that surfactant-rich phases other than

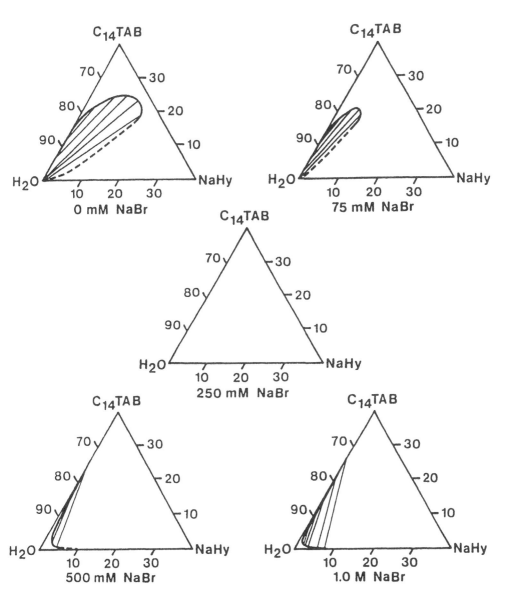

FIG. 10 Pseudo three-component phase diagrams for aqueous mixtures of sodium hyaluronate (NaHy) and tetradecyltrimethylammonium bromide. From ref. 36 with permission.

the isotropic micellar solution may form, notably hexagonal, cubic or lamellar liquid crystalline phases [27]. Such phases have been noted particularly in dilute mixtures of polymer and surfactant as concentrated solids or "pastes," as opposed to the liquid coacervates formed at higher solute concentrations or in the presence of external salt. Three-phase equilibria must therefore exist in the associative region, involving two phases concentrated in the complex salt (polyion + surfactant ion) in equilibrium with a dilute solution containing primarily simple salt. The normal quasi-ternary representation is quite ill-suited to illustrate this situation. An alternative quasi-ternary representation would appear to be more appropriate, referring to a mixing plane defined by complex salt, simple salt and water. This mixing plane represents, in the three-dimensional phase diagram (Figure 8), a plane perpendicular to that corresponding to the conventional quasi-ternary description. The alternative representation emphasizes the, intuitively expected, tendency of the complex salt to separate out as a concentrated phase in the mixture.

These points were addressed in recent simple experiments on stoichiometric (i.e., the equivalent charge concentrations of polyelectrolyte and surfactant were equal) aqueous mixtures of NaPA and C_{16}TABr. Note that such stoichiometric mixtures may equally correctly be described as mixtures of C_{16}TAPA with NaBr in water; their global compositions belong to both mixing planes. The mixtures were diluted, by successive additions of pure water, and the phase behavior was observed [44]. At sufficiently high concentrations, the mixtures were monophasic, because of the high concentration of simple salt, which screened the interactions between the polyions and the surfactant micelles. On dilution with water, an associative phase separation occurred which initially involved two isotropic solutions: one dilute and one concentrated. As the system was further diluted, the concentrated phase became more concentrated and, eventually, a second (hexagonal) concentrated phase was formed. The surfactant had now formed infinite aggregates. Finally, if the dilute phase was replaced by large quantities of pure water, the concentrated phase was transformed into a more concentrated lamellar phase of the complex C_{16}TAPA salt. This sequence of events is clearly best described as a consequence of the decreasing concentration of screening simple salt as the system is diluted.

The phase behavior of the above stoichiometric mixtures is shown in Figure 11. Since the pyramidal phase diagram tends to be complicated, the diagrams shown give the distributions of the *ions* in the separating phases. (This corresponds to side views of the pyramid.) The cations (Na^+ and CTA^+) are shown in Figure 11a, and the anions (Br^- and PA^-) in Figure 11b. Since both phases contain, in general, all four ions, each axis represents the sum of two salts. The global compositions all lie on the bisectors of both diagrams, since all investigated mixtures were stoichiometric. If the compositions of the

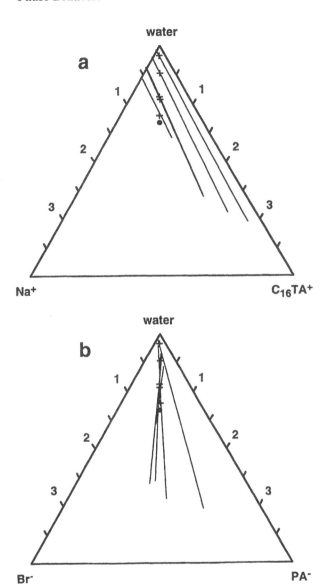

FIG. 11 Phase behavior of stoichiometric mixtures of NaPA and C_{16}TABr on dilution with water, represented as partitioning of the cations (a) and the anions (b) between the separating phases. Concentrations are in moles of charge/kg H_2O. The initial monophasic mixture (●) and the global compositions of phase-separating mixtures (+) are indicated. The concentrated phase was hexagonal in all cases except for the most concentrated biphasic system (the shortest tie-lines), where both phases were isotropic.

separating phases had all belonged to the C_{16}TABr, NaPA and water mixing plane--the normal quasi-ternary representation--then the C_{16}TA$^+$ axis in Figure 11a would obviously have been equivalent to the Br$^-$ axis in Figure 11b, and similarly for the Na$^+$ and PA$^-$ axes. The two diagrams would then have been mirror images of each other. Clearly, this is far from the real situation. Conversely, if all compositions had belonged to the C_{16}TAPA/NaBr/water mixing plane, the diagrams would have been identical. This is also not true. While essentially all of the C_{16}TA$^+$ ion is always in the concentrated phase, a significant concentration of the PA$^-$ ion is found in the dilute phase, and this concentration increases with increasing global concentration. Only at high global dilution does the distribution of PA$^-$ approach that of C_{16}TA$^+$. Two mechanisms should contribute to this difference. The first is that the polyion, being much smaller and flexible, is intrinsically much more soluble than the surfactant aggregate. Second, bromide is known to bind preferentially (over ions such as chloride) to cationic surfactants (*cf.* below). There is thus an element of chemical competition in the binding to the surfactant headgroups.

2. Some trends

We will now look briefly at how the phase behavior for oppositely charged surfactant and hydrophilic polyelectrolyte depends on the specific choices of polyelectrolyte, surfactant and simple salt. The diagrams in Figure 12 [31,39,40] illustrate the coacervation-redissolution behavior on addition of cationic surfactant to NaHy at a fixed concentration, in the presence of varying amounts of simple salt. The minimum concentration of surfactant required for phase separation decreases strongly as its chain length increases, *i.e.*, as the CMC decreases. This observation supports the notion that the associative phase separation involves the polyion and the self-associated surfactant. The redissolution concentration is less strongly dependent on surfactant chain length. As detailed above, an increase in the simple salt concentration leads to a shrinkage of the two-phase area. The amount of salt needed to give redissolution, the critical electrolyte concentration (CEC), increases with increasing surfactant chain length (Figure 12a) and is strongly dependent on surfactant concentration. Figure 12b shows that the CEC also depends appreciably on the nature of the added salt--the anion in particular. Increasing the valency of a simple ion lowers counterion entropy effects and decreases the entropic gain of polyion-micelle association. Due to nonelectrostatic contributions, bromide ions associate more strongly than chloride ions to micelles [26,45-48]. This reduces the tendency to associative phase separation. Both these trends (the dependence on surfactant chain length and on the

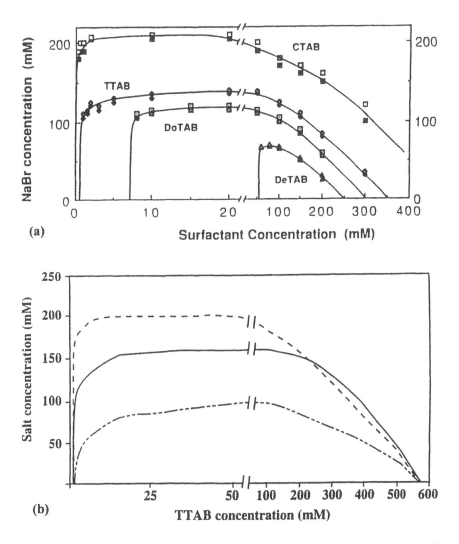

FIG. 12   a), top: critical electrolyte concentrations of NaBr in samples of 1.0 mM NaHy and different C_nTABr surfactants (n = 10, 12, 14 and 16; from bottom to top). From ref. 31 with permission; b), bottom: critical electrolyte concentrations of different sodium salts (NaCl, NaBr and Na_2SO_4; from top to bottom) in samples of 0.1 % NaHy and $C_{14}TABr$. From refs 39 and 40, with permission.

identity of the simple anion) have been confirmed also for mixtures of cationic surfactants with NaPA by Hansson and Almgren [49]. They suggested that the increasing CEC with increasing surfactant chain length could be attributed to a larger aggregation number of the micelles.

As might be expected, comparisons between polyelectrolytes with different charge densities show that the tendency to association strongly increases with increasing charge density. This is nicely illustrated in a study of poly(acrylate-co-acrylamide) where the fraction of ionized monomer units was varied systematically [43]. The reverse experiment of varying the charge density of the micelles, by mixing in nonionic surfactant, has also been made [41]. As expected, reducing the micellar charge density markedly reduces the extent of phase separation; at higher fractions of nonionic surfactant phase separation is completely eliminated.

III. HYDROPHOBICALLY ASSOCIATING POLYMERS

Hydrophobically associating water-soluble polymers (HAP) are essentially hydrophilic polymer molecules, modified by a small number of very hydrophobic moieties attached as side-groups or end-groups. We will reserve the term HAP for polymers were the hydrophobes are sufficiently hydrophobic so that they can self-associate into micelle-like aggregates even in the absence of surfactants. Mixtures of HAP with surfactants have been extensively studied for their interesting rheological properties, which vary widely depending on the surfactant content. However, the phase behavior of the mixtures also may be rich, as will be exemplified below. A qualitative understanding of this behavior has emerged only recently, and owing to the novelty of the subject we will slightly increase the level of detail in this section.

A. General Considerations

By now it is well established that added surfactant molecules join the HAP hydrophobes in mixed micelles [50-53]. A discussion of the implications of this mixed micellization for the binding of surfactants to HAP is given in Chapter 5 of this volume. As was realized early by Goddard [54], one would therefore expect that the phase behavior of HAP/surfactant mixtures should be understandable in terms of a quasi-species, the polymer-surfactant complex, which has a composition that varies with the composition of the mixture. This notion has been confirmed in recent studies [55,56]. We may thus think of our mixed HAP/surfactant solution as consisting of the solvent (water) plus the following species:

i)   the HAP-surfactant complex, with a variable composition.
ii)   excess monomeric surfactant (this is important especially for ionic surfactants, where the monomer concentration is often dominant and contributes to the ionic strength of the mixture).
iii) free surfactant micelles (at sufficiently high surfactant concentrations).
iv)   added simple salt.

To simplify, we will restrict the discussion in this section to the simple cases where the parent polymer (without the hydrophobes) is hydrophilic in the sense that it does not form mixed micelles with surfactants. (More complicated HAP are discussed briefly in section IV.C below.) Cases in point are, for nonionic parent polymers, hydrophobically modified HEC and PAm (HM-HEC and HM-PAm) and, for an ionic polymer, hydrophobically modified PA (HM-PA). Two stoichiometric variables are particularly important for surfactant complexes with such HAP: the overall charge (for ionic surfactants and/or polymers) and the ratio β_{sh} of surfactant to HAP hydrophobe in the mixed micelles:

$$\beta_{sh} \equiv C_{s,b} / C_{h,b} \tag{1}$$

Here C_s stands for the surfactant concentration and the subscript b denotes a species (HAP hydrophobe, h, or surfactant molecule, s) that is "bound" in a (mixed) micelle. It is important to realize that it is not safe to assume that the composition of the micelles is given by the global ratio of surfactant to hydrophobe, especially not if the surfactant is ionic. Analogies with mixed micellization in surfactant solutions [57] are useful to illustrate this point [55,56,58], as is explained further in Chapter 5 of this volume.

From the point of view of hydrophobe stoichiometry, we may identify the following composition regimes for the complex.

$\beta_{sh} \ll 1$: the mixed micelles are essentially HAP micelles, with some incorporated surfactant. The structure of the micelles should be only marginally perturbed by the surfactant. Still, ionic surfactants may contribute to the solubility of a nonionic HAP.

$\beta_{sh} \sim 1$: here both species contribute to the micelles in comparable amounts. This is a regime where, in certain systems, an associative phase separation occurs.

$\beta_{sh} \gg 1$: when the surfactant dominates, the mixed micelles should be similar to pure surfactant micelles. One consequence of this is that the surfactant monomer concentration should be close to the CMC of the surfactant in the same system (salt conditions, etc.) but without the HAP hydrophobes. In this composition regime, mixed micellar crosslinks will gradually disappear, and resolubilization often occurs in associatively phase-separating mixtures.

$\beta_{sh} > N_{agg,s}$: this is the "micellar excess regime," where the binding ratio exceeds the average aggregation number $N_{agg,s}$ of the surfactant in the micelles; pure surfactant micelles must therefore exist. The polymer-surfactant complex is saturated with respect to the HAP hydrophobes, mixed micellar crosslinks should have vanished, and the phase behavior should essentially depend on the interactions between the saturated polymer-surfactant complex and the pure surfactant micelles. This behavior may be either segregative or associative.

From the above it follows that two equivalent concentration units for the HAP have a special significance: the equivalent charge concentration, C_e, and the total concentration of HAP hydrophobes, C_h. While the importance of C_e is well recognized, surprisingly few authors discuss C_h or the hydrophobe ratio C_s/C_h in HAP/surfactant mixtures. In practice, C_h is typically of the order of 10^{-3} mole L^{-1} or less, since it is difficult to obtain homogeneous solutions where C_h is much larger than this value. ($C_h \sim 10^{-3}$ mole L^{-1} corresponds to a concentration of the order of 1% w/w of an HAP where a few per cent of the repeating units carry hydrophobes.) C_s, on the other hand, may easily be orders of magnitudes larger.

B. Nonionic Hydrophobically Associating Polymers

Studies have shown, as might be suspected, that water-soluble polymers modified by alkyl side-chains are miscible with various ionic and nonionic surfactants over a large composition range [59,60]. Indeed, an ionic surfactant may render soluble a water-insoluble HM-HEC [61,62]. The large miscibility is probably the reason why relatively little is done on phase-separation in mixtures of nonionic HAP with surfactants. Moreover, the few results that are available are not always entirely consistent. Our conclusions on this issue must therefore necessarily be tentative.

In some instances, phase separation has been observed at very high C_s/C_h ratios (of the order of 10^2) [11,62], where it may be suspected that the mixtures were in the micellar excess regime. Although the phase compositions were not determined, the overall behavior here strongly suggests a segregative phase separation. Figure 13 shows results on mixtures of a HEC, modified by a small fraction of fluorocarbon side-chains, mixed with a perfluorated surfactant. Not entirely in accordance with the discussion in the original paper [62], we would interpret these results in terms of a segregation between the complex and pure surfactant micelles, where the phase rich in pure surfactant micelles forms the *lower* phase (owing to the high density of the fluorocarbon component). This interpretation is supported by the observation (Figure 13) that the viscosity of the mixtures levels off at similar concentrations. The latter indicates that all

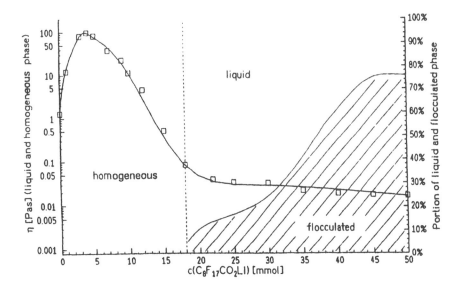

FIG. 13   Viscosity and phase behavior of 1% HEC, modified by fluorocarbon side-chains, in solutions of lithium perfluorononanoate at varying concentrations. From ref. 62, *Colloids Surf. A* 82, 279 (1994) with permission of Elsevier Science-NL, Sara Burgerhartstraat 25, 1055 KV Amsterdam, the Netherlands.

mixed micellar cross-links between the polymer chains have disappeared, implying that the polymer hydrophobes are, indeed, saturated with bound surfactant.

In contrast to the above studies, Goddard [54] found a phase behavior which has all the appearance of associative phase separation at much lower C_s/C_h ratios in mixtures of another type of HM-HEC with several different ionic and nonionic of surfactants. The reason for these conflicting results is not clear. The only obvious difference is that the HM-HEC studied by Goddard was modified with nonylphenol groups at a DS of 0.02, but it is difficult to imagine how this could account for the difference in phase behavior. Another possibility is the presence of salt impurities (no purification step is mentioned in the experimental protocol of Goddard [54]), which would enhance the tendency to associative phase separation with ionic surfactants. The latter was demonstrated by Effing *et al.* [63] for mixtures of HM-PAm with SDS. Phase separation was found only in the presence of salt, and studies of the phase compositions clearly confirmed its associative nature. Unfortunately, no results seem to be available for mixtures of HM-PAm with nonionic surfactants. Such results would have been interesting, since a segregative phase separation occurs in mixtures of non-modified PAm with $C_{12}E_5$ [21].

C. Ionic Hydrophobically Associating Polymers

In typical ionic HAP, the number of charges per polymer is at least equal to the number of hydrophobes; often the number of charges is very much larger. For mixtures with oppositely charged surfactants, a particularly instructive case is that where both these numbers are the same, *i.e.*, $C_e = C_h$. This holds for catHM-HEC, which is HEC modified by cationic hydrophobes of the quaternary ammonium type. When the hydrophobe carries one alkyl chain of sufficient length (typically C_{12}), it micellizes in solution at a C_h much smaller than the CMC of a comparable cationic surfactant [55]; thus, this polymer indeed behaves as an HAP. CatHM-HEC/surfactant mixtures were first studied by Goddard and Leung [64,65] and later by other groups [55,56,66]; the general features of all these studies were quite similar.

Figure 14 shows quasi-ternary phase diagrams from detailed phase studies [55] of mixtures of catHM-HEC with SDS, in pure water or in 10 mM aqueous NaCl. The phase separation is clearly associative with a concentrated liquid coacervate in equilibrium with a dilute solution of, mainly, monomeric surfactant. A notable feature (Figure 14a) is that the cloud point curve in the investigated composition range is given by two essentially straight lines; the "coacervation line" and the "redissolution line" at low and high surfactant contents, respectively. The coacervation line extrapolates to zero polymer and surfactant content, whereas the redissolution line extrapolates to a finite surfactant concentration close to the CMC of SDS in the polymer-free solvent (water or 10 mM NaCl). The latter feature was noted by Goddard [54] also for the redissolution of phase-separated mixtures of nonionic HM-HEC with several surfactants, as reported in section III.B above. The physical mechanism should undoubtedly be the same in these cases.

The phase behavior in Figure 14 could be rationalized if the HAP-surfactant complex was considered as a single quasi-component with a variable composition (and, thus, varying solubility), given by the surfactant binding isotherm [55]. The latter was assumed to be independent of the total polymer concentration in the investigated range. Figure 15 shows experimental binding isotherms in the presence and absence of added salt. The isotherms were obtained from phase-separated mixtures under the assumption that the surfactant concentration measured in the dilute phase was equal to the free surfactant concentration. As expected, virtually all added SDS enters into the (oppositely charged) mixed micelles until charge neutrality ($\beta_{sh} = 1$) of the

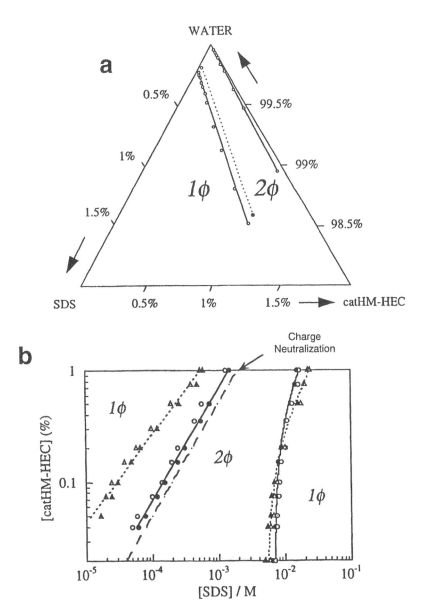

FIG. 14 Cloud-point curves for catM-HEC/SDS/water mixtures. ϕ_1 and ϕ_2 refer to one- and two-phase regions, respectively. (a) Quasiternary representation, with one tie-line indicated. (b) Log-log representation, showing boundaries in water (circles, solid lines) and in 10mM NaCl (triangles, dotted lines). Lines are fits to Eqs. (3) and (4). Data from ref. 55.

micelles is reached. Beyond charge neutrality, the binding isotherm displays an *anticooperative region*, where the added SDS is incorporated with increasing difficulty (the slope in the isotherm becomes progressively smaller) in the now increasingly negatively charged mixed micelles. Hence, the free surfactant concentration starts to build up when more surfactant is added. Ultimately, beyond the anticooperative binding region, a cooperative binding of the surfactant must set in as the free surfactant concentration approaches its maximum value, *i.e.*, the CMC of pure SDS. This region is only sketched in Figure 15, since it occurred largely in the one-phase region after redissolution.

The total surfactant concentration in the mixture may generally be written as

$$C_s = C_{s,f} + C_{s,b} = C_{s,f} + \beta_{sh}C_{h,b} \qquad (2)$$

where the subscript f denotes free (monomeric) surfactant. Coacervation occurred before charge neutrality, where the free surfactant concentration was found to be negligible. The reasonable assumption that all polymer hydrophobes were also incorporated in the mixed micelles thus leads to

$$C_{s,\text{coacervation}} = \beta_{sh,\text{coacervation}}C_h \qquad (3)$$

The total surfactant concentrations at the phase boundaries were found to be linear functions of the polymer concentration, as shown in Figure 14a. This implies that the complex phase separated at a fixed mixed micellar stoichiometry, independent of the total concentration. However, $\beta_{sh,\text{coacervation}}$ did depend on the concentration of added salt. Fits of Eq. (3) to the coacervation lines in Figure 14b gave $\beta_{sh,\text{coacervation}} = 0.73$ in the absence of salt, and 0.3 for the same ratio in the presence of 10 mM NaCl. Since the catHM-HEC sample alone derives its solubility in water from the dissociated counterions (phase separation occurred at 16 mM monovalent salt even without surfactant [55]), this behavior is as expected. Moreover, the mixed micellar crosslinking in itself may contribute to the associative phase separation with surfactant, as indicated by the results of Effing *et al.* [63] on mixtures of HM-PAm with SDS in the presence of salt.

The surfactant concentration at redissolution could similarly be written as

$$C_{s,\text{redissolution}} = C_{s,f,\text{redissolution}} + \beta_{sh,\text{redissolution}}C_h \qquad (4)$$

Again, the observed linearity implies that redissolution occurs at a given, but salt-dependent, stoichiometry. The fact that $C_{s,f,\text{redissolution}} \approx$ CMC was

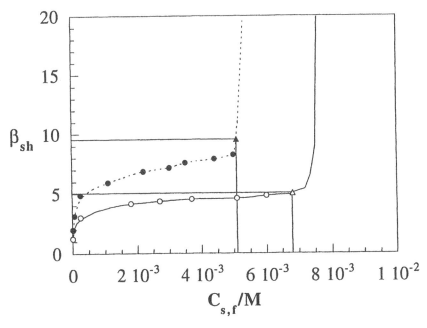

FIG. 15 Isotherm for the binding of SDS to catHM-HEC in water (open symbols) and in 10 mM NaCl (filled symbols). Triangles represent the values $C_{sf,redissolution}$ and $\beta_{sh,redissolution}$ at the phase boundaries; see text. Data from ref. 55.

rationalized by the following argument: it is likely that redissolution, which presumably requires a large net negative charge and/or a dissolution of mixed micellar crosslinks, should only occur when the surfactant dominates in the mixed micelles. For such micellar compositions, the free surfactant concentration must be close to the CMC. The redissolution coordinates $C_{sf,redissolution}$ and $\beta_{sh,redissolution}$ from a fit of Eq. (4) to the experimental phase boundaries are indicated in the binding isotherm in Figure 15, and connect smoothly to the points obtained in the two-phase region. This interpretation gives a simple explanation to the observation that the redissolution lines with and without added salt cross each other (Figure 14b): the CMC is lower in the presence of salt, but a larger binding ratio is needed in order to achieve redissolution. The comparatively strong sensitivity of $\beta_{sh,redissolution}$ to monovalent salt also provided a clue as to why the binding ratio had to be comparatively large (~5) in order to redissolve the complex also in the absence of added salt: the ionic strength from monomeric SDS alone was already 7 mM at redissolution.

A similar reasoning explains the phase behavior of mixtures of catHM-HEC with cationic surfactants shown in Figure 16. The data, obtained by Guillemet [56], are for C_nTABr surfactants of different chain lengths, including the reference tetramethyl ammonium bromide salt. For the long-chain surfactants with low CMC, the monomeric surfactant concentration never becomes high enough to cause coacervation. For $C_{12}TAB$ and $C_{10}TAB$, however, $C_{s,f}$ may become high; the CMC:s are 15 and 70 mM, respectively. Although there is a binding also of these surfactants to the HAP micelles at arbitrary low surfactant concentrations, the binding is low until the cooperative region of the isotherm is reached. This means that the surfactant concentration gradually builds up and, at coacervation, most of the added surfactant is actually monomeric and plays the same role as added simple salt. Redissolution occurs, as usual, when the binding ratio is high enough to dissolve mixed micellar crosslinks and/or to give the complex a sufficiently high net charge. The tetramethyl ammonium ion never joins in the micelles, hence, no redissolution occurs for C_1TABr.

Detailed phase studies are still lacking for surfactants mixed with charged HAP where $C_e > C_h$. Published reports concern mixtures with oppositely charged surfactant [52,66-68], in experiments where the surfactant concentration has been varied at a constant HAP concentration. The HAP concentration was not varied, and the focus was mainly on the conditions for coacervation; there is a scarcity of experiments extended to high C_s where redissolution should appear. A general observation seems to be that coacervation occurs before the complex is totally neutralized by the binding of oppositely charged surfactant.

Magny et al. [52,67] studied the effects of added $C_{12}TAC$ or $C_{12}TAB$ on 1% w/w solutions of fully neutralized HM-PA. Samples carrying one or three C_{12} or C_{18} hydrophobes per 100 repeating units were studied and compared with unmodified PA. The surfactant concentration required for coacervation increased dramatically on hydrophobic modification; phase separation occurred at about 2.5×10^{-2} M of surfactant for HM-PA compared to 3×10^{-3} M for unmodified PA. This effect was ascribed to the mixed micellization which occurs for HM-PA, but not for PA. For unmodified PA, coacervation occurs at surfactant concentrations far below the global charge neutrality, but not very far exceeding the CMC, where surfactant micelles form (cf. section II.B). In contrast, the surfactant molecules added to HM-PA form mixed micelles distributed on all the polymer molecules, and coacervation does not occur until a certain high binding ratio has been reached. The mechanism of this coacervation is not clear at the moment. We can postulate two possibilities. One is that the complex at some degree of surfactant binding simply becomes insoluble, owing to a shift in the balance between hydrophobic attraction (favoring coacervation) and the entropy of mixing of the counterions (favoring

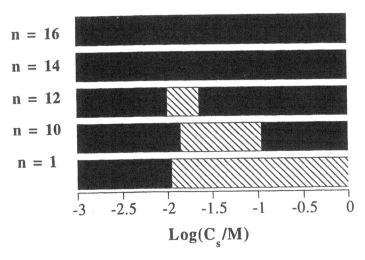

FIG. 16 Phase boundaries for mixtures of 1% catHM-HEC (C_h = 2 mM) with C_nTABr surfactants or with tetramethylammonium bromide. The filled and cross-hatched areas refer to one- and two-phase regions, respectively. Data from ref. *56, Adv. Coll. Int. Sci. 63*, 1 (1996), reprinted with permission from Elsevier Science-NL, Sara Burgerhartstraat 25, 1055 KV Amsterdam, the Netherlands.

miscibility). This is the mechanism invoked for the coacervation of catHM-HEC with oppositely charged SDS. The other possibility is that the HM-PA becomes saturated, so that pure surfactant micelles appear at those relatively high surfactant concentrations where phase separation occurs. The phase separation would then be analogous to the coacervation of unmodified PA, the difference being that the (still net negatively charged) HM-PA-surfactant complex now acts as the polyelectrolyte.

More systematic studies seem warranted, especially for mixtures of highly charged polyelectrolytes with low degrees of hydrophobe modification ($C_e \gg C_h$). The available results suggest that for hydrophilic polyelectrolytes, which do not form mixed micelles with the surfactants, the miscibility with oppositely charged surfactants can be significantly extended by a small degree of hydrophobic modification, owing to the formation of mixed micelles. This could be useful in applications.

IV. SLIGHTLY HYDROPHOBIC MAIN CHAIN POLYMERS

Here we will consider the case where the polymer (nonionic or ionic) has a more or less uniform hydrophobic character along its extension, a situation produced by polymerizing markedly amphiphilic monomers (*e.g.* PEO, PNIPAm or NaPSS) or by grafting a large number of small hydrophobic groups on a hydrophilic homopolymer (*e.g.* typical cellulose ethers). Such a slightly hydrophobic polymer (SHP) differs from an HAP, where a small number of strongly hydrophobic groups are grafted onto the polymer chain. With respect to their interactions with surfactants, the SHP represent a case intermediate between hydrophilic polymers and hydrophilic main chain HAP. Similar to the HAP, the SHP typically form molecularly mixed complexes with certain surfactants, but these complexes always involve self-associated surfactant aggregates, containing in the order of ten or more surfactant molecules. This is because the SHP, unlike the HAP, are not sufficiently hydrophobic to self-associate into hydrophobic domains with a capacity to solubilize single surfactant molecules. The complexation therefore only occurs above some critical surfactant concentration, often called the CAC, which may be regarded as the CMC of the mixed micelle. As for the oppositely charged pairs discussed in section II.B, the complexation between an SHP and a surfactant may thus be viewed as a polymer-induced surfactant micellization. This complexation is particularly well documented for monophasic mixtures of nonionic SHP and anionic surfactants [30].

Most nonionic SHP have another characteristic property, *i.e.* an inverse temperature dependence on solubility, described below. A polymer denoted here as an SHP has at least one of these properties: an inverse solubility in water, or a capacity to join surfactants in molecularly mixed micelles.

A. Nonionic Slightly Hydrophobic Main Chain Polymers

Nonionic SHP typically have an inverse temperature dependence of solubility, *i.e.*, the solubility decreases with increasing temperature. On heating, phase separation will eventually occur, which is first observed as an increased turbidity, hence the commonly used terms clouding polymers and cloud point (temperature). We will here use the more precise term lower consolute temperature (LCST) for this type of cloud point, to distinguish it from cloud points obtained by other infinitesimal changes (*e.g.*, in composition; cf. section II.B.1 above) in the system.

The phase diagram of a mixture of an LCST polymer with water consists of a closed two-phase region (Figure 17a) [69] that increases in extension with increasing molecular weight of the polymer. The nontrivial temperature

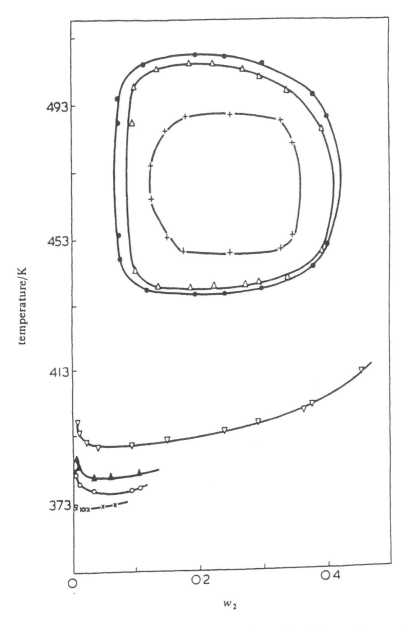

FIG. 17 (a) Phase diagrams for PEO/water mixtures for different polymer molecular weights. From the bottom the mean molecular weights are 10^6, 2×10^4, 1.4×10^4, 8×10^3, 2270, 2250 and 2160. From ref. 69 with permission.

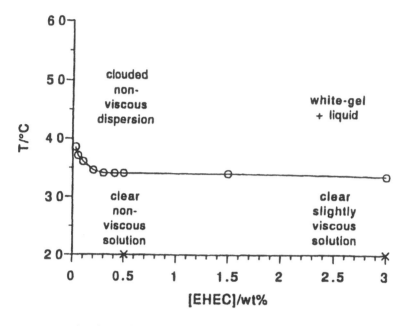

FIG. 17   (b) Phase diagram for a hydrophobic EHEC in water. From ref. 71 with permission.

dependence arises because at low temperature water is a good solvent for the ethylene oxide groups while it changes into a bad solvent at higher temperature. This can in turn be attributed to a change-over to less polar conformations of the EO chains as temperature is increased [70]. In practice, one is usually concerned with the dilute part of the phase diagram as illustrated for a nonionic cellulose ether, EHEC, in Figure 17b [71].

For an LCST polymer, a phase separation thus exists already in the binary polymer/water mixture, and most phase studies on mixtures with surfactants have been concerned with the influence of the surfactant on the LCST. This problem is significant from both a practical and a theoretical point of view:
i) nonionic LCST polymers are used extensively in various formulations, from paints to pharmaceuticals, and the control of the LCST is a condition for controlling performance.
ii) LCSTs offer an especially simple indicator of polymer-cosolute interactions. It should be noted, however, that the nature of the effective interaction (repulsive or attractive) usually has to be determined independently.

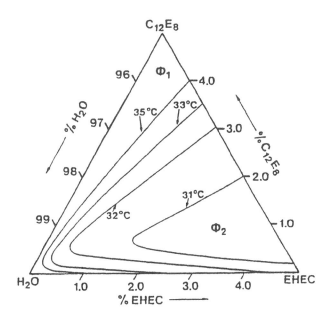

FIG. 18 (a) Experimental phase diagram for the EHEC/$C_{12}E_8$/water system at different temperatures. ϕ_1 and ϕ_2 refer to one- and two-phase regions, respectively. Tie-lines run almost parallel to the polymer-water axis. From ref 73, *Colloids Surf. 67*, 147 (1992) reprinted with permission of Elsevier Science-NL, Sara Burgerhartstraat 25, 1055 KV Amsterdam, the Netherlands.

We will consider nonionic polymers in mixtures with nonionic and ionic surfactants. In the latter case it will be especially important to investigate the influence of added electrolyte.

If, in mixtures of a nonionic polymer and a nonionic surfactant, we make the polymer progressively less polar, we expect a changeover from segregative to associative phase separation via conditions with no phase separation. The association should also increase with decreasing polarity of the surfactant. Although the matter has not been systematically investigated, these expectations are borne out by experiments on random copolymers of propylene oxide and ethylene oxide (UCON) and on different varieties of EHEC. For both types of polymer, an associative phase separation has been documented in mixtures with nonionic surfactants, as illustrated in Figure 18 [72-75]. The extent of phase separation increases with decreasing polarity of the nonionic surfactant or the polymer and, very appreciably, with increasing temperature. It is significant that an EHEC with a higher LCST than that of Figure 18a was

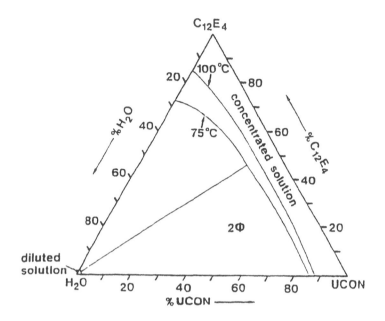

FIG. 18 (b) Experimental phase diagram for UCON/$C_{12}E_8$/water. From ref 75 with permission.

found to display a segregative phase separation with nonionic surfactants [76]; *cf.* Figure 22 below.

Systems of a nonionic SHP and an ionic surfactant are characterized by repulsive surfactant-surfactant interactions, while the polymer-polymer interactions are more or less attractive. For the cases considered here, the attraction is strong and the solution of the polymer alone is not far from the LCST curve. As mentioned above, LCST polymers typically form mixed micelles with (especially negatively charged) ionic surfactants, and this is the case which will concern us here. The general effect expected is that micellization of the surfactant along the polymer chains will raise the LCST of the complex compared to the bare polymer, since inter- (and intra-) chain repulsions are obtained. The effect is analogous to converting a nonionic polymer into a polyelectrolyte. This is indeed what is observed for many long-chain ionic surfactants [76-83] as illustrated in Figure 19. For short- or medium-chain surfactants the increase is smaller and is often preceded by a

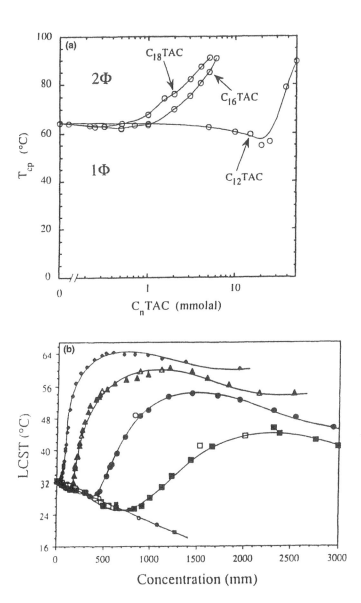

FIG. 19 (a, top): the effect of C_nTACl surfactants on the cloud point of EHEC (1 wt%). From ref. 76 with permission. (b, bottom): the effect of n-alkyl sulfates (C_4 to C_8; from bottom to top) on the cloud point of PNIPAm (0.04 wt%). From ref. 83 with permission.

shallow minimum in the curve of LCST as a function of surfactant concentration.

Simple salts also affect the LCST of SHP, but the changes are orders of magnitude smaller than those observed for ionic surfactants. The LCST may either increase or decrease on electrolyte addition, the effect being determined mainly by the anion [84]. The changes follow the classical Hofmeister or lyotropic series found to apply for a broad range of systems. I^- and SCN^- raise the LCST while decreases are observed with most anions like Cl^- and SO_4^{2-}. Short-chain surfactants also act as simple salts before the CMC, producing an initial decrease of the LCST, as illustrated for mixtures of PNIPAm with various alkyl sulfates in Figure 19b [83].

The effect of adding electrolyte becomes very different in the presence of an ionic surfactant [71,76,79-82]. In fact, addition of surfactant in the presence of electrolyte at concentrations which are 1/100 of those shifting the LCST appreciably by themselves causes a dramatic lowering of the LCST in a wide surfactant concentration range (Figure 20). The effect has been found to be very general; it appears for all electrolytes, for all ionic surfactants and for a large number of nonionic polymers (including PEO and several nonionic cellulose ethers). Still, there are important quantitative differences:

i)   the effect is larger for longer alkyl chains of the surfactant;
ii) cationic surfactants give smaller effect than anionics;
iii) divalent counterions give a two-phase region extending over a wider concentration region than monovalent ones [79];
iv) the ion specificity is small for the simple cations, but appreciable for the anions [79];
v)   the effect is quantitatively different for different polymers (*e.g.*, PEO gives much smaller effects than EHEC);
vi) the effect is not limited to water but is observed also in formamide [85,86]. Interestingly, cationic surfactants give larger effects than anionics in formamide.

Studies of the ternary phase diagrams of nonionic SHP/ionic surfactant/water systems are yet few and incomplete [72-75]. Partial phase diagrams in mixtures of SDS with three polymers, EHEC, UCON and PPO, have demonstrated an associative phase separation which becomes more pronounced as temperature increases [75]. Low concentrations of added electrolyte increase the extent of phase separation.

We know from the lowering of the CMC that there is an attractive interaction between these polymers and ionic surfactants, which could drive an associative phase separation. In the absence of electrolyte, however, associative phase separation is accompanied by a significant entropy loss due to the enrichment of the surfactant counterions in the concentrated phase. Therefore,

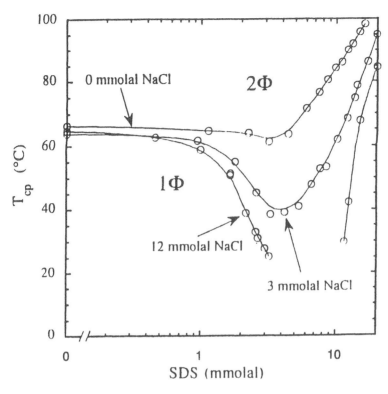

FIG. 20 The effect of SDS, with and without added NaCl, on the cloud point of EHEC (1 wt%). From ref. 76 with permission.

phase separation is strongly inhibited, as can be seen from the increase in LCST as small amounts of a long-chain ionic surfactant is added to a polymer solution. Adding salt eliminates the entropic penalty and phase separation becomes strongly favored as can be seen from the often quite dramatic decreases in LCST on surfactant addition in the presence of very low concentrations of salt. The non-micellized surfactant will have an effect analogous to that of simple salt, explaining the LCST decreases observed for surfactants with high CMC even in the absence of salt.

The degree of counterion association is generally higher for cationic than for anionic surfactants, due to nonelectrostatic attractive interactions [48]. This explains why the effects of added electrolyte are smaller for cationic surfactant systems. The strong counterion specificity for cationic, but not for anionic, surfactants, is likewise due to these other (dispersive) interactions.

With respect to macroscopic appearance, the aqueous nonionic polymer/ionic surfactant mixtures are often much more complex than indicated by the LCST curves presented. Increasing the temperature for a polymer/surfactant solution can lead to major rheological effects of significant technical interest, in particular a dramatic increase in viscosity and/or gelation [87]. The rheological effects depend on subtle features of polymer structure, such as the blockiness of the substitution. This thermal gelation is correlated with the phase behavior [88,89] and can best be seen as a microphase separation, according to the recent conclusions of Cabane *et al.* [71]. The polymer molecules become more strongly associating the higher the temperature, leading to polymer "lumps" of different sizes. An ionic surfactant will have a dispersing effect and cause a reduction in size of the lumps to the order of 500 Å, as well as a swelling of the gel network. The mechanical rigidity of the gels comes from the association of the polymer molecules through the lumps.

B. Oppositely Charged Systems

Hansson and Almgren [49] compared the associative phase behavior of a hydrophilic and a hydrophobic main chain polyelectrolyte (NaPA and NaPSS, respectively) in their mixtures with oppositely charged surfactants. The results are shown in Figure 21. For NaPSS, the phase separation is suppressed, especially at low concentration, when the polyelectrolyte is in excess with respect to charge equivalence. The explanation of this behavior should be quite similar to that given for oppositely charged HAP/surfactant mixtures in section III.C: a quasicomponent, *i.e.*, the molecularly mixed polymer-surfactant complex, is formed, and phase separation does not occur until the solubility limit for this complex has been reached, close to charge neutrality. Figure 21b shows the dilute region in more detail, and also shows that the complexes, *i.e.*, the polymer-bound surfactant micelles with PSS, are formed at much lower surfactant concentrations. In contrast, the phase separation with NaPA occurs almost immediately on micellization of the surfactant, except at very low polymer concentrations, where the phase behavior reflects the finite solubility of the complex in the presence of 1 mM of simple salt.

We know of no results showing segregation in oppositely charged mixtures involving an SHP. Even large concentrations of salt do not seem to inhibit the formation of a mixed micellar complex [90]. A segregation between complexes and free micelles, as for HAP/surfactant mixtures, might be possible at conditions of high salt and large excesses of surfactant, but does not seem to have been observed.

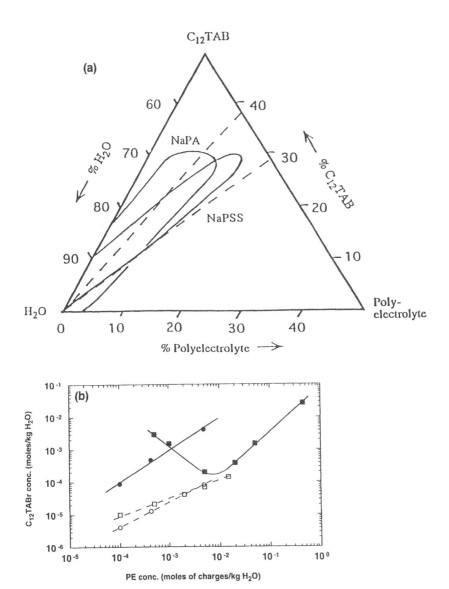

FIG. 21 Comparison of phase behavior for mixtures of $C_{12}TABr$ with NaPA or NaPSS. a) (top): quasi-ternary representation; dotted lines indicate charge equivalence; b) (bottom): log-log representation of the dilute region. Solid lines are cloud-point curves, dotted lines show the CMC values for $C_{12}TABr$ in the respective polyelectrolyte solutions. Circles are for NaPSS, squares for NaPA. Data from refs. 49, 126 and P. Hansson (unpublished).

C. Hydrophobically Associating Polymers

Thuresson *et al.* [76] compared the effects of added surfactants on the LCST of a comparatively hydrophilic EHEC with that of its hydrophocially modified variety, HM-EHEC, the latter containing 1.7 nonylphenol groups per 100 sugar units. Both nonionic and ionic surfactants were investigated. Results for SDS were obtained also in the presence of salt. Compared to the results for unmodified EHEC (Figure 19a), the hydrophobe modification led to a lowering of the LCST at low SDS concentrations, and a broadening and deepening of the minimum; the qualitative behavior was, however, the same for HM-EHEC as for EHEC.

More significant differences were found with the nonionic surfactant $C_{12}E_8$, which did not form mixed micelles with the unmodified EHEC (Figure 22). The LCST of EHEC was completely unaffected by $C_{12}E_8$, up to quite high concentrations where, presumably, a segregative phase separation occurred. In contrast, the curve for HM-EHEC with the same surfactant showed a non-monotonic behavior, quite similar to that found for ionic surfactants with both types of EHEC. The initial decrease on additions of small amounts of surfactant was interpreted in terms of an additional hydrophobic attraction between the mixed HM-EHEC-surfactant complexes, conferred by the surfactant molecules. Eventually, this effect was overtaken by the solublilizing power of the added surfactant which, at sufficiently high concentrations, gradually dissolved the mixed micellar crosslinks between the polymer chains, making the complex more and more soluble. At high surfactant concentrations, the one-phase area extended to higher temperatures for HM-EHEC (which formed a complex with the surfactant) than for EHEC (where no complex was formed). Interestingly, the minimum in the LCST for the HM-EHEC-$C_{12}E_8$ complex occurred at a surfactant-to-hydrophobe ratio of ~5. This is a regime where the surfactant begins to dominate in the mixed micelles, and consequences in terms of dissolution of crosslinks are expected. This interpretation is given independent support by the onset of a decrease in viscosity with added surfactant in HAP/surfactant mixtures at similar mixed micellar stoichiometries [58].

Alami *et al.* [91] obtained results similar to those of Thuresson *et al.* on mixtures of hydrophobically end-capped PEO ($C_{12}EO_{460}C_{12}$, M_W= 20300) with $C_{12}E_8$. The mixtures displayed an LCST that was lower (by up to 40°C) than that of either of the solutes on its own. A well-defined minimum in the LCST appeared when the concentration of either the HAP or the surfactant was varied, that of the other solute being constant. The stoichiometry of the mixed complex and, indeed, the composition of the mixed micelles at the LCST minimum, were found to be quite insensitive to the overall concentration in the studied range (0.5 - 4 % w/w of HAP). With HM-EHEC, the mixed micelles at the minimum had a roughly five-fold excess of the surfactant over the HAP

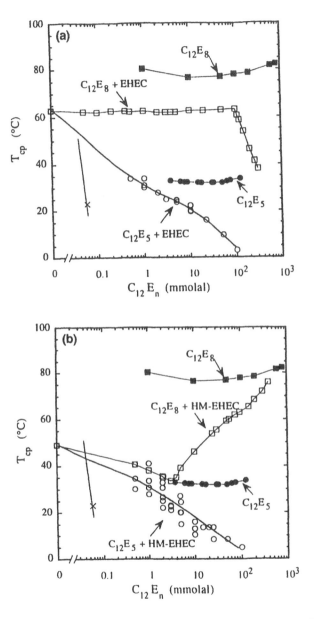

FIG. 22 The influence of $C_{12}E_5$ and $C_{12}E_8$ surfactants on the LCST of EHEC (top) and HM-EHEC (bottom). Data from ref. 76.

hydrophobes. However, the viscosity maximum for $C_{12}EO_{460}C_{12}/C_{12}E_8$ mixtures occurred at a significantly higher surfactant content ($\beta_{sh} \approx 16$). No segregative phase separation was found in these mixtures, but this--if it indeed occurs--would only be expected at much higher concentrations than those investigated [12].

V. PROTEINS

A large literature exists on interactions between proteins and surfactants in dilute solution [92]. Of particular relevance for this chapter are reports on precipitation and redissolution phenomena occurring on the successive addition of charged surfactants to solutions of oppositely charged proteins [93-98]. However, it is only very recently that such phase studies were extended to concentrated mixtures [99,100].

Apart from their intrinsic importance, globular proteins in particular have the distinct advantage over those polymers whose phase behavior we have discussed so far, in that they are monodisperse. A complicating factor, on the other hand, is that the association of certain surfactants, ionics in particular, may give rise to a denaturation of the protein, so that the mixtures with surfactants involve proteins of different conformational states. The native globular protein is quite different from a flexible polymer in that it is compact and essentially lacks conformational degrees of freedom. Complexes involving a denatured protein, on the other hand, should be more similar to other complexes of surfactants and partially hydrophobic, flexible polymers.

Both similarities and differences may thus be expected when the phase behavior of protein/surfactant mixtures are compared with those involving flexible linear polymers. These expectations have, indeed, been confirmed by the studies of Morén et al. Figure 23 shows the quasi-ternary phase diagram [99] for mixtures of lysozyme and SDS in water. It is interesting to compare this phase diagram to that for mixtures of SDS with catHM-HEC, Figure 14a above. Both diagrams refer to mixtures of SDS with positively charged, partially hydrophobic polymers, and both display an associative phase separation at low surfactant contents and a redissolution to a one-phase solution at higher surfactant contents. Moreover, the redissolution phase boundaries are in both cases straight lines. The latter indicates that the condition for redissolution is that some fixed number of surfactant molecules (the actual number depending on pH and on the ionic strength [96]) have joined the polymer-surfactant complex, and that this number is reasonably independent of the complex concentration. Studies on more dilute mixtures of lysozyme/SDS [96], as well as phase diagrams of other oppositely charged protein/surfactant

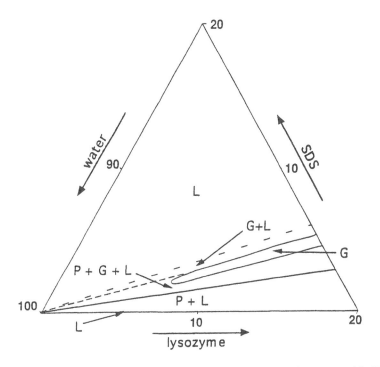

FIG. 23 Quasi-ternary phase diagram for mixtures of lysozyme with SDS, showing single-phase and multiphase areas. L, G, and P denote solution, gel, and neutral precipitate, respectively. From ref. 99 with permission.

mixtures [100] have shown that the redissolution of a protein typically occurs in the cooperative part of the binding isotherm, when the free surfactant concentration approaches the CMC. The protein is here measurably denatured by the bound surfactant [96].

   Despite the above similarities in the redissolution behavior, the nature of the phase separation of studied protein/surfactant mixtures differs in important respects from that of other oppositely charged polymer/surfactant pairs. The early binding to the protein results in the formation of a neutral complex that precipitates out at very low surfactant concentrations [96,99]; *cf.* Figure 24. This complex may be regarded as a compound of a fixed composition, which is in equilibrium with a solution containing the excess protein. This interpretation is supported by the analysis of the lysozyme-SDS precipitates [99]. Below charge neutrality, the tie-lines were found to extrapolate to the neutral complex, independently of the global SDS/protein ratio. This behavior is quite different from the phase separation observed for catHM-HEC + SDS, where a liquid

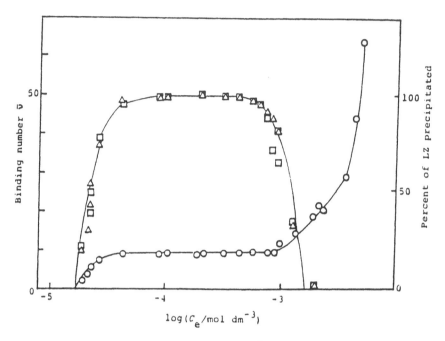

FIG. 24   Dilute mixtures of lysozyme (3.76×10^{-5} M) and SDS in water: Number of bound SDS molecules per lysozyme molecule, and the fraction of the protein recovered in the precipitate phase, as functions of the free SDS concentration. From ref. 96 with permission.

coacervate was formed, which always contained virtually all of the polymer, but where the surfactant/polymer ratio varied with the global composition. The solid-like precipitates formed in dilute mixtures of NaPA and cationic surfactants (section II.B.1) were also found to have a variable composition [44], unlike the protein-surfactant precipitates.

At surfactant concentrations exceeding charge neutrality, Morén and Khan discovered a new "gel" phase, distinct from the solution phase (Figure 23). While not a true compound, this phase appeared only in a narrow range around a well-defined SDS/lysozyme ratio. Both gels and neutral precipitates were found also in mixtures of β-lactoglobulin with C_{12}TAC [100]. The overall features of the phase diagram in Figure 23 thus appears to be quite general.

It seems natural to ascribe the above differences between proteins and linear polymers, which occur at relative low levels of surfactant binding, to the globular nature of the protein. A well-defined tertiary structure should generally favor the formation of well-defined surfactant complexes. Moreover, the water content of a neutral protein-surfactant precipitate should be quite low, as it is formed by rigid particles. This is in contrast to the gel-like coacervates formed in stoichiometric mixtures of SDS and catHM-HEC, which contained 85 - 90 % water even at charge neutrality [55]. The large water content of the latter phase should be attributable to the osmotic pressure generated by the intrinsically water-soluble, semi-flexible parent polymer (HEC).

It should be noted that the occurrence of compounds of more-or-less well-defined stoichiometries has consequences for the interpretation of the binding isotherms observed in protein/surfactant mixtures [94-97,101]. As is shown in Figure 24, these isotherms have many similarities with those shown for SDS and catHM-HEC above (Figure 15), displaying a high affinity binding before charge neutrality, followed by an anti-cooperative plateau and, finally, a cooperative binding as the CMC is approached. However, in the multiphase regions, the average binding to the protein does not generally reflect the stoichiometry of the complex; rather, it reflects the coexistence, at varying proportions, of more than one of three invariant species: the free protein, the neutral precipitate, and the complex forming the gel. This point does not seem to be generally appreciated.

As noted above, the onset of surfactant binding to an oppositely charged protein is accompanied by the precipitation of a neutral complex; cf. Figure 24. A simple interpretation of the "cooperative binding" observed at this critical (low) surfactant concentration is in terms of the solubility product, K_s, of the neutral compound, according to the following simple equilibrium,

$$PS_n(s) \leftrightarrow P^{n+} + nS^- \; ; K_s = [P^{n+}][S^-]^n \qquad (5)$$

given here for the explicit case where the complex is formed by a negatively charged surfactant, S^-, and a net positively charged protein, P^{n+}. The critical surfactant concentration for binding is that where the solubility product is just exceeded. Since the complex involves many surfactant molecules ($n \approx 10$), the critical surfactant concentration depends only weakly on the protein concentration; moreover, the binding becomes quite cooperative.

VI. INFINITE SURFACTANT SELF-ASSEMBLIES

As we have seen above, the extent of phase separation increases with increasing molecular weight of either cosolute in mixed polymer/surfactant systems. Specifically, micellar growth augments phase separation. Strong tendencies to phase separation are thus expected in systems where the surfactants form non-discrete aggregates, connected in 1, 2 or 3 dimensions. The field of mixtures of polymers with infinite surfactant self-assemblies is vast but experimental phase diagram work is still in its infancy. We will here mainly outline general expectations as put forward in recent work by Olsson and coworkers [102-108] and confront them with the experimental observations of these authors. Two infinite surfactant structures will be considered: the lamellar liquid crystalline phase, which is infinite in 2 dimensions, and the bicontinuous microemulsion, which is connected in 3 dimensions. The principles discussed are basically the same for both structures. However, it is easier to picture lamellar systems, because of the periodicity in one dimension; earlier studies on polymers in lamellar phases can be found in [109-111].

We can view these infinite surfactant phases in terms of connected surfactant self-assembly structures separated by water channels of different shapes. In the lamellar phase there are water layers, which are planar with a thickness that increases in proportion to the water content. The water domain in a bicontinuous microemulsion can be regarded as a tubular network, with a certain mean thickness of the water tube. In either case, the thickness (D) of the water domain may be of the same order of magnitude as the size of a polymer molecule, which we can approximate by twice its radius of gyration (R_g). This leads to new effects, compared to the situations discussed in previous sections. If $R_g \ll D$, size does not influence the ability of the polymer to enter the surfactant structure. If $D < 2R_g$, however, the unperturbed polymer cannot enter into the unperturbed surfactant structure. This conflict leads to different results, depending on the polymer-surfactant interactions. In a case where there is no strong attraction between the polymer and the surfactant assembly, the polymer will simply be excluded from the water channel, since penetration would lead to a quite unfavorable reduction in its conformational entropy. For a strongly attractive situation, on the other hand, penetration will occur and it will affect the swelling of the lamellar phase. If the polymer is charged, the incorporation will strongly affect the swelling due to electrostatic interactions. The solubility of an adsorbing polymer has been found to be much larger in bicontinuous microemulsions than in the lamellar phase [108], demonstrating that surfactant film topology is an important factor in controlling polymer compatibility with these surfactant phases.

Studies of the lamellar phase of $C_{12}E_5$ in hydrocarbon/water systems mixed with non-modified or hydrophobically modified sodium poly (acrylate) (PA or HM-PA) [102] have shown that unmodified PA (nonadsorbing) tends to keep out of the lamellar phase; there is a segregation, leading to a surfactant phase with a reduced water content and thus a lower degree of swelling. With HMPA, the strong association to the bilayers gives rise to an increased swelling owing to electrostatic repulsions. Due to entropic effects resulting mainly from the counterions (but also from restricting the conformational freedom of the polymer chains), the polymer-containing lamellar phase becomes unstable at higher surfactant concentrations (lower D values). Therefore, there will be a

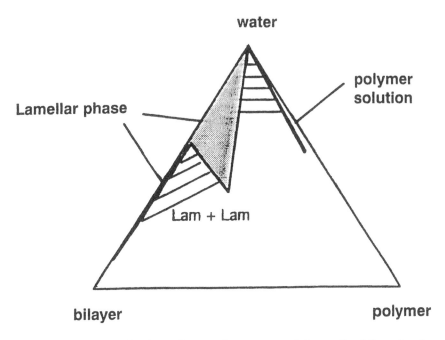

FIG. 25   A schematic phase diagram for aqueous mixtures of a bilayer-forming surfactant and an adsorbing polymer. (U. Olsson, unpublished.)

phase separation into one polymer-rich and one polymer-lean lamellar phase [102]. Adding salt will dramatically decrease the tendency to swelling. This will stabilize the lamellar phase at higher surfactant concentrations and also create a phase separation into a concentrated lamellar phase and a dilute solution [102].

A generic ternary phase diagram for a "bilayer"/polymer/water system can be sketched as in Figure 25. We note that an adsorbing polymer is incorporated to a considerable extent into the lamellar phase, before a polymer-rich solution forms, at low bilayers concentrations, but that incorporation is very limited at high bilayer concentrations resulting in the phase separation into two lamellar phases. This was confirmed by observations of two X-ray diffraction patterns with different spacings [112].

Because of the technical and theoretical interest, a number of studies have concerned water-soluble polymers in microemulsions; the technical applications have concerned, for example, enhanced oil recovery, cleaning- and cosmetics formulations. Microemulsions exhibit a rich structural variation, the structure changing with surfactant(s), nonelectrolyte and electrolyte cosolutes and temperature. The principal extreme structures are the water-in-oil droplet, bicontinuous, and oil-in-water droplet structures. Effects of added polymer will be quite different for these different structures. The considerations for oil droplet structures are analogous to those for the micellar systems treated above. Addition of a water-soluble polymer to surfactant phases of different microstructure illustrates the above-mentioned role of connectivity on polymer-surfactant compatibility: while sodium poly(acrylate) is effectively insoluble in the lamellar and sponge phases of a nonionic surfactant, it is highly soluble in oil-in-water microemulsions [102].

We will here focus on the challenging bicontinuous systems; the first study of these was accomplished only recently [104]. We will consider a three-component so-called balanced surfactant system, where the spontaneous curvature of the surfactant films is zero and the microemulsion forms at equal volume fractions of oil and water. The phase diagram is given in Figure 26. A convenient way of investigating the polymer-surfactant interactions is to

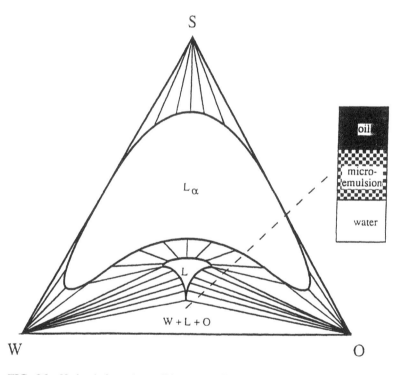

FIG. 26 Under balanced conditions a surfactant(S)/water(W)/oil(O) system gives a microemulsion phase (L) and a lamellar liquid crystalline phase (L_α), in addition to the phases dominated by the three components. From ref. 106 with permission.

observe the change in relative phase volumes on addition of polymer to a sample in the three-phase region of microemulsion+oil phase+water phase [104]. Experiments on the system $C_{12}E_5$/dextran (shown to be segregative in section II.A.1 above) confirm our expectation that a polymer which is not attracted to the surfactant aggregate will enter the bicontinuous phase extensively if it is small compared to the width of the water channels, but not at all if it is too large (Figure 27) [105]. The distribution of polymer between the three phases will affect the phase volumes (Figure 28), which can be understood in terms of the effect of the polymer on the osmotic pressure of the aqueous domains (water activity) [105]. A polymer that is excluded from the microemulsion phase will exert an osmotic stress on the latter phase, hence its water content will decrease. From arguments based on the Gibbs-Duhem relation it can be shown that there should be a concomitant expulsion of oil into the oil phase. Addition of high molecular weight oil-soluble polymers has a completely analogous effect on the phase volumes in a middle-phase microemulsion system.

The behavior will be very different for an attractive polymer-surfactant interaction, where the polymer will adsorb onto the surfactant films and form mixed aggregates. Here a lowering of the water activity in the microemulsion results and there is an increased volume of the microemulsion. However, the initial swelling of the microemulsion is reversed when the surfactant films are saturated with polymer. Excess polymer is dissolved in the lower aqueous phase where the osmotic pressure results in a contraction of the microemulsion phase (Figure 28b).

The dramatic effect of hydrophobic "stickers" on the polymer is further illustrated by the addition of PEO and hydrophobically end-capped PEO (HM-PEO) to nonionic surfactant systems [103]. The two polymers destabilize both the lamellar and the oil-in-water microemulsion phases. However, with PEO a segregative phase separation is seen but with HM-PEO, phase separation is associative.

Many significant questions are still open in the field of polymers in infinite surfactant self-assemblies. One important issue relates to the effect of an adsorbed polymer on the spontaneous curvature of the surfactant film. Changes in curvature may induce transitions between phases, for example from one of infinite surfactant self-assemblies to one of discrete aggregates. Under good solvent conditions, the adsorption of a polymer onto the surfactant film from the aqueous side results in an increase in the spontaneous curvature away from water [106]; the reverse would apply for adsorption from the oil side. For aqueous systems of PEO derivatives, the spontaneous curvature effects and the swelling can be controlled by temperature because of the temperature-dependent solvency conditions.

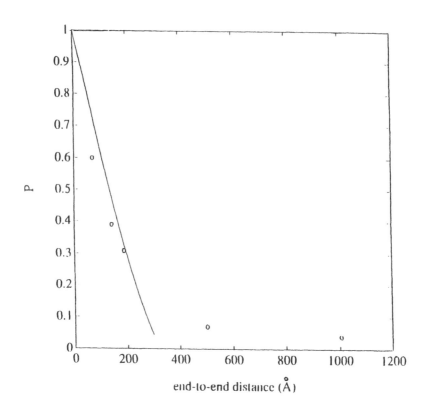

FIG. 27   Variation of the partition coefficient of dextran between the balanced microemulsion and the excess aqueous phase for the $C_{12}E_5$/water/decane system with the unperturbed end-to-end distance of the polymer. The solid line is a theoretical prediction. From ref. 106, with permission.

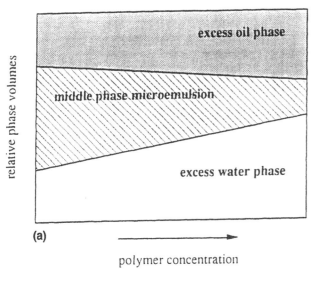

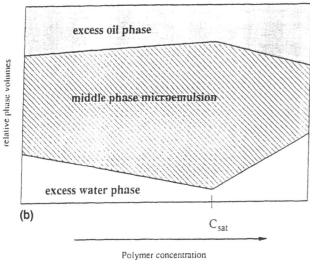

FIG. 28   Typical variation of the relative phase volumes of a three-phase microemulsion system with the polymer concentration for a non-adsorbing (a, top) and an adsorbing high molecular weight polymer (b, bottom). C_{sat} denotes the saturation concentration for association of the polymer to the surfactant films. From ref .106 with permission.

VII. MODELS OF PHASE EQUILIBRIA IN MIXTURES OF HOMOPOLYMERS AND MICELLES

The phase behavior of a mixture is a manifestation of the interactions between the molecules in the studied system, and this section will deal with the modelling of the phase equilibria of aqueous polymer/surfactant systems in terms of such interactions. This is useful for several reasons. If a theoretical model is capable of reproducing the experimentally observed phase behavior, then one can trust the intermolecular interactions used in the model with some confidence. These interactions may then be used to make meaningful predictions of how the system can be modified in a systematic way, in order to obtain a desired behavior. They can also be used to indicate interesting features in unexplored regions of phase space, or to predict the equilibrium phase behavior in (concentrated) regions that are not easily accessible experimentally.

A. Approaches Taken

Theoretical modelling of chemical systems can be performed at various levels of sophistication. The most sophisticated approach is to use theoretically derived potentials and to simulate the properties of the system using all-atom Monte Carlo or molecular dynamics procedures. To model a phase diagram by such methods is, however, a difficult task even for simple two-component liquids, and at present quite beyond reach for systems as complex as aqueous polymer/surfactant solutions. Thus, one is still forced to rely on simple models, capable, at best, to capture the gross features of the phase diagram. In the literature, one finds two different approaches to this problem. Both are based on polymer molecules and micellar surfactant aggregates, and both ignore the equilibrium between free surfactant molecules and micellar aggregates. The differences between the models have their origin in the description of the micelles and, also, in the description of the micelle-polymer interactions.

The oldest approach is based on the concept of depletion flocculation: An impenetrable surface (such as a particle surface) restricts the number of possible conformations of a polymer molecule close to that surface. Therefore, polymer molecules avoid the surface, and there will be a depletion of polymer segments in the vicinity of the surface. This increases the chemical potential of the micellar aggregates and, ultimately, leads to a segregative phase separation. Models to describe this phenomenon have been derived by Hall [113,114]. The analysis is based on the following general mechanism. If the minor component in a mixed solvent avoids an added colloidal particle, for enthalpic or entropic reasons, then the solubility of that colloidal particle is reduced [113-116]. In our case, the minor component of the mixed solvent is the polymer molecule, and the added particle is the surfactant micelle.

In order to treat the entropic repulsion between the particle and the polymer solution, one has considered the particle as an uncharged sphere, which cannot be accessed by the polymer molecule [10,117]. The effective interaction between the sphere and the solvent is then calculated using a statistical mechanical integral equation technique [10,11,117]. The model has been used to model the experimental phase behavior of nonionic polymers mixed with nonionic surfactants of the C_nE_m type. To describe the effective polymer-micelle repulsion a parameter, the depletion layer thickness, is introduced into the model. There are, however, several problems with this approach. Micelles of the C_nE_m type are "hairy," and thus there is no well-defined size of the impenetrable sphere. Moreover, the micelles are often smaller than the size of a polymer coil. This means that the usual simple estimate of the depletion layer in terms of the radius of gyration of the polymer, which holds for a large particle, is certainly wrong. Lastly, the depletion model as such is based on the assumption that the micelles do not interact enthalpically with the components in the solvent. In the real system, however, enthalpic effects must affect the concentration of polymer segments close to the micelle. Because of these complications, it is difficult to predict the depletion layer thickness. It is therefore usually treated as a parameter, which is fitted to get agreement between theory and experiment [11].

For the rest of this section, we will concentrate on an entirely different approach. This approach is based on the Flory-Huggins (FH) lattice model [25], as developed for solutions of two flexible polymers in a common solvent. The surfactant micelle is regarded simply as one of the two polymers. Since random mixing within each single phase is a basic assumption in the FH model, entropic depletion effects are neglected entirely. On the other hand, short-range (enthalpic or entropic) interactions are included explicitly via interaction parameters. The FH model has been used to describe both nonionic and ionic mixtures, and the arguments for using the model differ for the two types of mixture. As noted above, a nonionic micelle of the C_nE_m type looks like a hairy ball in aqueous solution. It thus carries a certain resemblance to the flexible polymer coil for which the FH model was derived. We may also incorporate the temperature-dependent interactions of the EO groups, which are important both for EO type polymers and for C_nE_m micelles in the model. Still, as was pointed out in section II.A.3. above, there are important differences between polymer/polymer and polymer/surfactant mixtures, so the model should only be used to investigate general trends. On the other hand, owing to the severe approximations inherent to the FH model, the same may be said about applying it to mixed polymer solutions.

Arguments for the use of a generalized FH model for mixtures of involving ionic micelles and/or polymers are based on the observation that for such systems, the mixing entropy is totally dominated by the counterions to the micelle and polyions. The contribution from the micelle and polyions are of little importance. Consequently, it is enough that the model is qualitatively correct with respect to the concentration of the micelles and polymers, as long as it models the influence of the counterions in a more realistic way. The easiest way to include the latter along the lines of the FH model is to simply treat each ionic species as a monomeric species with its own set of interaction parameters, and to impose the constraint that each phase in the model must be electrically neutral.

B. LCST Polymers and Surfactants

Before we give the basic equations necessary for the description of polymer surfactant systems according to the FH model, there is one additional complication that must be discussed. A common feature of both polymers containing EO groups and nonionic surfactants of the EO-type is that their solubility in water decreases with increasing temperature, as discussed above. Such a behavior can never be explained within a standard FH model. In practice, it means that the effective interaction between the solvent and the solute must be repulsive, and that the interaction strength must increase stronger than linearly with increasing temperature.

Three different explanations, on a molecular level, have been given to this phenomenon. The oldest one is due to Kjellander and Florin. They noted that the structuring of water around a hydrophobic solute results in a temperature dependence for its solubility in water [118], similar to what is observed for EO-containing solutes. Based on this observation they argued that this mechanism is active for the latter type of compounds. Later Lang [119] applied the old ideas of Hirschfelder, Stevensson and Eyring [120], (HSE) that a reversed temperature dependence of the solubility may arise from the properties of the interaction potential between the solvent and the solute. HSE showed that if there were small strongly attractive regions and large repulsive regions in this potential, then a reversed temperature dependence of the solubility could be obtained. Goldstein used these assumptions and developed a thermodynamic model for the phase behavior of PEO in water [121]. Somewhat later one of us suggested, on the basis of quantum chemical and statistical-mechanical calculations, that the reversed solubility behavior was linked to the conformational equilibrium of the EO chain, and a thermodynamic model was developed to describe this effect [70,122]. The essential idea in the model is that polar conformations are favored by a polar surrounding, and non-polar conformations are favored by a non-polar surrounding. It is further assumed that the polar conformations are energetically stabilized and that the non-polar ones are entropically favored. Recently, all-atom statistical-mechanical simulations have shown that the conformational equilibrium is coupled both to the structuring of water and to hydrogen bonds between water and the EO system to produce the anticipated thermodynamics [123].

An interesting consequence of the reversed solubility of these systems is that they show a rich phase behavior and that they can serve as probes of the effective intermolecular interactions between molecules.

C. Basic Equations of the Flory-Huggins Model

We will consider a system built up from m components, where each has the possibility to assume $L(i)$ different conformations. The relative probability of each conformation at infinite temperature is given by F_{ij}, and the degree of polymerization of component i is M_i. Using this notation, one may write for the entropy S of a phase

$$S = -nR \left[\sum_{i=1}^{m} \left(\frac{\phi_i}{M_i} \ln \phi_i + \phi_i \sum_{j=1}^{L(i)} P_{ij} \ln \frac{P_{ij}}{F_{ij}} \right) \right] \qquad (6)$$

In this equation, ϕ_i is the volume fraction of component i, P_{ij} is the probability for conformation j of component i, n is the total number of moles of monomers or the systems, and R is the gas constant. Using the same notation, we may write the internal energy of the phase as

$$U = 0.5n \sum_{i=1}^{m} \phi_i \sum_{j=1}^{L(i)} P_{ij} \sum_{k=1}^{m} \phi_k \sum_{l=1}^{L(k)} P_{kl} W_{ij,kl} \qquad (7)$$

In Eq. (7), $w_{ij,kl}$ is an interaction parameter describing the effective interaction between conformation j of component i and conformation l of component k. Eqs. (6) and (7) contain some unnecessary parameters. Without loss of generality, F_{il} can be set equal to 1 and the diagonal interaction parameters $w_{il,il}$ can be set equal to 0. The free energy for the phase can be obtained from

$$A = U - TS \qquad (8)$$

Here T is the absolute temperature and A the free energy. The total free energy for a given composition is minimized with respect to the probabilities P_{ij}. If two or more of the components in the system are charged, then one must impose the condition that only such variations are allowed that yield phases that are electrically neutral, as was pointed out above.

In order to calculate a phase diagram one proceeds as follows. 1) Specify the maximum number of phases possible for the considered system. 2) Distribute the different components among the phases in some way and calculate the free energy of the system for this composition of the different phases. 3) Redistribute the components among the phases slightly; if the free energy is lowered, then accept this new composition of the phases; otherwise, go back to the previous best composition and redo point 3 again until some convergence criterion is fulfilled. For a more thorough derivation and discussion see [75]. In the way described above one obtains the equilibrium composition of the phases, the binodal lines and the critical points of the system. Gerharz and Horst have also used the method to calculate the spinodal lines [14].

There may naturally be many objections to a simple minded theory as the one described above. First of all it can not be derived in the same way as the ordinary FH theory when one of the components is of micellar type. Second, the equilibrium between monomers and micelles is completely neglected. This also implies that it is assumed in the model that the size and shape of a micelle is independent of all solute concentrations (surfactant, polymer and, possibly, salt). On the other hand, one may argue that the limiting behavior at low concentration is reasonable, that the total mixing entropy is small (as it should be), and that a large part of the errors in the description of the entropy can probably be described through the interaction parameters. It may also be argued that the problems indicated are probably not larger than those associated with assumptions of the FH model as applied to polymer solutions, *i.e.*, that a polymer segment has the same size and volume as a solvent molecule, that the interactions are non-directional, that there is random mixing in the system and, finally, that the interaction parameters are temperature independent. Considering the problems indicated above, it is clear that we cannot expect quantitative agreement between the model predictions and experimental observations, but it seems reasonable to assume that the qualitative picture emerging from the calculations is realistic. The main advantage of the modelling compared to experimental investigations is that it requires much less work, and that large parts of both the concentration and temperature regime can easily be investigated.

D. Sample Calculations

The FH model has been used to model the phase behavior of aqueous solutions of nonionic polymers with nonionic surfactants [75,124,125], mixtures of a nonionic polymers with an ionic surfactant, [75,124] and mixtures of similarly [21] and oppositely [32,35,37] charged ionic polymers and surfactants. We will show some representative results here. Since the calculations make no distinction between a surfactant micelle and a polymer, both species will here be referred to as "polymers."

1. Nonionic mixtures

The FH model predictions for simple ternary polymer/polymer/solvent mixtures have been thoroughly investigated in the literature. Most of the older calculations have focussed on segregating systems, showing the roles of the lengths of the polymers and the interaction parameters, including the (important) interactions with the solvent [3-5]. Recently, calculations have also been performed for associating systems [6,75]. Here, we will show one example of a calculation of the temperature dependent phase diagram for a mixture of an LCST polymer with a surfactant with ordinary solution behavior. In the model there is a weak effective attraction between the surfactant and the polymer. As a consequence addition of the surfactant slightly lowers the LCST and a phase separation of associative nature is observed. The calculated phase diagram is presented in Figure 29 [127]. The interaction parameters used are shown in the figure caption. The LCST for the pure polymer/solvent system is 30 °C. This means that the two-phase region seen in Figure 29a is induced by the surfactant and that the polymer is completely soluble in pure water at this temperature. The fact that the two-phase region appears so close to the water polymer axis is a consequence of the high degree of polymerization used for the polymer and the surfactant in the modelling. In practice, this means that only a very small amount of surfactant is needed to induce a phase separation at temperatures that are not too far below the clouding temperature. At the higher temperatures the two phase region reaches the water/polymer side of the phase diagram since they are calculated at temperatures above the LCST.

2. Mixtures of similar low charge

Figure 30 shows calculations by Nilsson [21] for an intrinsically segregating polymer pair where small fractions of similarly charged ionic groups are introduced on both polymers, as in the experiment described in Figure 6 above.

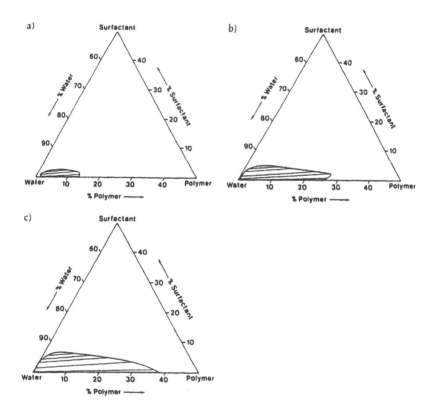

FIG. 29 Phase diagram calculated for an LCST polymer, a surfactant and a solvent at a)17°C, b) 33°C, and c) 49°C. The following interaction parameters have been used. w_{1p} = 562.6 J/mol, w_{1u} = 9209.9 J/mol, w_{pu} = 5491.0 J/mol, w_{uu} = 8793.0 J/mol, w_{13} = -4927.1 J/mol, w_{p3} = -8039 J/mol, w_{u3} = -907.6 J/mol. Degree of polymerization: polymer, 2000; surfactant: 100. Tie lines and phase boundaries shown. From ref 127 with permission.

The dashed reference phase diagram shows the segregation for the uncharged polymers; this has purposely been made asymmetric through differences in the polymer-solvent interactions. The solid diagrams are for the corresponding polyelectrolyte mixtures, where the fraction of ionic segments of PE1 is varied but is kept constant for PE2.

When the fractions of charged residues are the same on both polyelectrolytes (Figure 30a) the two-phase area is smaller, compared to the nonionic mixture. On increasing the fraction of charges on PE1, the extension of the two-phase area changes in a non-monotonic fashion. A maximum two-phase area occurs when the slope of the tie-lines in the polyelectrolyte system is the same as the slope in the reference system (Figure 30c). At this point, the phase diagrams of the nonionic and polyelectrolytic mixtures are almost identical. This agrees with the conclusion drawn in section II.A.2 above: the segregation becomes maximal in the balanced case, when the fractions of charged residues on the polyelectrolytes are chosen such that the dissociating counterions do not perturb the distribution of the polymers from that which would have been obtained in the absence of charges. A consequence of this is that the polymer that has the highest affinity for the solvent in the neutral mixture must also have the highest charge density in the balanced polyelectrolyte mixture.

3. Oppositely charged mixtures

Here we will present some new calculations, illustrating the complete phase behavior of mixtures of two oppositely charged polyions (P^+ and P^-) and two simple ions (A^+ and A^-) in water. The results are relevant to the experimental diagrams shown in section II.B above, except for those that involve anisotropic phases, since it is assumed in the theoretical modelling that all phases are liquid and isotropic. For simplicity, we will assume that the phase diagram is symmetric, i.e., that both polyions are described with the same interaction parameters, but that they have opposite charges. Similarly, the simple ions will be assumed to be equivalent apart from their opposite charges. The interaction parameters and the degree of polymerization are given in Table 1. These parameters are chosen in such a way that the parts of the phase behavior that can be experimentally observed are in agreement with what is frequently observed for oppositely charged polyelectrolyte/surfactant mixtures. The choice of a low degree of polymerization (20) may perhaps deserve special justification, since a larger value would have been more relevant for most systems studied experimentally. The advantage of this relatively small number

Piculell, Lindman, and Karlström

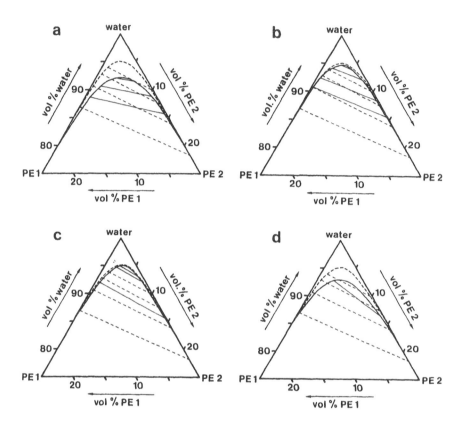

FIG. 30 Calculated phase diagrams for aqueous mixtures of intrinsically segregating similarly charged polyelectrolytes. The degree of polymerization is 200, for both polymers, and the interaction between polymer 2 and water is slightly repulsive. Dashed diagram for uncharged polymers, solid diagrams for 10% charged units on PE1 and a) 10, b) 15, c) 20, or d) 30% of charged units on PE2. From ref. 21 with permission.

TABLE 1 Interaction parameters (in units of kT) used to calculate the phase diagrams in Figures 31-32. The degree of polymerization was 20 for the polyions (P^+ and P^-) and 1 for the counterions (A^+ and A^-) and for the solvent.

Species	Species				
	solvent	polyion+	polyion–	ion+	ion–
solvent	0.0	0.1353	0.1353	–0.3007	–0.3007
polyion+	0.1353	0.0	–3.1874	2.1801	–4.9164
polyion–	0.1353	–3.1874	0.0	–4.9164	2.1801
ion+	–0.3007	2.1801	–4.9164	0.0	–2.1049
ion–	–0.3007	–4.9164	2.1801	–2.1049	0.0

is that the phase boundaries are not positioned at very high or very low concentrations. This means that trends occurring when the different parameters are changed are more clearly seen.

The complete phase diagram can be visualized in the pyramidal representation (cf. Figures 8 and 9 above). In the four corners of the square bottom the four possible ionic pairs (P^+P^-, P^+A^-, A^+P^-, and A^+A^-) are placed, and water is at the top. Consequently, each side of the bottom square corresponds to one of the ionic components, and components with the same charge are placed on opposite sides of the square. In Figure 31a we present the phase behavior for the water-free system. The observed behavior is a consequence of the chosen interaction parameters, where the P^+P^- and A^+A^- attractions are stronger than the P^+A^- or P^-A^+ attractions. The parameters have been chosen in this way to model the fact that the polyions contain hydrophobic material, whereas the simple ions are assumed not to do so. In Figure 31b and c we present cuts parallel to the bottom square, at 45 and 60% water. Here, as in the following, the position in the pyramid of the illustrated plane is indicated in a small insert in each figure. Figures 32a and b show the $P^+P^-/P^+A^-/H_2O$ and $A^+P^-/A^+A^-/H_2O$ surfaces of the pyramid, and Figure 32c the cut corresponding to the $P^+P^-/A^+A^-/H_2O$ mixing plane.

From Figures 31 and 32 it is clear that the phase behavior is characterized by two three-phase regions and two two-phase regions. For the present set of interaction parameters the upper (dilute) two-phase region is separated from the three-phase regions by a one-phase region. In the dilute two-phase region the phase separation is associative, and thus the P^+ and P^- ions are enriched in the same phase. With the present interaction parameters the simple salt (A^+A^-) has a slight preference for the water-rich phase. The origin of the associative two-phase region is the strong attraction between P^+ and P^- which results in a limited solubility of the complex salt (P^+P^-) in water.

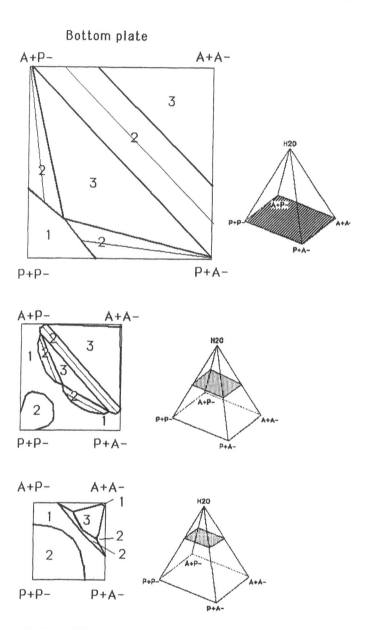

FIG. 31 Different cuts parallel to the bottom plane in the P$^+$/P$^-$/A$^+$/A$^-$/H$_2$O phase diagram. The cuts are made at 0, 45, and 60 % H$_2$O, as indicated in the inserts. Parameters from Table 1. Thicker lines indicate phase boundaries, thinner lines are tie-lines.

The concentrated two-phase region is clearly segregative; the tie-lines are perpendicular to the $P^+P^-/A^+A^-/H_2O$ mixing plane. This behavior is a result of the A^+P^+ and A^-P^- interactions together with the P^+P^- and A^+A^- interactions. The essential feature is that there is a stronger repulsion between A^+ and P^+ (or A^- and P^-) than between the A^+ ions (or P^+) ions among themselves. This means that it is unfavorable to locate A^+ and P^+ (or P^- and A^-) in the same phase. This is the force driving the phase separation into one phase with A^+P^- and one phase with A^-P^+. The phase separation where the A^+P^- and A^-P^+ pairs phase separate from the A^+A^- and the P^+P^- pairs originates from a larger effective attraction between A^+ and P^- than between A^+ and A^- or P^+ and P^-.

The two three-phase regions seem not to have been observed experimentally but, as we see, their existence is predicted from reasonable interaction parameters. Moreover it is only in the symmetry plane $A^+A^-/P^+P^-/H_2O$ that it is possible to come direct from a one-phase region into the three phase region. For all other projections the three-phase regions are hidden behind two phase regions. A consequence of this is that, for non-symmetric systems, one will always enter into a two-phase region before entering into the three-phase regions; the three-phase regions will be observed only at even higher polymer and surfactant concentrations where experimental observations are difficult.

An advantage of the theoretical modelling is that it is relatively easy to investigate how changes in the interaction parameters and/or the degrees of polymerization affect the phase behavior. In Figure 33a and b we illustrate the influence of the degree of polymerization by showing the correspondence to Figures 31a and 32c when the degree of polymerization is changed from 20 to 10 in the model. In a similar way we show in Figures 33c and d the effect of making the interaction between the salt and the polyion with opposite charge less attractive. If these latter figures are compared with Figures 31a and 32c it is clearly seen that a lower degree of polymerization decreases the extension of all two and three phase regions, whereas decreasing the attraction between A^+ and P^-, or P^+ and A^- leads to an increase of the upper (dilute) two phase region of associative nature and a decrease in the extension of the lower (more concentrated) two- and three-phase regions.

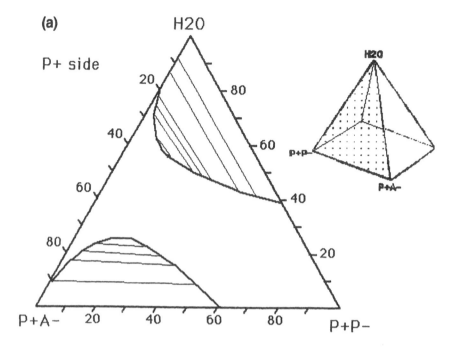

(a)

H2O

P+ side

20

80

40

60

60

40

80

20

P+A−   20   40   60   80   P+P−

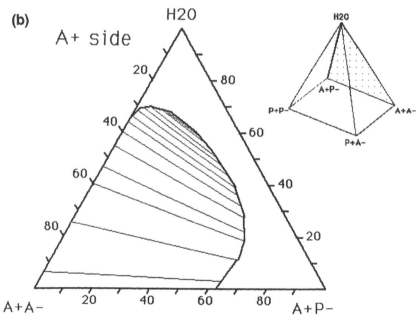

(b)

A+ side

H2O

20

80

40

60

60

40

80

20

A+A−   20   40   60   80   A+P−

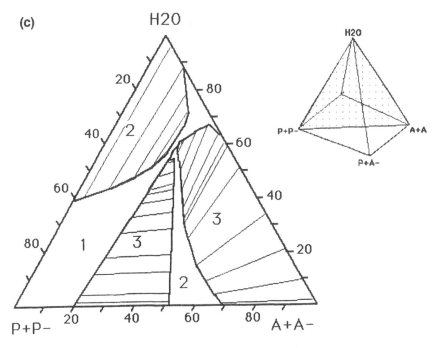

FIG. 32 The $P^+P^-/P^+A^-/H_2O$ (top, previous page) and $A^+P^-/A^+A^-/H_2O$ surfaces (bottom, previous page) of the pyramid, and the cut corresponding to the $P^+P^-/A^+A^-/H_2O$ mixing plane (above). The locations of the surfaces are shown in the inserts. Parameters from Table 1. Thicker lines indicate phase boundaries, thinner lines are tie lines (2-phase regions) or projections in the plane of three-phase triangles (3-phase regions).

VIII. CONCLUSIONS AND OUTLOOK

It may be useful at this stage to summarize some of our main conclusions in terms of general principles that govern the phase behavior of polymer/surfactant mixtures.

Charges on one of the components (polymer or surfactant) generally increases not only the solubility of that component, but also the miscibility of the system. This effect, which is due to the entropy of mixing of the small counterions, holds both for associating and segregating mixtures. In segregating similarly charged mixtures, the increased miscibility may be largely balanced out by introducing the appropriate amount of charges on the

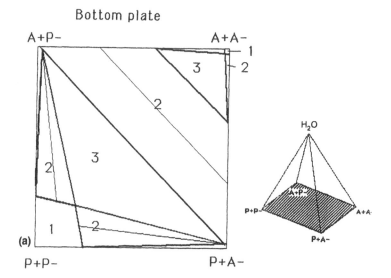

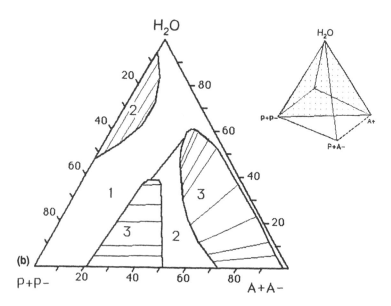

FIG. 33 Cuts corresponding to Figures 31a and 32c. **(a,b)**: the effect of lowering the degree of polymerization from 20 (as in Figs. 31 and 32) to 10. Thicker lines indicate phase boundaries, thinner lines are tie lines (2-phase regions) or projections in the plane of three-phase triangles (3-phase regions).

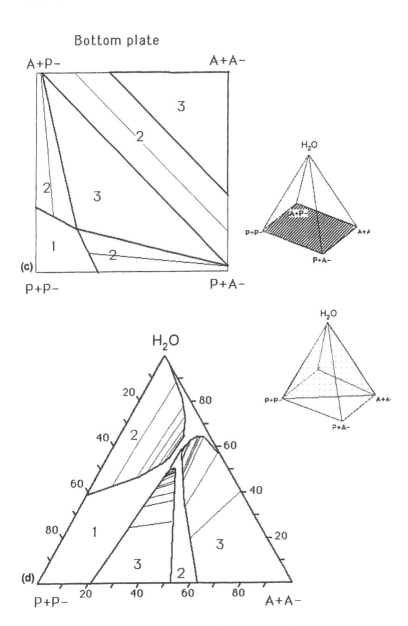

FIG. 33 (c,d) : Effect of increasing the P^+A^- and the A^+P^- interactions with 0.3007 kT to -4.6157 kT.

other component. Another way of counteracting the increased miscibility is to add simple salt.

For segregating mixtures, an increased miscibility can also be obtained by hydrophobic modification of the polymer (HAP or SHP) so that mixed aggregates are formed. On the other hand, if the polymer is made too hydrophobic, the complex may become insoluble and an associative phase separation occurs. Conversely, for slightly modified polymers, segregation may still occur between a saturated complex and pure surfactant aggregates at a large excess of surfactant.

The concentration of simple salt has different effects in oppositely charged mixtures, depending on whether the polymer and surfactant are intrinsically (*i.e.*, in the absence of charge effects [1]) associating or segregating. For intrinisically segregating mixtures, large concentrations of simple salt screens the attraction between polions and surfactant aggregates. Initially this results in an increased miscibility and, at high concentrations of salt, segregation. For intrinsically associating mixtures, however, salt generally results in an increased phase separation (a salting out of net charged complexes). Both the small counterions (of the polelectrolyte and the surfactant), and free surfactant ions contribute to the screening effect. The salt emerging from the counterions is thus a "hidden variable" in oppositely charged systems; its concentration obviously increases at increasing global concentrations of polymer and surfactant. Removing the counterions (by dialysis of a neutral concentrated phase) results in a complex salt with a very low solubility. This complex may, nevertheless, absorb large quantities of water, if it contains a flexible polymer. For infinite surfactant aggregates, confinement effects may be important, if the polymer coil is larger than the smallest dimension of the water pores.

Finally, we should stress that the discussion in this chapter has only dealt with the locations of phase boundaries, and the phase compositions, at equilibrium. In many instances, however, the phase separation is not macroscopic, but results in more or less stable dispersions, or connected particulate gels. Often, such metastable systems can be switched over to other states (solutions, macroscopically phase-separated samples) by small variations in temperature or in the concentrations of the components. The factors controlling the metastability are still quite poorly understood, although some patterns and empirical rules are suggested by available experiments. Such partial or frustrated phase separations are challenging topics for future research. Other important issues concern the phase behavior of more complex systems: Here, we may think of star copolymers, HAP with moieties that are associating, but only partly miscible with surfactant aggregates, and oil soluble polymers, modified by hydrophilic groups.

ACKNOWLEDGEMENTS

We thank Per Hansson, Philippe Ilekti, Anna Karin Morén and Krister Thuresson for contributing data and figures to this chapter. We are also indebted to these and other colleagues and collaborators, whose work we have cited above, for countless stimulating discussions on the subject of this chapter. LP gratefully acknowledges The Swedish Research Council for Engineering Sciences (TFR) for their support.

ABBREVIATIONS

C_nE_m	$CH_3(CH_2)_{n-1}(OCH_2CH_2)_mOH$
C_nTA	$CH_3(CH_2)_{n-1}N^+(CH_3)_3$
EHEC	ethyl hydroxyethyl cellulose
HEC	hydroxyethyl cellulose
HPC	hydroxypropyl cellulose
HM	hydrophobically modified
NaHy	sodium hyaluronate
PAm	poly (acrylamide)
PEO	polyoxyethylene
PNIPAm	poly (n-isopropylacrylamide)
PPO	polyoxypropylene
SDS	sodium dodecyl sulfate
UCON	a random copolymer of oxyethylene and oxypropylene

REFERENCES

1. L. Piculell and B. Lindman, *Adv. Colloid Interface Sci. 41*: 149 (1992).
2. H.G. Bungenberg de Jong, in *Colloid Science* (H.R. Kruyt, ed.), Elsevier, New York, Volume II, 1949, pp. 232-258.
3. L. Zeman and D. Patterson, *Macromolecules 5*: 513 (1972).
4. C.C. Hsu and J.M. Prausnitz, *Macromolecules 7*: 320 (1974).
5. D. Patterson, *Polymer Eng. Sci. 22*: 64 (1982).
6. K. Bergfeldt, L. Piculell, and P. Linse, *J. Phys. Chem. 100*: 3680 (1996).
7. B. Lindman and K. Thalberg, in *Interactions of Surfactants with Polymers and Proteins* (E. Goddard and K. P. Ananthapadmanabhan, eds.), CRC Press, Boca Raton, 1993, pp. 203-276.
8. K. R. Wormuth, *Langmuir 7*: 1622 (1991).
9. K. Thalberg and B. Lindman, *Colloids Surf. A 76*:283 (1993).
10. S.M. Clegg, P.A. Williams, P. Warren, and I.D. Robb, *Langmuir 10*: 3390 (1994).

11. I. D. Robb, P. A. Williams, P. Warren, and R. Tanaka, *J. Chem. Soc., Faraday Trans. 91*: 3901 (1995).
12. L. Piculell, K. Bergfeldt, and S. Gerdes, *J. Phys. Chem. 100*: 3675 (1996).
13. K. Bergfeldt and L. Piculell, *J. Phys. Chem. 100*: 5935 (1996).
14. B. Gerharz and R. Horst, *Colloid. Polym. Sci. 274*: 439 (1996).
15. N.K. Pandit, J. Kanjia, K. Patel, and D.G. Pontikes, *Int. J. Pharm. 122*:27 (1996).
16. P.-Å. Albertsson, *Partition of Cell Particles and Macromolecules*, 3rd ed, Wiley-Interscience, New York, 1986.
17. P.-G. Nilsson, H. Wennerström and B. Lindman, *J. Phys. Chem. 87*: 1377.
18. R. Zana and C. Weill, *J. Physique Lett. 46*: L-953 (1985).
19. B. Lindman and M. Jonströmer, in *Physics of Amphiphilic Layers* (J. Meunier, D. Langevin and N. Boccara, eds.), Springer Proceedings in Physics, Vol. 21, 1987, pp. 235-240.
20. O. Anthony and R. Zana, *Langmuir 10*: 4048 (1994).
21. L. Piculell, I. Iliopoulos, P. Linse, S. Nilsson, T. Turquois, C. Viebke, and W. Zhang, in *Gums and Stabilisers for the Food Industry Vol. 7* (G.O. Phillips, P.A. Williams, and D.J. Wedlock, eds.), Oxford University Press, Oxford, 1994, pp. 309-322.
22. I. Iliopoulos, D. Frugier and R. Audebert, *Polymer Prepr. 30*: 371 (1989).
23. A. Khokhlov and I.A. Nyrkova, *Macromolecules 25*: 1493 (1992).
24. U. Sivars, K. Bergfeldt, L. Piculell, and F. Tjerneld, *J. Cromatogr. B 680*: 43 (1996).
25. P.J. Flory, *Principles of Polymer Chemistry*, Cornell University Press, Ithaca, N.Y., 1953.
26. B. Lindman and H. Wennerström. *Topics in Current Chemistry 87*: 1 (1978).
27. D.F. Evans and H. Wennerström, *The Colloidal Domain*, VCH Publishers, New York, 1994.
28. K. Hayakawa and J.C.T. Kwak, in *Cationic Surfactants: Physical Chemistry* (D.N. Rubingh and P.M. Holland, eds.), Dekker, New York, 1991.
29. Y.-C. Wei, and S.M. Hudson, *Rev. Macromol. Chem. Phys. C(35)1*: 15 (1995).
30. E. D. Goddard and K. P Ananthapadamabhan (eds.), *Interactions of Surfactants with Polymers and Proteins*, CRC Press; Boca Raton: 1993.
31. K. Thalberg and B. Lindman, *J. Phys. Chem. 93*: 1478 (1989).
32. K. Thalberg, B. Lindman, and G. Karlström, *J. Phys. Chem. 94*: 4289 (1990).
33. K. Thalberg, *Polyelectrolyte - Surfactant Interactions*. Ph.D. Thesis, Lund, 1990.
34. K. Thalberg, B. Lindman, and G. Karlström, *Prog. Colloid Polym. Sci. 84*: 8 (1991).
35. K. Thalberg, B. Lindman, and G. Karlström. J. Phys. Chem. 95:6004 (1991).
36. K. Thalberg and B. Lindman, in *Surfactants in Solution*, (K. L. Mittal and D. O. Shah, eds), Plenum Press, New York, Vol. 11, 1991, p. 243-260.
37. K. Thalberg, B. Lindman, and G. Karlström, *J. Phys. Chem. 95*: 3370 (1991).
38. J. O. Carnali, *Langmuir 9*: 2933 (1993).
39. Å. Herslöf-Björling, *An Experimental Macromolecular Study of Sodium Hyaluronate and Amphiphile Interaction in Electrolyte/Water Solutions*, Ph.D. Thesis, Uppsala 1995.

40. Å. Herslöf Björling, M. Björling, and L.-O. Sundelöf, in ref. 39.
41. Y. Li and P. L. Dubin, *Langmuir 11*: 2486 (1995).
42. J. Merta, and P. Stenius, *Colloid Polym. Sci. 273*: 974 (1995).
43. S. Ranganathan and J. C. T. Kwak, *Langmuir 12*: 1381 (1996).
44. P. Ilekti, L. Piculell, B. Cabane and F. Tournilhac, *J. Phys. Chem. B 102*: 344 (1998).
45. B. Lindman and S. Forsén, *Chlorine, Bromine and Iodine NMR. Physico-Chemical and Biological Applications*, Springer Verlag, Heidelberg, 1976.
46. H. Fabre, N. Kamenka, A. Khan, G. Lindblom, B. Lindman, and G.J.T. Tiddy, *J. Phys. Chem. 84*, 3428 (1980).
47. B. Lindman, H. Wennerström and S. Forsén, *J. Phys.Chem. 96*:5669 (1992).
48. B. W. Ninham and V. Yaminsky, *Langmuir, 13*: 2097 (1997).
49. P. Hansson and M. Almgren, *Langmuir 10*: 2115 (1994).
50. R. Tanaka, J. Meadows, G.O. Phillips and P.A. Williams, *Carbohydr. Polymers 12*: 443 (1990).
51. I. Iliopoulos, T.K. Wang and R. Audebert, *Langmuir 7*: 617 (1991).
52. B. Magny, I. Iliopoulos, R. Zana, and R. Audebert, *Langmuir 10*: 3180 (1994).
53. K. Thuresson, O. Söderman, P. Hansson and G. Wang, *J. Phys. Chem. 100*: 4909 (1996).
54. E.D. Goddard, *J. Colloid Interface. Sci., 152*: 578 (1992).
55. F. Guillemet and L. Piculell, *J. Phys. Chem. 99*: 9201 (1995).
56. L. Piculell, F. Guillemet, K. Thuresson, V. Shubin, and O. Ericsson, *Adv. Colloid. Interface Sci., 63*: 1 (1996).
57. Ogino, K. and Abe, M. eds., *Mixed Surfactant Systems*; Marcel Dekker, New York: 1993.
58. L. Piculell, K. Thuresson, and O. Ericsson, *Faraday Discuss. 101*: 307 (1995).
59. R.A. Gelman, in *TAPPI Proceedings of the International Dissolving Pulps Conference*, Geneva, 1987, pp.159-165.
60. R. Tanaka, J. Meadows, P.A. Williams, and G.O. Phillips, *Macromolecules 25*: 1304 (1992).
61. A.J. Dualeh and C.A. Steiner, *Macromolecules 23*: 251 (1990).
62. U. Kästner, H. Hoffman, R. Dönges, and R. Ehrler, *Colloids Surf. A 82*: 279 (1994).
63. J.J. Effing, I.J. McLennan, and J.C.T. Kwak, *J. Phys. Chem. 98*: 2499 (1994).
64. E.D. Goddard and P.S. Leung, *Colloids Surf. 65*: 211 (1992).
65. E.D. Goddard and P.S. Leung, *Langmuir 8*: 1499 (1992).
66. U. Kästner, H. Hoffman, R. Dönges, and R. Ehrler, *Coll. Surf. A 112*: 209 (1996).
67. B. Magny, I. Iliopoulos, R. Audebert, L. Piculell and B. Lindman, *Prog. Colloid Polymer Sci. 89*:118 (1992).
68. O. Anthony and R. Zana, *Langmuir 12*: 3590 (1996).
69. S. Saeki, N. Kuwahara, M. Nakata, and M. Kaneko, *Polymer 17*: 685 (1976).
70. G. Karlström, *J. Phys. Chem. 89*, 4962 (1985).
71. B. Cabane, K. Lindell, S. Engström, and B. Lindman, *Macromolecules 29*: 3188 (1996).

72. K. Zhang, G. Karlström, and B. Lindman, *Prog. Colloid Polymer Sci. 88*: 1 (1992).
73. K. Zhang, G. Karlström, and B. Lindman, *Colloids Surf. 67*: 147 (1992).
74. K. Zhang, *Ternary Aqueous Mixtures of a Nonionic Polymer with a Surfactant or a Second Polymer*, Ph.D. Thesis, Lund 1994.
75. K. Zhang, G. Karlström, and B. Lindman, *J. Phys. Chem. 98*: 4411 (1994).
76. K. Thuresson, and B. Lindman, *J. Phys. Chem. 101*: 6460 (1997).
77. T.F. Tadros. *J. Colloid Interface Sci. 46*: 528 (1974).
78. M.Y. Pletnev and A.A. Trapeznikov, *Kolloidn. Zh. 40*: 948 (1978).
79. A. Carlsson, G. Karlström, and B. Lindman, *Langmuir 2*:536 (1986).
80. A. Carlsson, G. Karlström, B. Lindman, and O. Stenberg, in *Surfactants in Solution* (K.L. Mittal, ed.), Plenum New York, Vol. 8, 1989, pp 25-34.
81. A. Carlsson, *Nonionic Cellulose Ethers. Interactions with Surfactants. Solubility and Other Aspects*, Ph.D. Thesis, Lund 1989.
82. C.J. Drummond, S. Albers, and D.N. Furlong, *Colloids Surf. 62*: 75 (1992).
83. H.G. Schild and D.A. Tirrell, *Langmuir 7*: 665 (1991).
84. F.E. Bailey Jr., and J.V. Koleske, *Poly(ethylene oxide)*, Academic Press, New York, 1976.
85. A.A. Samii, B. Lindman, and G. Karlström, *Prog. Colloid Polym. Sci. 82*:280 (1990).
86. A.A. Samii, G. Karlström, and B. Lindman, *Langmuir 7*: 653 (1991).
87. A. Carlsson, G. Karlström, and B.Lindman, *Colloids. Surf. 47*: 147 (1990).
88. K. Lindell and S. Engström, *Int. J. Pharmaceutics 124*, 117 (1995).
89. K. Lindell, *An Investigation of Thermogelling Aqueous Systems of Ethyl(HydroxyEthyl) Cellulose and Ionic Surfactants*. Ph.D. Thesis, Lund 1996.
90. P. Hansson and M. Almgren, *J. Phys. Chem. 100*: 9038 (1996).
91. E. Alami, M. Almgren, and W. Brown, *Macromolecules 29*:5026 (1996).
92. K.P. Ananthapadmanabhan, in *Interactions of Surfactants with Polymers and Proteins* (E. Goddard and K. P. Ananthapadmanabhan, eds.), CRC Press, Boca Raton, Fla., 1993, pp. 319-365.
93. E.D. Goddard and B.A. Pethica, *J. Chem. Soc.* 2659 (1951).
94. M. N. Jones and P. Manley. *J. Chem. Soc., Faraday Trans. I* 75:1736 (1979).
95. M. N. Jones and P. Manley. *J. Chem. Soc., Faraday Trans. I* 76:654 (1980).
96. K. Fukushima, Y. Murata, N. Nishikido, G. Sugihara, and M. Tanaka, *Bull. Chem. Soc. Jpn. 54*:3122 (1981).
97. K. Fukushima, Y. Murata, G. Sugihara, and M. Tanaka, *Bull. Chem. Soc. Jpn. 55*:1376 (1982).
98. J. Chen and E. Dickinson, *Colloids. Surf. A 100*: 255 (1995).
99. A.K. Morén and A. Khan, *Langmuir 11*: 3636 (1995).
100. A.K. Morén, K. Eskilsson, and A. Khan, *Colloids Surf. B 9*: 305 (1997).
101. M.N. Jones and A. Brass, in *Food Polymers, Gels, and Colloids* (E. Dickinson, ed.), Special Publication No. 82 The Royal Society of Chemistry, Cambridge, 1991, pp. 65-80.
102. H. Bagger-Jörgensen, U. Olsson, and I. Iliopoulos, *Langmuir, 11*, 1934 (1995).

103. H. Bagger-Jörgensen, L. Coppola, K. Thuresson, U. Olsson, and K. Mortensen, *Langmuir 13*: 4204 (1997).
104. A. Kabalnov, U. Olsson, and H. Wennerström, *Langmuir 10*, 2159 (1994).
105. A. Kabalnov, U. Olsson, K. Thuresson, and H. Wennerström, *Langmuir 10*, 4509 (1994).
106. A. Kabalnov, B. Lindman, U. Olsson, L. Piculell, K. Thuresson and H. Wennerström, *Colloid Polym. Sci. 274*, 297 (1996).
107. I. Iliopoulos and U. Olsson, *J.Phys. Chem. 98*: 1500 (1994).
108. V. Rajagopalan, U. Olsson, and I. Iliopoulos, *Langmuir 12*, 4378 (1996).
109. P. Kékicheff, B. Cabane, and M. Rawiso, *J. Colloid Interface Sci. 102*, 51 (1984).
110. C. Ligoure, G. Bouglet, and G. Porte, *Phys. Rev. Lett. 71*, 3600 (1993).
111. M. Singh, R. Ober, and M. Kleman, *J. Phys. Chem. 97*, 11108 (1993).
112. H. Bagger-Jörgensen, U. Olsson, I. Iliopoulos, and K. Mortensen, *Langmuir,* in press.
113. D.G. Hall, *Trans. Faraday Soc. 67*: 1516 (1971).
114. D.G. Hall, *J.Chem.Soc. Faraday Trans. 2 70*: 1526 (1974).
115. G. Karlström, H.- O. Johansson , F. Tjerneld, *On the use of clouding polymers for purifying chemical systems.Cellulose and Cellulose Derivatives: Physico-chemical aspects and ndustrial applications.* (J.F. Kennedy, G.O. Phillips, P.A. Williams and L. Piculell, eds), Woodhead Publishing Ltd., Cambridge, England, 1995, pp. 417-423.
116. H.- O. Johansson, G. Karlström, and F. Tjerneld, *Macromolecules 26*: 4478 (1993).
117. H.N.W. Lekkerkerker, W.C.-K. Poon, P.N. Pusey, S. Stroobants, and P.W. Warren, *Europhys. Letters, 20*, 559, (1992).
118. R. Kjellander and E. Florin-Robertsson, *J. Chem. Soc. Faraday Trans. 1, 77*: 2053 (1981).
119. J.C. Lang and R.D. Morgan, *J. Chem. Phys. 73*: 5849 (1980).
120. J. Hirschfelder, D. Stevenson, and H. Eyring, *J. Chem. Phys. 5*: 896 (1937).
121. R.E. Goldstein, *J. Chem. Phys 80*: 5340 (1984).
122. M. Andersson and G. Karlström, *J. Phys. Chem. 89*: 4957 (1985).
123. O. Engkvist and G. Karlström, *J. Phys. Chem. 106*:2411 (1996).
124. G. Karlström, A. Carlsson, and B.Lindman, *J. Phys. Chem. 94*: 5005 (1990).
125. K.Zhang, M. Jonströmer, and B. Lindman, *Colloids Surf. A 87*: 133 (1994).
126. K. Thalberg, B. Lindman, and K. Bergfeldt, *Langmuir 7*: 2893 (1991).
127. K. Zhang, G. Karlström, and B. Lindman, *Prog. Colloid Polymer Sci. 88*: 1 (1992).

4
The Nature of Polymer-Surfactant Interactions

KEISHIRO SHIRAHAMA   Department of Chemistry and Applied Chemistry, Faculty of Science, Saga University, Saga, Japan

I.   INTRODUCTION

The presence of both hydrophilic and large hydrophobic groups in a surfactant molecule leads to many intriguing properties which are academically interesting and practically useful. These aspects have been well documented by many authors [1-5]. The second molecule of interest for the current topic is the polymer, a molecule with a high molecular weight and multiple modes of internal motion. The physical and chemical properties of polymers, including amphiphilicity, are also diverse because of the many possible variations of monomers and molecular size.   Thus the study of polymers forms an independent discipline [6-8]. The combination of these two classes of molecules has brought us a rich area of investigation, the results of which have over the years already been summarized in a number of worthwhile reviews [9-17], as well as in other chapters of this- book. Each of these reviews often represents a particular view of the system of interest, or a particular application. The purpose of this chapter is to provide some historical context and personal observations to one of the topics often treated in reviews of polymer-surfactant interactions, i.e. binding isotherms and their theoretical interpretation. As a result, this article does not necessarily aim at an extensive review of the topic, for which the reader is referred to other chapters of this book and the reviews already cited. Rather, we will describe the development of experimental evidence for cooperative binding in dilute polymer-surfactant solutions through the direct determination of binding isotherms, as well as the theoretical description of this cooperative effect. This discussion will lead us to recent descriptions by the author based on small system thermodynamics and experimental verification of some of the conclusions, based on electrophoretic light-scattering experiments on the polymer-surfactant aggregates.

The first examples of binding isotherms for polymer surfactant systems may have been the results for poly(vinylpyrrolidone)-sodium alkyl sulfates systems (PVP-SAS) presented by Arai *et al.* at the 24th Annual Meeting of the Chemical Society of Japan in 1970, which were later published as a full paper [18]. The realization that the steep increase of binding within a small change in equilibrium concentration is characteristic of a cooperative effect led the present author to perform surfactant binding measurements on poly(ethylene oxide)(PEO)-SAS systems, using PEO as the polymer because it has a much simpler chemical structure than PVP [19,20]. The method used in these experiments was equilibrium dialysis. These experiments were based on the need to determine binding isotherms with direct analytical methods, in order to provide the reliable experimental data essential to the development of theoretical treatments using the tools of thermodynamics and statistical mechanics. In addition to the binding isotherms so determined, physical techniques including especially spectroscopy also serve the theoretician in setting up models which resemble the actual physical information available on these systems. Section II presents a brief summary of binding isotherms typical for polymer-surfactant systems, together with a full description of two of the major experimental techniques, i.e. dialysis and surfactant selective electrodes. Section III is devoted to review of some theoretical treatments of polymer-surfactant interactions with special reference to the statistical mechanical theory originally presented by T. L. Hill [21,22]. All the data referred to are limited to solutions dilute with respect to both polymer and surfactant, for which this theoretical treatment is valid. More concentrated regimes are of both academic and practical interest. Such systems are the topic of many recent investigations [23] and are covered in detail for instance in Chapters 2, 3 and 10 of this book. For the present discussion, it is realized that dilute solutions are important as a first step in considering concentrated systems, since a concentrated polymer-surfactant aqueous mixture can be regarded as a system where polymer-surfactant complexes known to be present in dilute solution interact with each other.

In addition to the cooperative nature of binding, another feature of polymer-surfactant systems is the small size of the polymer-bound surfactant aggregates. Accordingly, we need to focus our attention on the aggregate size on polymer strand as obtained by various experimental techniques. Section IV deals with electrophoresis of polymer-surfactant complexes, a method which can reveal some unusual aspects of polymer-surfactant interactions. During the early binding studies by the present author on PEO-SAS systems [19,20], work by Takagi *et al.* et Osaka University was aimed at explaining the physical basis of sodium dodecyl sulfate-polyacrylamide electrophoresis (SDS-PAGE) [104-106,110]. A fruitful collaboration between these two groups led to the recognition of common elements in the application of the SDS-PAGE technique in biochemistry and the understanding of the basic elements of polymer-surfactant interactions. These studies ultimately resulted in the proposal of a "necklace" model for polymer surfactant complexes [110], as recalled in section IVA. In section IVB, experimental evidence, based on electrophoretic light-scattering measurements, for the phenomenon of bimodality of aggregate sizes in polymer-surfactant solutions is demonstrated. This bimodality, although unexpected, is in fact deeply rooted in the nature of polymer-surfactant complexes described in the previous sections. Finally, section V presents a number of models for polymer-surfactant complexes, collected together with the comments from the original authors. In considering such models, it should be pointed out that although "seeing is believing," and indeed visualization of a model is an important part of the process leading to the theoretical description of a system, at the same time such visualizations include a number of possibilities for pitfalls which could lead to incorrect descriptions and conclusions. Accordingly, it should be emphasized that such models can be considered only as approximations of reality, and may exclude much refinement at the microscopic level.

II. BINDING ISOTHERMS

There are many methods to study interaction between polymers and surfactants in solution. One of the fundamental and necessary methods is the determination and analysis of the binding isotherm. A binding isotherm is an equation of state. Binding isotherms will obviously tell us about the interactions between polymer and surfactant. However, theoretical approaches are required to understand what they tell us. Such approaches will be discussed in more detail in Section III of this Chapter.

The experimental determination of a binding isotherm requires a procedure to separate surfactant molecules (ions) in the whole system into bound and free species. Conveniently, equilibrium dialysis is a standard method: the amounts of surfactant in a polymer solution and in the polymer-free equilibrium solution are determined after dialysis equilibrium has been established. This usually takes a long time because of a partial uptake of surfactant ion by the dialysis membrane and resultant electrostatic repulsion, and has to be followed by tedious chemical analyses. Surfactant selective electrodes have been developed which make it possible to determine the amount of free surfactant in a polymer-surfactant solution, and thereby the binding isotherm.   Various types of spectroscopic methods may allow *in-situ* determination of the two states, but systems in which such methods are applicable are limited.   Other indirect methods include more approximate results derived from viscosity and surface tension measurements.

Binding isotherms have been reported for a wide variety of systems, well documented in review articles [9-17]. In the present section, only some selected binding isotherms will be shown which are typical and necessary for further discussion in the following sections.

Polymer-surfactant interactions are conveniently classified by the electric charges of the two components.   Since there are three types of interacting molecules as classified from their electric charge: positive, negative, and neutral, there are nine possible charge combinations. The two cases with the equal charge signs are excluded because of strong electrostatic repulsion, although there are reports of interactions between polymers and surfactants carrying charges of the same signs [24,25,26]. Interactions between nonionic species are usually very weak. Such systems have been reviewed [27], and a current review is presented by Saito in Chapter 9 of this book. This leaves the binding of ionic surfactants to neutral polymers or to polyions of opposite charge.

Surfactant ion binding to a neutral polymer actually leads to an increasingly charged complex, a charging process similar to micelle formation. On the other hand, surfactant ion binding to a polyion with opposite charge is a discharging process, and is more favorable than binding to a neutral polymer.

The work (Gibbs energy) involved in the discharge process may be expressed as $-e\phi_d$ and that of the charging process as $e\phi_c$, where ϕ_d and ϕ_c are electrostatic potentials in the respective processes and e is the unit electronic charge. $\phi_c\phi_d$ < 0 for an ionic surfactant. If the discharge and charge potentials are assumed to be 50mV and -50mV, reasonable approximations for many systems, the energy difference between the discharging and charging processes equals $0.1e$ or 4 kT units in Gibbs energy. This factor (= $\exp(0.1e/kT)$ = 55) explains the concentration difference between surfactant binding to a neutral polymer (close to the cmc) and binding to a polyelectrolyte of opposite sign, which may start as much as two orders of magnitude below the cmc.

A.   Methods: Equilibrium Dialysis and Surfactant-Selective Electrodes

Before turning to the experimental results, two experimental techniques to obtain binding isotherms will be discussed briefly, i.e. dialysis equilibrium and surfactant selective electrodes.

Dialysis is one of the simplest standard experiments in binding studies. Dialysis cells are commercially available, or may be custom made. One example is to use an assembly of pairs of cylindrical Teflon block cells where each block has a cell cavity of 2cm in diameter and 1cm depth. A dialysis membrane is inserted between each pair of cells. The cell pairs are placed in a row inside a thermostat, and rotated by means of a mechanical device to allow stirring the solutions. The membrane is usually cellulose, a cellulose derivative, or a synthetic polymeric material. Porosity of the membrane is a crucial factor for efficient experiments, since surface-active ions are apt to be trapped in the membrane, which obstructs further passage of surfactant ions resulting in a prolonged equilibration time. Pore size is controlled effectively by immersing the membrane into alkaline solution, whereby alkalinity, immersion time and temperature may be adjusted for the maximum performance. A small cylindrical Teflon disk placed in each cell cavity is effective for stirring solution and for making the dialysis membrane surface to solution volume ratio larger. Teflon is one of the best materials for dialysis cells. It is so hydrophobic and water-tight that no sealing gasket has to be used and the cell inside is easily rinsed. The only drawback is that one cannot see through inside the cells. When visual observation is necessary, methacrylate resin cell material may be used.

Another small but useful device that has led to great progress in the study of polymer-surfactant interactions is the surfactant-selective electrode [12]. Many investigators have employed surfactant-selective electrodes to study a wide variety of surfactant solutions. An early example was to employ a mercury electrode in contact with mercury salts of surfactant carboxylate and sulfate [28]. Gilany and Wolfram [29] used an Au/Hg/Hg$_t$DS amalgam electrode sensitive to dodecyl sulfate (DS) to study polymer-surfactant binding, but more convenient systems have now become available. Also less frequently used are liquid membrane electrodes. These electrodes contain a hydrophobic salt of surfactant ion, or a mobile ion-exchanger, dissolved in water-immiscible organic solvent with rather high dielectric constant like nitrobenzene with the cell formula:

reference electrode / agar bridge / surfactant solution, C$_1$ / ion-exchanger in nitrobenzene / surfactant solution, C$_2$ / agar bridge / reference electrode   (1)

The electromotive force (emf) of this symmetrical cell is expressed by the Nernst equation (neglecting activity coefficient differences),

$$E = (RT/zF)\ln(C_2/C_1)$$ (2)

The electrode performance is judged by the following criteria:
(1) Is the plot of E vs. log C$_2$ a straight line with RT/zF = 59.1mV at 25° C?
(2) Is E at C$_1$ = C$_2$, i.e. the asymmetry potential, zero?
(3) Does the straight line extend to the lowest surfactant concentrations needed, e.g. 10^{-6}M?
(4) Do the data correctly reflect solution prpoperties such as the cmc ?
(5) Does the electrode have a sufficient selectivity against other ions in the system?

Liquid membrane electrodes usually function very well, but they are inconvenient to handle because of the labile liquid membrane. The latter drawback was partly solved by using a cell with a stopcock at the center of the U-shaped cell which is very useful to stabilize the liquid levels when solutions are exchanged [30]. However this kind of electrode is still disliked because of the solubility of the organic liquid especially in micellar solutions.

Currently, the most commonly employed electrodes are based on a polymer membrane containing a mobile ion-exchanger. The electrode membrane consists of poly(vinylchloride) with a high plasticizer content (usually more than two third of the whole membrane mass), which forms actually a polymer "gel" electrode rather than a polymer "solid" electrode. Polymer gel electrodes have been successfully employed in constructing binding isotherms of polymers and surfactants with opposite electric charges [31,32]. Excellent and useful as they are, there is still a problem: dissolution (solubilization) of mobile ion-exchanger and plasticizer into polymer-surfactant complexes and subsequent aging of electrode performance. Use of polymeric plasticizer and partially charged PVC has meant a breakthrough for this difficulty [33,34]. The polymeric plasticizer is a commercially available copolymer commonly called an "engineering plastic," such as Elvaloy (Du Pont). Charged groups are introduced as a monomer before [33], or by chemical modification after [35] polymerization. It was found that introduction of a charged group such as acrylamide propanesulfonate into PVC also gave an excellent electrode membrane usable not only in micellar and polymer-surfactant complex solutions, but also in aqueous organic mixed solvents [34,35]. The polymeric membrane components cannot dissolve, and thus the electrode lifetime may extend as long as a year. It should be noted that electrodes for cationic surfactants have been successfully developed, but it has been difficult to obtain a good sensor for anionic surfactants. The only exception is the Salford school which has been successful in constructing SDS electrodes [33,36]. These small but highly useful experimental setups may be expected to survive even in these days of large instruments armed with sophisticated electronics.

B. Binding to Neutral Polymers

1. Binding of anionic surfactants

The first study of binding isotherms by Arai *et al.* showed a set of binding isotherms for PVP-SAS systems as obtained by a surface tension method [18].

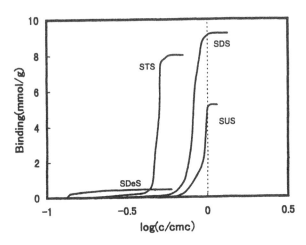

FIG. 1 Binding isotherms for SAS-PEO systems. The abscissa is a reduced equilibrium concentration, C_f/cmc.

The sudden appearance of binding and subsequent saturation with a relatively small increase in the surfactant equilibrium concentration is a clear sign of a cooperative phenomenon; once one surfactant molecule is bound, the following surfactant molecules are more favorably bound aided by the already bound surfactant molecules. The three binding isotherms reported by Arai *et al.* are nearly equally spaced on the equilibrium concentration axis suggesting the contribution of the alkyl chain to the binding affinity. All these features are based upon the assumption that neither polymer nor polymer-surfactant complex are surface active, which is not assured by theory or experiment. Binding isotherms obtained by more reliable experimental techniques were needed. Equilibrium dialysis yielded binding isotherms for PEO-SAS systems as shown in Figure 1 [19,20]. The binding features are similar to those of the PVP-SAS systems, but show a slight difference in the chain-length dependence on the critical aggregation concentration (cac), C_c, i.e. the concentration where binding starts. In Figure 2, log C_c is plotted vs. the number of carbon atoms in the alkyl chain, together with log(cmc). The slope of the log C_c vs n straight line for the PVP-SAS system is the same as that for log(cmc) vs. n, while the same relation in the PEO-SAS system has a lower slope resulting in a crossing with the cmc plot. This crossing corresponds with the fact that binding to PEO is not more favorable than micelle formation for surfactants with shorter alkyl chain lengths, in agreement with the observation that there is no binding with decyl sulfate. It is not known whether this difference is due to the different nature of PVP and PEO, or to the different experimental methods.

The importance of binding isotherms has been well recognized and many cases with similar binding characteristics have been reported in more recent

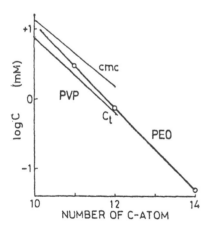

FIG. 2 log C_c and log(cmc) vs. number of carbon atoms in an AS molecule.

articles, often in combination with various kinds of physical measurements. Homberg *et al.* obtained a binding isotherm for the SDS-EHEC (ethyl(hydroxyethyl)cellulose) system with the dialysis method using a Donnan correction.   In this study the amount bound is plotted against total SDS concentration which makes interpretation more difficult [41].   Anad *et al.* carried out dialysis experiments for PVP-NaDBS (Na-dodecylbenzene sulfonate) systems [42], however, there is no mention of added salt or Donnan correction.   Wan-Badhi *et al.*[39] reported cooperative binding of SDS to PVP as measured by an SDS-electrode.   There is no difference in binding to PVP with various molecuar weights (10,000 - 70,000). This study is complemented by gel-chromatography, counterion activity, and ultrasonic relaxation experiments. Kamenka *et al.* studied the interaction of SDS with EHEC by electric conductivity, self-diffusion, and fluorescence methods.   Binding isotherms were constructed based on the diffusion measurements,   and the cluster size was estimated by the fluorescence method

2. Binding of cationic surfactants

It is well known that cationic surfactants have low or even negligible affinity to water-soluble neutral polymers, in marked contrast to the strong affinity shown by anionic surfactants [11].   According to Saito this is closely related to the water structure in the medium, evidenced by the observation that the interaction is controlled by adding structure breaking agents such as iodide and thiocyanate [41, 42].

In the absence of these additives, Saito presented an affinity table for neutral polymer-cationic surfactant systems [43]. According to this table, alkylammonium surfactants show more affinity than alkyltrimethylammonium, or alkylpyridinium surfactants. These tendencies are also confirmed in the binding isotherms as seen in Figure 3 [46].   Actually binding does occur, although micelle formation impedes further increase of binding. This situation can be compared to the binding of undecyl sulfate to PEO as seen in Figure 1, where binding of undecyl sulfate is interrupted before saturation by the onset of normal micelle formation. Other than in the binding strength, there seems to be no essential difference between anionic and cationic surfactants.

Saito also pointed out that the molecular structure of neutral polymers has a great influence on the affinity which decreases in the order [43]:

poly(partially acetylated vinylalcohol) > poly(vinylalcohol)
> poly(ethylene oxide) > poly(vinylpyrrolidone)

This trend is also reflected in the binding isotherms of C_{16}TABr to various neutral   polymers [45].   It   was   noted   also   the   binding   strength   of

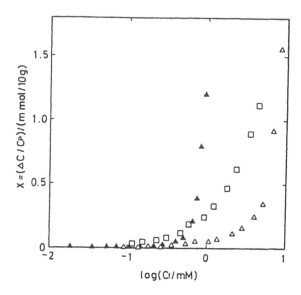

FIG. 3   Binding isotherms of cationic surfactant-PVA90 systems in 5mM NaCl at 30°C. ▲C₁₄PCl, ☐C₁₂PCl, △C₁₂ACl. From ref. 46 From ref. 46, *Colloids Surf. 40*, 261 (1989), with permission from Elsevier Science-NL, Sara Burgerhartstraat 25, 1055 KV Amsterdam, the Netherlands.

tetradecylpyridinium bromide(C_{14}PBr) with acetylated PVA increases with the degree of acetylation [45] as seen in Figure 4, suggesting the importance of hydrophobic regions as possible binding sites and the role of hydrogen bonding (i.e., hydroxyl groups) in the polymer coil. However it is not clear if the hydrophobicity is the only factor which enhances binding to neutral polymers, since the most hydrophobic polymer, poly(vinylpyrrolidone), does not bind cationic surfactants at all. The affinity order for neutral polymers shown above is reversed for the binding of anionic surfactants [11,47]. This may imply an electrostatic effect. If it is assumed that these neutral polymers are protonated, even if very minute, to the reversed order of (3), then the affinity order would be reasonably explained for both anionic and cationic surfactants. Such an assumption is partially verified by experiments in high pH media where the neutral polymer would be completely deprotonated.   Figure 5a shows potentiograms of tetradecylpyidinium cation in the absence and in the presence of PVP, where the pH of the medium is neutral or alkaline (pH 11) [48]. The potentiogram in the presence of PVP coincides with the calibration curve at neutral pH, but a deviation from the Nernstian response is apparent in a high

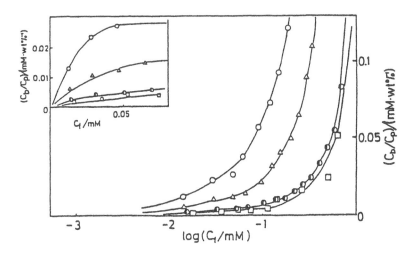

FIG. 4 Binding isotherms of PVA at 25°C: ○ PVA70, △ PVA80, ● PVA90, and □ PVA100. the numbers after PVA are the percentage of hydrolysis. The insert is for a lower concentration region showing a Langmuir type saturating tendency. From ref. 45, *Colloids Surf. 66*, 275 (1992), with permission from Elsevier Science-NL, Sara Burgerhartstraat 25, 1055 KV Amsterdam, the Netherlands

pH medium where PVP would be completely deprotonated. From this deviation, a binding isotherm can be constructed as seen in Figure 5b, more or less resembling those in Figures 3 and 4.

Binding isotherms of different surfactants coincide when a reduced concentration scale, $C_f/$cmc, is used as seen in Figure 6, where an apparent binding constant, i.e., amount of binding divided by the equilibrium concentration, is plotted. This result strongly suggests that binding to polymer and micelle formation share mostly the same mechanism. Only a slight effect may trigger surfactant binding prior to ordinary micelle formation.

Brackman and Engberts measured the "cmc" of n-dodecyldimethylamine oxide in the presence of poly(vinyl methylether)(PVME), PPO, and PEO by their "pH method" [49]. They observed a decrease in cmc in the presence of PVME and PPO, but not with PEO. The decrease is more marked with the more charged (protonated) surfactant. These authors suggest that the polymer brings about a relative stabilization of the "micelles" by means of an electrostatic effect, reducing the mutual headgroup repulsion. This work shows how the interaction with neutral polymers changes with the transition from non-ionic surfactant to ionic surfactant.

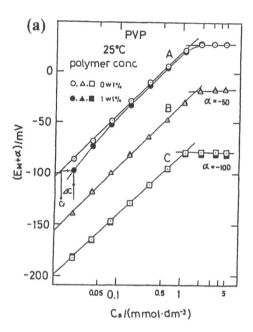

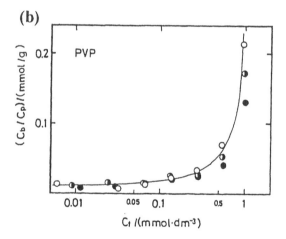

FIG. 5   Interaction between PVP and C_{14}PBr   (a); EMF of surfacatant-selective electrode. Open marks refer to the systems without PVP, and filled marks to those with PVP. Medium A: 2.5mM Na_2CO_3 (pH=11.3), B: 2.5mM Na_2CO_3, pH adjusted to 7 by adding HCl, and C: distilled water pH =6-7. (b) Binding isotherms of C_{14}PBr to PVP in alkaline media.   Open and half-filled circles; 2.5mM Na_2CO_3 (pH=11.3), and filled circles; pH=11 by adding NaOH. From ref. 48 with permission from Steinkopf Verlag.

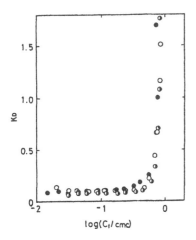

FIG. 6 The apparent equilibrium constant, K_D vs. the reduced equilibrium concentration, C_f/cmc plot for C_{14}PBr-PVA90 systems at 30°C. K_D is the amount of binding divided by the equilibrium concentration. NaBr concentration:● 5mM, O 10mM, half-filled circle: 20mM. From ref. 46, *Colloids Surf.* 40, 261 (1989), with permission from Elsevier Science-NL, Sara Burgerhartstraat 25, 1055 KV Amsterdam, the Netherlands.

Of particular interest are the factors making binding to polymer more favorable than micelle formation in the absence of polymer. Even a minute effect can contribute an appreciable enhancement because of the cooperative effects commonly observed with polymer-surfactant interactions. There is an informative experimental result in Figure 7 showing the partial molar adiabatic compressibilities of both anionic and cationic surfactants in solutions [50]. The partial molar adiabatic compressibility is obtained by combining sound velocity and density measurements, and is thought to be a good measure of hydration strength [51]. The authors conclude that hydrophobic hydration is stronger for cationic than for anionic surfactants for molecules with alkyl chains shorter than the pentyl group. This means that hydrophobic hydration around methylene groups adjacent to charged group is differently affected by the presence of positive or negative charges. In connection with the hydration problem, Anthony and Zana [52] found that PEO and PVP show signs of binding C_{14}TABr at 60°, but not at 25°. They interpreted this as being due to reduced polarity of the polymer at higher temperature as a result of dehydration. Moreover, decreasing hydrophobic hydration around the methylene groups adjacent to the cationic group of C_{14}TABr at higher temperature may also play a role. Enhanced affinity at higher temperature is

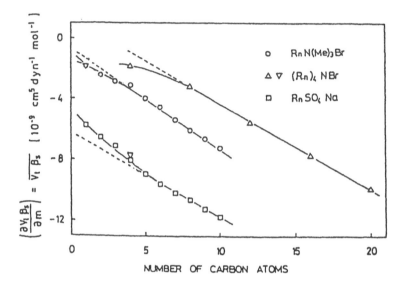

FIG. 7 The partial molar adiabatic compressibility of surfactant vs. the number of Carbon atoms in a surfactant molecule From ref. 50 with permission.

clearly manifested by the binding isotherms [53]. Veggeland and Nilsson reported that branched alkylethoxysulfonates [$(C_6)_2CE_x SO_3^-Na^+$] lose affinity to POE when x>3[54], showing the importance of the nature of the binding site. The differences in the interaction of neutral polymers with anionic or cationic surfactants and neutral polymers seems to be of more general significance: there is a report that anionic alkylorange dyes are bound to PVP, while the cationic analogues do not [55,56].

C. Binding to Polyelectrolytes

The determination of binding isotherms in these systems was initiated by the work of Satake and Yang [57] who studied the binding of SAS to poly(peptides) using a liquid membrane electrode. The experimental results were analyzed in terms of what can be called the Satake-Yang equation, based on a one-dimensional Ising model. Later, by employing polymer gel electrodes, it was well established that ionic surfactants are bound strongly and cooperatively to polyelectrolytes of opposite charge (synthetic or natural), as comprehensively reviewed by Goddard [11,13], Hayakawa and Kwak [12], and Wei and Hudson [16].

With the development and gain in popularity of reliable surfactant electrodes, many investigators employed this technique to determine binding isotherms, often in combination with other experimental techniques. Hanssen and Almgren [58] studied the binding of C_nTABr (n = 12, 14) to sodium carboxymethyl celluloses of different linear charge density, together with time-resolved fluorescence quenching measurements to estimate aggregation numbers of the surfactants bound to the polymer. Their analysis will be described in more detail in Section III. Shimizu et al. reported a number of binding isotherms of dodecylpyridinium chloride to various polycarboxylates, including alternating copolymers of maleic acid with styrene (MASt) and with ethylene (MAE) [59]. The MASt system shows a two-step binding process, while binding to MAE occurs in a single step. The difference was attributed to the role of hydrophobic interactions in the MASt system. In a second paper [60], the behavior of counterions (H^+ and Na^+) was examined as a function of surfactant binding by using counterion responsive electrodes in combination with the surfactant electrode. It was found that pH passes through a maximum on binding surfactant. The maximum is much more enhanced with MASt than with MAE. Shimizu ascribed this to the lowered dielectric constant around charged groups brought about be the surfactants. The sodium ion concentration in a salt-free system is shown to be approximately proportional to the bound surfactant concentration. In the following paper [61], Shimizu compared binding behaviors of four poly(carboxylic acids): poly(fumaric acid)(PFA), poly(maleic acid)(PMA), poly(acrylic acid), and MAE at degrees of neutralization 0.5 and 1.0. He tentatively identified four types of binding sites depending on the neighboring chemical groups. Satake et al. reported binding isotherms of $C_{12}TACl$ to poly(vinylsulfate) in aqueous mixtures of dioxane, ethanol, propanol, and butanol, and compared their findings to the solvent effect in ordinary micelle formation [34]. By using an electrode membrane composed of polymeric materials only this study avoids complications due to dissolution of membrane materials into the mixed solvents [35].

Mel'nikov et al. made a direct fluorescence microscopic observation of DNA(T4) in $C_{16}TAB$ solutions, where a globule-coil transition was seen with increase in $C_{16}TAB$ concentration [62,63]. This transition takes place at the same concentration as binding of the cationic surfactant as shown in Figure 8. It should be pointed out that DNA undergoes an all-or-none type transition. No DNA molecule occurs in an intermediate state with one polymer molecule having both a globular and a coil part. Wei and Hudson [64] studied binding of SDS to chitosan with various degree of acetylation (da) where the polymer charge density decreases with increasing da. The SDS binding is cooperative in character and the binding affinity is reduced with increasing da. Shirahama et al.[65] reported on the interaction between SDS and N-methylglycolchitosan.

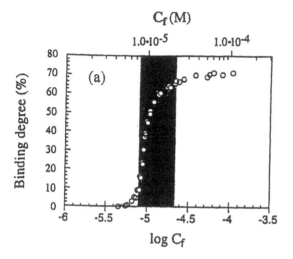

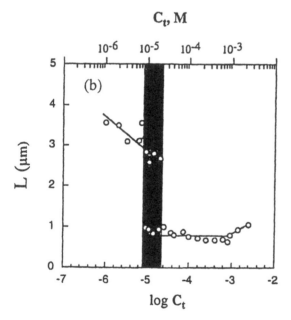

FIG. 8 Interaction of $C_{16}TABr$ with T4DNA(0.5 M TBE buffer, 20°C). (a) Binding isotherm, (b) Long-axis lengths of T4DNA molecules as a function of total $C_{16}TABr$ concentration. The shaded region corresponds to the coexistence of two species. From ref. 63 with permission from the American Chemical Society.

D. Binding to Polymer Gels

Since Tanaka's statistical mechanical description of the volume phase transitions found in polymer gels [66], research on these materials has multiplied. A number of studies have appeared describing the influence of surfactants on the transition behavior [67-80]. Polymer gels are made of normally linear polymers, cross-linked to make a network in which water and ions can penetrate. Surfactant binding to polymer gels can be compared to polymer-surfactant interactions in solution. Binding can be determined relatively easily, since the gels form a separate phase. Polymer gels are also classified by their electric charge. Among many polymer gels, the neutral N-isopropylacrylamide (NIPAM) gel is the most notable. Binding of surfactants has been reported [71]. The binding features are much like the ones for neutral polymers: anionic surfactants have a high affinity, while cationic and nonionic surfactants do not bind. Binding of SDS is less cooperative than for solution phase binding. The reasons for this lower cooperativity should be of interest. Surfactant binding to ionic gels also is similar to the case of binding to polyelectrolytes in solution: both cationic and anionic surfactants bind to polymeric gels of opposite polymer charge, but binding is less cooperative [72-74]. Okuzaki and Osada showed that for dodecylpyridiniumchloride (C12PyCl) -poly(2-(acrylamido)-2-methylpropanesulonic acid) systems initiation of binding is at lower surfactant concentration, but the cooperative nature is reduced as the degree of cross-linking increases [73]. Gong and Osada's [74] theoretical analysis ascribes the reduced cooperativity to a strong osmotic pressure effect which is absent in linear polyelectrolyte systems (Figure 9).

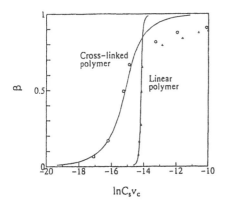

FIG. 9 Binding isotherms of $C_{12}PCl$ to linear polymer and cross-linked polymer. From ref. 74 with permission from the American Chemical Society.

Shirahama *et al.* also studied the binding of $C_{12}PyCl$ to 1 mole % cross-linked gels of co-polymers of sodium 2-(acrylamido)-2-methyl-propanesulonate and NIPAM of various compositions together with the corresponding linear co-polyelectrolytes [75]. They also found that the binding to gels is less cooperative than to the corresponding linear polyions, and ascribed this effect to the interruption of binding at the cross-linking sites leading to a more limited extent of interaction. Surfactant binding is closely related with the volume transition of the gel. Tanaka [76] pointed out that the gel phase transition is a macroscopic manifestation of a coil-to-globular transition. Clearly it is informative to compare binding isotherms with volume behavior.

III. THEORETICAL TREATMENTS

In order to extract information about the interactions between polymer and surfactant from binding isotherms, it is necessary to model these interactions resulting in mathematical expressions which allow comparison with experimental binding isotherms. If the fit is not good enough, the model is modified and a comparison is repeated until satisfactory agreement is obtained, which may then be useful in our interpretation of the molecular events. A number of such modeling procedures are described in this section.

A. Multiple Binding Equilibria

It is reasonable to assume that binding occurs in multiple steps, even if many surfactant molecules or ions are involved in the binding process.

$$P + L = PL \qquad\qquad k_1 = [PL]/[P][L] \qquad\qquad\qquad (4)$$

$$PL + L = PL_2 \qquad\qquad k_2 = [PL_2]/[PL][L] \qquad\qquad\qquad (5)$$

and in general,

$$PL_{i-1} + L = PL_i \qquad\qquad k_i = [PL_i]/[PL_{i-1}][L] \qquad\qquad\qquad (6)$$

where P, L, and PL_i refer to the equilibrium concentrations of polymer, free ligand (surfactant), and polymer-ligand complex carrying i-ligands. Brackets [ ] stand for the equilibrium concentration, k_i's are the equilibrium constants of relevant steps.

By using this notation, the average number of ligands per polymer $\bar{\nu}$ is:

$$\bar{v} = \sum_{i=0}^{B} i[\text{PL}_i] \Bigg/ \sum_{i=0}^{B} [\text{PL}_i] = \sum_{i=0}^{B} iK_i[\text{L}]^i \Bigg/ \sum_{i=0}^{B} K_i[\text{L}]^i \qquad (7)$$

In Eq. (7),

$$[\text{PL}_i] = K_i\,[\text{L}]^i, \text{ with } i = 0,1,2,\,...\text{B}, \qquad (8)$$

where $K_i = \prod_i k_i$ and B is the total number of binding sites. Thus, with the knowledge of $\bar{v}$ as a function of [L], a binding isotherm is constructed and the pertinent parameters, K_i (the equilibrium constants) are obtained. This expression has been frequently employed in analyzing, for example, the binding of ligand to protein [77,78]. The number of parameters obtainable is usually limited to five or less because of limited accuracy of the data. In the case of polymer-surfactant interaction, B is usually much larger (say, 30), and some other analytical method has to be sought. It is informative at this stage, however, to consider a case where all K_i's are equal,

$$K_i = K_0, \text{ for } i = 1,2,3\,...\text{B}. \qquad (9)$$

Then Eq.(7) becomes

$$\bar{v} = \sum_{i=0}^{B} iK_i[\text{L}]^i \Bigg/ \sum_{i=0}^{B} K_i[\text{L}]^i = \sum_{i=0}^{B} i\cdot\,_B C_i \{K_0[\text{L}]\}^i \Bigg/ \sum_{i=0}^{B} {}_B C_i \{K_0[\text{L}]\}^i$$

$$(10)$$

where the combinatory factor, $_B C_i$ arises from the multiplicity of placing i-ligands to B homogeneous binding sites as assumed in Eq.(9). It will be easily recognized that Eq.(7) together with the binomial theorem leads to the Langmuir binding isotherm,

$$\bar{v} = BK_0[\text{L}]\big/\big(1 + K_0[\text{L}]\big) \text{ or } \beta = \bar{v}/B = K_0[\text{L}]\big/\big(1 + K_0[\text{L}]\big) \qquad (11)$$

It is noted in Eq. (11) that the degree of binding, β is no longer a function of B, since binding sites are assumed to be independent from each other. There should not be a dependence on the size of a system.

    Polymer-surfactant interactions are characterized by a high cooperativity as seen in the preceding section. Thus the Langmuir isotherm cannot be expected to apply but is a conceptually important starting point. Cooperativity can be introduced by using a β-dependent parameter, K, in place of K_0 in Eq.(11).

$$K = K_0 \exp(-\beta\varepsilon/kT) \tag{12}$$

$$\beta = K_0\exp(-\beta\varepsilon/kT)[L]/(1 + K_0\exp(-\beta\varepsilon/kT)[L]) \tag{13}$$

where the interaction energy term, $\beta\varepsilon$ shows that when the number of neighbors increases, the interaction term becomes larger as well. Actually the K value increases as binding proceeds when $\varepsilon < 0$, *i.e.* when the interaction is attractive.

Expression (12) is called the Bragg-Williams approximation in statistical mechanics, which will be introduced later in a different way. It can be used to analyze the case of cooperative adsorption. This theory predicts a first order phase transition when $\varepsilon < -2kT$ as seen in Figure 10. However, this treatment fails in polymer-surfactant systems where $-\varepsilon$ is much larger than $2kT$, since no first order phase transition is observed in polymer-surfactant systems. Still some other method is needed.

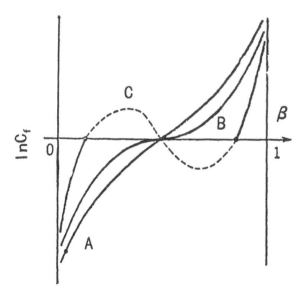

FIG. 10 Model calculation of binding isotherms from Eq(13). $-\varepsilon/kT$ values: A, 0; B, 2; C, >2, the case where binding should show a phase-transitional jump.

B.   Statistical Thermodynamics for Small Systems

Suppose that there is a solution system where there are polymer molecules binding 0, 1, 2, 3,..., B surfactant molecules (ions) in equilibrium with free surfactant in the bulk. This situation is depicted in Figure 11, where a polymer molecule carrying surfactants is thought to be in equilibrium with other polymers through a reservoir (the bulk solution) characterized by the free surfactant concentration C_f (or the chemical potential $\mu = \mu^0 + RTlnC_f$) and the temperature, T.   Each polymer molecule is viewed as surrounded by a hypothetical membrane permeable only to surfactant. Within the membrane, a polymer molecule and a small number (small relative to N_{avo}) of bound surfactants form a "small system." Hill developed a statistical thermodynamic theory for such small systems, and applied it to polymer-ligand binding problem [22]. In order to express an average value of a given property (e.g., binding), he used a (semi)grand partition function, Ξ.

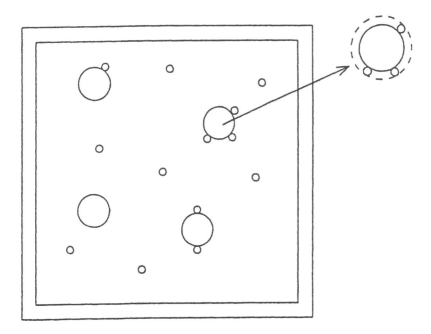

FIG. 11 Schematic display of grand canonical ensemble. One big circle and a few small circles constitute a small system. Such small systems are equilibrated in a bath with temperature, T, and chemical potential of ligand, μ. A small system is thought to be surrounded by a hypothetical membrane permeable to the ligand.

$$\Xi = \sum_{i=0}^{B} Q_i \exp(-i\varepsilon / BkT)^i \lambda_i \tag{14}$$

with

$$Q_i = {}_BC_i\, q_p q_L{}^i \tag{15}$$

in which Q_i is a partition function for a small system containing i ligands and the binomial coefficient, ${}_BC_i$ again takes care of the multiplicity arising from the assumption of homogeneous binding sites on the polymer; q_p and q_L are the molecular partition functions for polymer and ligand, respectively, containing all the molecular information, which, however, does not have to be considered in detail for the moment. λ is the absolute chemical potential proportional to the free ligand concentration, C_f in the bulk,

$$\lambda = \exp(\mu/RT) \quad \propto\ C_f \tag{16}$$

Equation (15) is related to the free energy of a small system (a polymer molecule plus i ligands) through

$$G_i = -RT\ln Q_i \tag{17}$$

The larger each term in Eq.(14), the lower the free energy of a small system is, showing the *relative contribution* from each small system to the whole system. In this sense, it is understood that these terms correspond to the respective terms in Eq.(7); a direct and intuitive relationship between chemical stoichiometry and statistical thermodynamics.

In order to explicitly introduce the interaction between the ligands in a small system, the Bragg-Williams approximation in the form, $\exp(-i\varepsilon/BkT)$ is applied; the interaction is proportional to the number of ligands in a subsytem. Then the semigrand partition function takes the form

$$\Xi = \sum_{i=0}^{B} Q_i \exp(-i\varepsilon / BkT)^i \lambda_i \tag{18}$$

$$= \sum_{i=0}^{B} {}_BC_i q_p q_L^i \exp(-i\varepsilon / BkT)^i \lambda^i = \sum_{i=0}^{B} {}_BC_i q_p [q_L \exp(-i\varepsilon / BkT)\lambda]^i \tag{19}$$

Once a partition function is obtained, the average number of bound ligands is calculated by an equation similar to Eq.(7).

$$\overline{v} = \sum_{i=0}^{B} i \cdot {}_B C_i q_p [q_L \exp(-i\varepsilon / BkT)\lambda]^i \Bigg/ \sum_{i=0}^{B} {}_B C_i q_p [q_L \exp(-i\varepsilon / BkT)\lambda]^i \qquad (20)$$

$$= d\ln\Xi / d\ln\lambda \qquad (21)$$

Statistical thermodynamics for large systems (B in the order of N_{Avo}) takes the maximum term (after differentiation) in place of the summing, and equates it with Ξ which is quite accurate enough for such large systems, and in turn, substitution into Eq.(21) leads to an equation similar to Eq.(13) with Eq.(9). Therefore such a method for large systems is not applicable to polymer-surfactant systems where no transitional binding is observed.

Hill, instead, takes the sum over all the terms for small systems with a definite number of bound ligands, and numerically calculates binding isotherms for some cases as shown in Figure12 where it is seen that large systems would yield transitional isotherms. Hill explained this difference by the fact that many terms still contribute to the total partition function because of the difference by

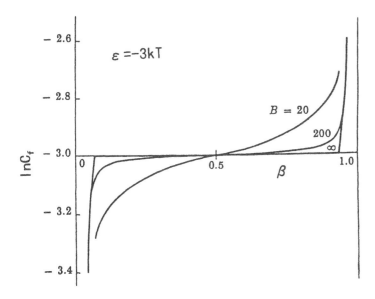

FIG. 12 Model calculation for small systems. Even for a large interaction energy $(-\varepsilon/kT= 3)$, smooth but steep binding isotherms are obtained in marked contrast with the B = ∞ case. Smooth binding isotherms are obtained even with $\varepsilon < -2kT$. Reprinted from ref. 22 with permission from Benjamin, Inc.

the fact of large statistical fluctuation in small systems.

    Shirahama met a similar difficulty with a transitional change appearing for $\varepsilon < -2kT$ when Eq.(13) was applied to the binding isotherm for a PEO-SDS system [19]. Adopting Hill's method, Eq.(20) with $\varepsilon = -10kT$ ($-13kT$ for hydrophobic interaction between dodecyl groups plus $+3kT$ from electrostatic repulsion), and changing the size of the system B as a parameter, resulted in a smooth fit for B = 25. The observed maximum amount of binding is 9.4 mmole/g PEO, or about 145 SDS molecules bound to a single POE molecule with the molecular weight = 1.54×10^4. Accordingly, small system analysis finds that there are only 25 SDS molecules in a cluster formed, and about 6 clusters are dispersed along a single polymer strand somewhat reminiscent of a necklace. This was the first numerical estimate of cluster-like aggregation, or a discrete binding model which until that time had been advocated only verbally [18, 79, 80]. In the following year Smith and Muller [81] estimated the cluster size from the analysis of their binding isotherm by the Hill plot, Eq.(23). Although initially uncertain about these unexpectedly small aggregation numbers, the results were soon reinforced by the realization that a similar model might be applied to protein polypeptide-SDS complexes in SDS PAGE, as described in section IV. Only several years later were such rather small aggregation numbers reported from direct measurements by fluorescense methods [82].

    It is interesting to point out that if the partition function is approximated by the most significant m-th term in Eq.(18), then the A.V. Hill Equation (23) is recovered [83].

$$\Xi = 1 + Q_m\lambda^m \tag{22}$$

$$\beta = d\ln\Xi/d\ln\lambda = Q_m\lambda^m/(1 + Q_m\lambda^m) \tag{23}$$

This equation has been adopted in many theoretical treatments [29, 84-86]. Actually this approximation leads to results not widely different from the full form of the T.L. Hill small systems theory: the aggregation number of surfactant in a cluster for a PEO-SDS system (in 0.1 M NaCl) was estimated by Eq.(23) as m = 20, while Eq.(20) including all the terms gave B = 25. Also, it has been known that the (A.V.) Hill coefficient, m, is 2.4 for the number of interacting O_2-binding sites in a hemoglobin molecule containing 4 heme groups. As an aside, it is confusing but interesting as well to note that there are two Hill equations, with A.V. Hill and T.L. Hill both working on descriptions of cooperative systems but some 50 years apart.

    Ruckenstein et al.[85] formulated equations that explain why and when binding to polymer is more favorable than ordinary micelle formation, and pointed out the importance of surface energy term between polymer and surfactants in clusters. They incorporated an equation similar to Eq.(23) to introduce cooperative nature to binding. Nikas and Blanckschtein [86] worked out a theory of surfactant binding to neutral water-soluble polymer based upon a necklace model. They elaborated on calculating the free energy of each hypothetical step in forming a polymer-surfactant complex and showed how the critical aggregation concentration (cac) is affected by such factors as the hydrophobicity of polymer and the size of surfactant head group.

It would be desirable to account explicitly for the interaction between clusters on a polymer in terms of electrostatic repulsion and entropy of elasticity of polymer chain connecting clusters. Theories so far published have been concerned only with the binding process, the effects of inter-cluster interactions have been neglected or implicitly included into various binding parameters. In this sense these are largely theories for adsorption processes.

Hanssen and Almgren [58] made an interesting modification of the Hill equation (23) by adding a term corresponding to binding to an isolated site (noncooperative) which may increase to some extent the otherwise underestimated configurational entropy of binding.

C. Ising Model

A linear polyelectrolyte is regarded as one-dimensional array of B-binding sites where ligands (surfactant ions) are bound. For such a system, the partition function, Z is,

$$Z = \begin{pmatrix} 1 & 1 \end{pmatrix} \begin{pmatrix} 1 & 1 \\ s/u & s \end{pmatrix}^B \begin{pmatrix} 1 \\ 0 \end{pmatrix}$$

$$(24)$$

where $s = uK_0C_f$ is proportional to the binding probability but is also a reduced concentration defined to yield $s = 1$ when C_f is equal to half saturation (B/2), K_0 is the binding constant for a ligand binding to an isolated binding site where two adjacent sites are vacant, and u is a cooperativity parameter, or a magnifying factor of K_0 when a ligand is bound to a site adjacent to sites already occupied and interacting with them. Eq. (24) was originally proposed to describe magnetization phenomena of analogous logic, i.e., the parallel/antiparallel alignment of magnets vs. binding models with bound/vacant sites. Schwarz applied this method to the binding of dyes to polypeptides [88]. To become familiar with the method, we start with expanding the matrix in Eq.(24). When B = 1

$$Z = \begin{pmatrix} 1 & 1 \end{pmatrix} \begin{pmatrix} 1 & 1 \\ s/u & s \end{pmatrix} \begin{pmatrix} 1 \\ 0 \end{pmatrix} = 1 + s/u \qquad (25)$$

This result shows that Z contains a term for a non-binding state, 1, and another, s/u, for a binding state which is proportional to K_0 and C_f. The fraction of binding, β (degree of binding) is

$$\beta = (s/u)/[1 + s/u] = KC_f/[1 + KC_f] \qquad (26)$$

This is exactly the same form as the Langmuir binding isotherm Eq.(11). There is no inter-ligand interaction for B = 1. For B = 2,

$$Z = 1 + 2(s/u) + s^2/u \qquad (27)$$

Each term in Eq.(27) corresponds in turn to the bound or unbound states schematically represented by

⊥⊥ ,   ⟨⊥⊥⟩ , and   ⟨⟨

where O stands for a bound ligand. The degree of binding, β is.

$$\beta = 2[s/u + s^2/u]/Z /B = (1/B)d \ln Z/d \ln s \qquad (28)$$

In the same way, we have for B = 3,

$$Z = 1 + 3(s/u) + (s/u)^2 + 2(s^2/u) + s^3/u \qquad (29)$$

which corresponds to the following binding states, respectively;

⊥⊥⊥   1   ⟨⊥⟨ s^2/u^2

⟨⊥⊥

⊥⟨⊥   ⟩ s/u ⟨⟨⊥   ⟩ s^2/u
         ⊥⟨⟨
⊥⊥⟨

⊥⊥⟨   ⟨⟨⟨ s^3/u

The degree of binding, β equals:

$$\beta = [3(s/u) + 2(s/u)^2 + 2\times 2(s^2/u) + 3s^3/u]/Z/B = (1/B)d(\ln Z)/d(\ln s) \qquad (30)$$

Once we accept the validity of Eq.(24), values for Z can be calculated for each value of B, and in turn the degree of binding, β is calculated from

$$\beta = (1/B)d \ln Z/d \ln s \tag{31}$$

The most useful case is for $B \to \infty$, or a "sufficiently long" chain. After a long matrix manupilation [89,90],

$$Z = (1-D)\lambda_0^{B+1} [1 + DR^{B+1} /(1-D)] \tag{32}$$

is obtained with $D = (\lambda_0-1)/(\lambda_0 -\lambda_1)$ and $R = \lambda_1/\lambda_0$, where λ_0 and λ_1 are the eigenvalues of matrix in Eq.(24). As $B \to \infty$, Eq.(32) gives [89,90]:

$$\beta = \frac{1}{2}\left\{1-(1-s)/\sqrt{(1-s)^2 +4s/u}\right\} \tag{33}$$

Schneider et al. successfully applied the Ising model to a binding study of iodine to amylose [90]. This informative paper is also an excellent reference for the matrix expression. It is also interesting to note that the interaction between iodine molecules bound in an amylose helix is as strong ($u = 86$) as the interaction between bound surfactants ($10 - 10^2$) as documented by Hayakawa and Kwak [12]. From Eq.(33),

$$d \beta/d \ln s = \sqrt{u/4} \qquad \text{at} \quad \beta = 0.5 \tag{34}$$

This is used to conveniently determine u, and K_0 together with $1/uK = C_f$ at $\beta = 0.5$. From Eq. (33), it is seen that a binding isotherm should be symmetrical with respect to the point ($s = 1$, $\beta = 0.5$). Experimentally obtained isotherms are rarely symmetrical because of the decreased electrostatic potential as a result of surfactant binding. Ion-condensation theory [91] predicts a relatively constant electrostatic potential at low degrees of binding where the linear charge density is kept higher than a critical value, $\xi = 1$. If this is the case it is better to estimate u-values by fitting the lower part of binding isotherm only (usually, $\beta < 0.5$) to Eq.(33). Delville [92] attributed the asymmetry to overcrowding of alkylchains and a reduced electrostatic potential at high degree of binding, and utilized a 4 x 4 matrix and the Poisson-Boltzmann equation for a cell model. He used an iteration procedure to derive the degree of binding from the matrix [89,93].
        Satake and Yang [50] reached Eq.(33) by another method corresponding to the regular solution theory [94] with the coordination number, $z = 2$, i.e., a linear array model. Their paper on the binding of alkyl sulfates to a

polypeptide presented the first application of the Ising model to polymer-surfactant systems.

The studies on polyelectrolyte-ionic surfactant interactions are well documented by Hayakawa and Kwak [12] and more recently by Wei and Hudson [16]. Wei and Hudson [64] employed two theoretical methods to analyze binding of SDS to chitosans with various degree of deacetylation (da): one was the Satake-Yang equation, and another a sequence generating function originally devised by Lifson for the analysis of conformational transitions of biopolymers [89]. This method models surfactant bound polymers as comprising bound sequences, unbound sequences, and flanking sequences. There is no binding in flanking sequences because of acetylation, i.e., lack of cationic sites. This method may be useful for analyzing binding behaviors in copolymers in general.

The effect of polymer molecular weight on binding behavior is important in model considerations. Shirahama et al. [96] reported binding isotherms of dodecylpyridinium bromide to sodium dextransulfate of various molecular weights as shown in Figure 13, where apparent decreases in binding affinity as measured by C_f at $\beta = 0.5$, and in the steepness of the isotherm are seen with reducing the length and thereby the number of binding sites of the polymer.

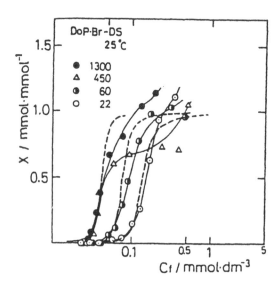

FIG. 13 The dependence of binding of DTAB on number of binding sites (degree of polymerization of Dextransulfate). Full lines: Eq. (33) with interaction parameter, u = 100. From ref. 96 with permission from Kluwer Academic Publishers.

TABLE 1 Dependence of binding affinity on polymer size

Number of binding sites	K_0 / (mmol dm^{-3})$^{-1}$
1300	0.23
450	0.23
60	0.15
22	0.08

Data from Figure 13 (ref. 96), u = 100

The lower cooperativity reflects the fact that the end-effect becomes more marked for the shorter polymers. Binding parameters K_0 and u are obtained for the largest polymer (number of binding sites = 1300) first, and then by using the same u-value, binding isotherms were directly calculated by Eqs. (24) and (31) fitting the experimental isotherms for the shorter polymers for the best K_0 value. We note (Table 1) that for the lower molecular weights K_0 starts to decrease, possibly reflecting the lower electrostatic potential in the vicinity of the shorter polymer ion.

Anthony and Zana [52] studied binding of $C_{12}TAC$ to hydrophobically modified polyelectrolytes. They found especially for highly hydrophobic polymers that the slopes of binding isotherms, $d\beta/dC_f$ at $\beta = 0.5$ are usually very low; ca. 0.065 which is far below the expected value of 0.25 for Langmuir-type binding. This low slope is indicative of anticooperativity, probably originating from electrostatically unfavorable partition of cationic surfactants in hydrophobic regions of the polymer. This is reminiscent of the binding of cationic surfactants to nonionic surfactant [97]. Hayakawa et al. also observed anticooperative binding of 12-hydroxydodecyltrimethylammonium bromide to PSS (98), and found that the anticooperative nature decreased with added NaCl concentration. However, the origin of the anticooperativity may be quite different for the two cases.

D.   The Size of Surfactant Aggregates on Polymers

The aggregation number is a variable in the theories of surfactant binding described above. This quantity however is subject to independent measurement, especially using time-resolved fluorescence quenching (TRFQ) as described in detail in chapters 7 and 10 of this volume. For the purposes of the present discussion only, Table 2 summarizes some typical results for neutral polymers and ionic surfactants. It is noted that the aggregate sizes are always smaller than the micellar aggregation numbers under the same solvent conditions. This may be due to the fact that the amphiphilic polymer chains are meandering

TABLE 2 Surfactant aggregation numbers in aqueous polymer solutions polymers compared to those in free micellar solution (all data at 25 °C).

System	Method	Aggregation number	Ref.
SDS-PEO	fluorescence	35	100
SDS-PEO	fluorescence	35 ± 5	102
SDS-PEO (0.1 M NaCl)	fluorescence	45 (90)	102
SDS-PEO	neutron scattering	60 (70)	99
TrifluoroSDS-PEO	Hill coefficient	15	81
SDS-PDXL	fluorescence	13	100
SDS-PPG	fluorescence	30	101
SDS-PVP	fluorescence	28 ± 4	102
SDS-EHEC	fluorescence	20 (70)	40
C16TAB-EHEC	fluorescence	37 (95)	103

Numbers in brackets denote micellar aggregation numbers in the absence of polymer

through the methylene groups adjacent to the ionic groups of aggregated surfactants. The polymer chain protects these methylene groups from the aqueous solvent, favoring aggregation in the presence of polymer chain and allowing smaller aggregation numbers. Kamenka *et al.* [40] observed a decrease in aggregation number with increasing temperature for the SDS-EHEC system, with the aggregation numbers as low as 12 at 40°C . Zana *et al.* reported similar temperature dependence for C_{16}TAX-EHEC systems [103]. The apparent higher hydrophobicity of EHEC at higher temperature leads to smaller surfactant aggregates. Roden and Sierra measured the concentration dependence of aggregation number for SDS-PPG systems [101].

IV. ELECTROPHORESIS OF POLYMER-SURFACTANT COMPLEXES

A. SDS-Polyacrylamide Gel Electrophoresis (SDS-PAGE)

SDS-PAGE is a compact and inexpensive technique to estimate molecular weights of water-soluble proteins, and its impact to biochemical studies cannot be overestimated. The experiment consists of electrophoresis in a polyacrylamide gel medium of a mixture of SDS and protein-polypeptide which is a reduced (if any disulfide linkage) and denatured protein. Standard experimental methods may be found in reference 104.

In Figure 14, relative electrophoretic mobilities are plotted against molecular weights of 37 proteins [105]. The linearity of the relationship between Log MW and electrophoretic mobility in the presence of SDS is remarkable taking into account that the data are for proteins with a variety of bioactivities and of physical and chemical properties. This striking result attracted the attention of many research groups.

Reynolds and Tanford [106] proposed a theory which may be summarized as follows. SDS is bound to protein-polypeptide in constant ratios of 1.4 g SDS per g protein-polypeptide or 0.6 g/g. The actual experimental condition of SDS-PAGE satisfies 1.4 g/g binding. The bound SDS brings a definite amount of negative electric charge per gram protein, overwhelming the original charges

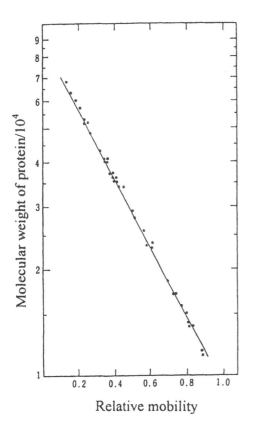

FIG. 14 SDS-PAGE plot. Gel concentration: 10%, cross linkage 2.6% From ref. 105 with permission.

of the protein. The SDS-protein-polypeptide complex thus formed assumes a helix-rich rodlike structure somewhat resembling myoglobin. The length of the rod is proportional to the molecular weight of protein in question. To quote: "This comparison (with myosin) shows that a model consisting of a helical polypeptide chain, folded back upon itself near its middle to give a double helical rod, with the SDS forming a shell about the rod would be consistent with the hydrodynamic data. Of course there is no reason to believe that this is the actual structure of the complexes." [106].   SDS-protein-polypeptide complex electrophoresis is driven by a constant force per gram protein, since the complex carries a constant negative charge per gram at saturation while experiencing the frictional force exerted by gel matrix.   This friction is proportional to the length of the rod-shaped complex, or the molecular weight of the protein. Thus there would be a unique relationship between electrophoretic mobility and molecular weight.   Since that explanation, a number of authors have voiced objections to this rod model.   One such objection is seen in the electric birefringence of SDS-protein polypeptide systems [107, 108]. This experimental technique measures rotational relaxation times yielding hydrodynamic information of polymers. The results support a model of a long extended polymer with some flexibility, but not a rigid rod. This conclusion equally applies to the SDS-PVP system, which does not have a helical structure [109].   In order to investigate this question, Tagaki et al. carried out electrophoresis experiments on SDS-polypeptide complexes with a free boundary method in a a conventional Tiselius apparatus at the "1.4 g/g" condition where polypeptide was supposed to be saturated with bound SDS [110]. The experimental result was striking: there is practically no molecular-weight dependence of the electrophoretic mobility as seen in Figure 15.

In fact the constancy was recently verified by a more refined capillary electrophoretic method [111] as shown in Figure 16. Shirahama et al. [110] compared this observation to the similar constancy noted for the electrophoretic mobility of the polyelectrolyte poly(acrylate) studied by Noda et al. [112]. The latter had explained the constancy in terms of free drainage model in which polyelectrolytes are viewed as surrounded by their counterions which tend to move in opposite direction from the polyion during electrophoresis.   The friction caused by the counterions is not spread out over the whole polyion, but local around charged sites on the polymer, because of electrostatic attraction. Then the counterions are draining through a polymer coil , leading to an electrophoretic mobility independent of polymer molecular weight.   The charged groups in the SDS-polypeptide complex and their counterions would be in a similar position. This analogy led to the proposition of the cluster binding model, or "necklace model" [110].

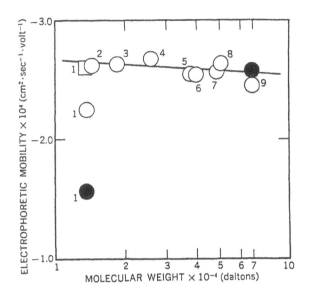

FIG. 15 The free-boundary electrophoretic mobility vs. molecular weight of protein. "1.4g/g complex" with free SDS concentration. Proteins: 1; bovine pancreatic RNase (□ RCAM 0.1%SDS, O RCAM, ● RCM), 2; RCAM hen's egg-white lysozyme, 3; RCM bovine β-lactoglobulin, 4; RCM bovine pancreatic chymotripsinogen A, 5; reduced rabbit-muscle lactate dehydrogenase, 6; RCM rabbit-muscle aldolase, 7; denatured *B. subtilus* α-amylase, 8; RCM Taka-amylase, and 9: bovine serum albumin (● RCM, O RCAM), where RCAM means reduced and carboxyamidomethylated, and RCM reduced and carboxymethylated). From ref. 110, with permission from the Biochemical Society, Japan).

The SDS cluster on a protein-polypeptide strand experiences electrophoretic friction from the counterion atmosphere migrating in opposite direction from the SDS-polypeptide complex. Indeed the electrophoretic mobility is comparable to that of an ordinary SDS micelle as seen in Table 3. Once this model is accepted, electrophoretic behavior of SDS-protein polypeptide complex in a gel medium is easy to understand. A rather expanded complex as a result of electrostatic repulsion between SDS-clusters experiences a frictional force from the gel network, which is related to the molecular weight of the protein as demonstrated by Figure 14.

The constant electrophoretic mobility in a medium free from gel matrix is not limited to the SDS-protein polypeptide systems. It has been shown that SDS-PVP complexes also have a constant electrophoretic mobility, independent

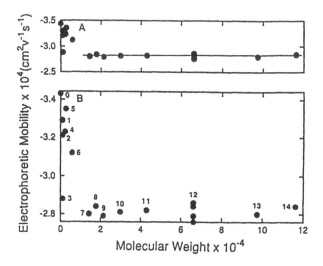

FIG. 16 Electrophoretic mobility of protein in the presence of 0.1% SDS as measured by capillary electrophoresis (50 mM sodium phosphate buffer, pH 7.0, 20°C) O; SDS micelle, proteins: 1: peptide III, 2: vassopressin, 3: peptide II, 4: peptide I, 6: insulin, 7: lysozyme, 8: myoglobin, 9: trypsin inhibitor, 10: carbonic anhydrase, 11: ovalbumin, 12: BSA, 13: phosphrylase *b*, 14: β-galactosidase.   All proteins are reduced and chemically modified except for BSA, the bottom data are for intact BSA. From ref. 111, with permission.

TABLE 3 Electrophoretic mobilities of polymer-SDS complexes

System	polym. mol.wt./10^3	mobility (μm.cm/V.s)	Ref.
PVP/SDS	24.5	-2,73	110
	40	-2.69	
	700	-2.66	
protein polypeptide/SDS	20 - 120	-2.83	111
SDS micelle		-3.43	111
		-3.4	113
		-3.5	114
POE/SDS		-2.6	115
PVNO/SDS		-2.1	116

of molecular weight [110]. Their mobilities are similar to that of ordinary micelles as seen in Table 3. Xia also reported an electrophoretic mobility independent of polymer molecular weight for the POE-LiDS system [115]. More recently Bahadur reported electrophoretic mobilities in the poly(vinylpyridine oxide) (PVNO)-SDS system which are very close and independent of polymer molecular weight [116]. Samso *et al.* reported "seeing" directly the necklace structure of SDS-protein polypeptides by cryo-electron microscopy complemented by small-angle X-ray scattering [117]. The SDS-protein polypeptide complexes have a constant diameter of 6.2 nm, slightly larger than that of SDS micelles, and the polypeptide chains are mostly situated at the interface of the sulfate groups and hydrocarbon core. The distance between clusters is 7-12 nm. Lundahl proposed an alternative model for protein-polypeptide -SDS complexes: a cylindrical rod composed of SDS is surrounded by a polypeptide chain, held together by hydrophobic and electrostatic forces and hydrogen bonds [118]. Chen and co-workers report that their SANS study of a protein-SDS complex supports a necklace model with an SDS cluster size of 29-44 [119,120]. They further propose a fractal structure for packing SDS clusters on polypeptide chain: the fractal dimension obtained is 1.5-2.0 depending on the protein/SDS ratio. This dimension suggests a relatively loose packing of clusters on polypetide chain. Turro *et al.* concentrated on the BSA-SDS system employing steady state and time-resolved fluorescence, ESR, and ^{2}H NMR [121]. The results of this multitechnique approach are consistent with a structure dominantly of the "necklace and beads" type where unfolded protein wraps around the SDS micelles . It is noted in this work that even a protein with the disulfide linkages intact shows a cluster structure similar to the one found in protein polypeptides which are devoid of the disulfide bonds. The model thus obtained from electrophoretic experiments is not in contradiction with those obtained by other methods such as NMR and scattering techniques. Apparently cluster binding allows the surfactant to minimize the aqueous contact surface when forming a complex with polymers.

B. Bimodality of Polymer-Surfactant Complexes

The question can be asked if in polymer-surfactant systems with strong inter-surfactant interactions we should expect a random distribution of surfactants among polymers. An example may be found for the case of the cooperative binding of dodecylpyridinium chloride (C_{12}PCl) to sodium dextran sulfate (NaDx) [122,123]. Based on the known binding isotherm, experimental conditions are set for electrophoresis with quasi-elastic light scattering detection which allows *in situ* observation of migrating species. The resulting

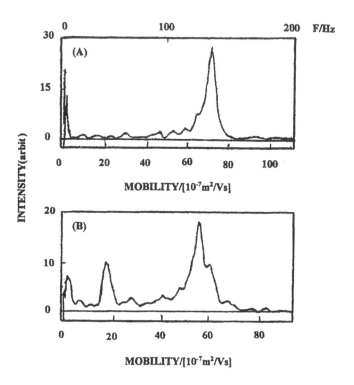

FIG. 17 Power spectra as obtained from electrophoretic measurement with quasi-elastic light scattering detection (20 mMNaBr, 25°C). (A) sodium dextran sulfate only, (B) C_{12}PBr-DS, β= 0.23. From ref. 122 with permission from the American Chemical Society.

electropherograms show that two species are migrating in the NaDx-C_{12}PCl system but only one in the absence of the cationic surfactant, as seen in Figure 17. The electrophoretic mobilities at the plane of zero electroosmotic flow [124] are plotted as a function of the degree of surfactant binding in Figure 18. Figs. 17 and 18 show that a fast migrating species coexists with a slow one, indicating bimodality in the charge of the migrating species. Kato *et al.* also observed two peaks in their electrophoretic experiments [125]. Skerjanc and Kogel came to a similar conclusion in interpreting the osmotic, calorimetric, and electric conductivity results for the PSS-CTABr system [126]. Xia *et al.* used the term "bimodality" to refer to the coexistence of a PEO-LiDS complex and ordinary LiDA micelles, although this usage of the term is different from the present context [115].

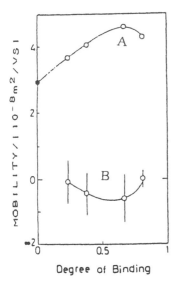

FIG. 18 Electrophoretic mobility vs. degree of surfactant binding. Two modes coexist for a range of equilibrium concentrations, characteristic of strongly interacting small systems. From ref. 122 with permission from the American Chemical Society.

The bimodality results when there is a strong cooperative interaction in a small system like a polymer-surfactant complex. The free energy of the total system becomes smaller when surfactant molecules are bound fully onto limited number of polymer molecules strongly interacting with each other, while leaving other polymer molecules nearly vacant, at the expense of a lower entropy. The above observation and notion are in accordance with the theory by Hill presented some 40 years ago [22]. Hill calculated each term in Eqs.(4-6), which is proportional to the concentration of polymer carrying a definite number of ligands. The result of this calculation (Figure 19) shows that two preponderant species coexist when $\alpha < -2$ (or $\varepsilon < -2kT$), i.e. for the case of cooperative interaction between bound ligands (surfactant ions). It is noted that the compositions of the two species nearly remain the same even when β equals 0.8 or 0.9, and that the bimodal distribution becomes broader as the number of binding sites (in our case the molecular weight of host polymer) is reduced: each term in Eq.(14) contributes even more to total binding because of the increased statistical fluctuation. It is satisfying to find that this experimental study of polymer-surfactant complexes confirms the statistical mechanics theory. Computer simulation also predicts bimodality [127]. The theoretical treatment of the effect of finite size in cooperative systems remains of current interest [128,129].

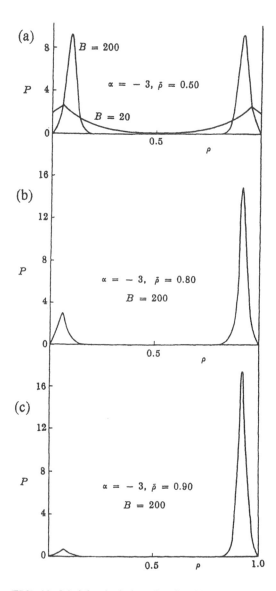

FIG. 19 Model calculation showing bimodality in strongly interacting small systems. $\rho = i/B$, $B = 200$, α (= ε/kT) = -3, (a) $\beta = 0.50$ (also shown the even broader distribution for B = 20), (b) $\beta = 0.80$, and (c) $\beta = 0.90$ From Ref. 22, with permission from Benjamin Publishers, Inc.

V.   MODEL REPRESENTATIONS

Simplified depictions and models of polymer-surfactant complexes have been used ever since the early studies of these systems, and indeed they serve a useful purpose.   In this section, such descriptions are displayed together and in chronological order, accompanied by the comments as given by the original authors and/or the present author.

Takagi and Shirahama, following the publication of their papers on SDS-PAGE, wrote two illustrative articles on the problem, one for biochemists to make them well acquainted with the colloid chemical background [130], and another for colloid chemists to remind them of colloid chemical problems in biochemistry [131]. In Section IV we already discussed the covered helix model for the SDS-proptein polypeptide complex as proposed by Reynolds and Tanford [106], where a folded-helix is covered by SDS.   Figure 20 shows a necklace model   depicted for the first time in the article by Takagi and Shirahama.   Typically, at low SDS concentration, cationic and hydrophobic sites are the initial binding sites, around which micelle-like aggregates are clustered with increasing SDS concentration.   Goddard sketched the multi-equilibrium phase diagram shown in Figure 21, including depiction of the polymer-surfactant complexes [11, 132].

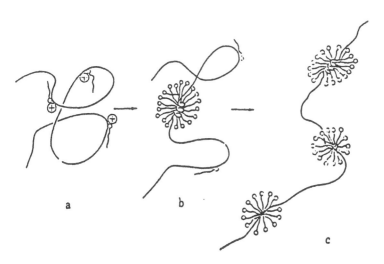

FIG. 20 Necklace model for SDS-protein polypeptide complex. From ref. 130 with permission.

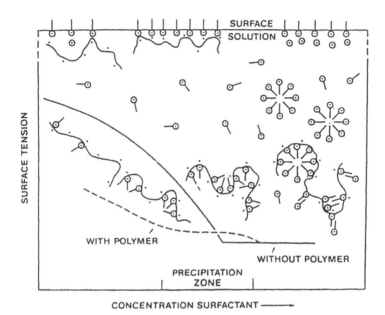

FIG. 21 Coexisting species in a polymer-surfactant system. From ref. 11, *Colloids Surf. 19*: 301 (1986), with permission from Elsevier Science-NL, Sara Burgerhartstraat 25, 1055 KV Amsterdam, the Netherlands.

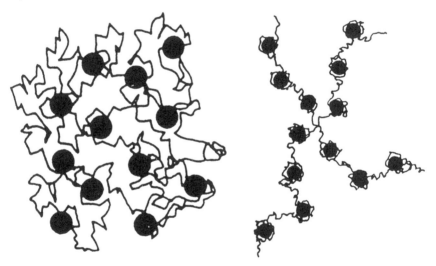

FIG. 22 Necklace model by Cabane and Duplessix. From refs. 133 and 134, with permission.

Some years later, Cabane and Duplessix represented a model based on the results of their NMR and neutron scattering studies as shown in Figure 22 [99, 133,134]. They also applied the term "necklace" to model the complex formed with SDS and synthetic polymer. Nagarajan and Kalpakcir depicted more details of the surfactant aggregate as shown in Figure 23 [135], where Cabane had not speculated on the specific binding sites. This last picture was meant to represent the binding between PEO and SDS, where the binding mechanism is not necessarily via the sulfate groups as perhaps assumed by Nagarajan and Kalpakcir. A more realistic system with specific binding sites is the case of a polyion and oppositely charged surfactant ion, as drawn by Abuin and Scaiano in Figure 24 [136]. This model clearly suggests the oversimplification of an Ising-type binding model. Instead, flexibility of the polyelectrolyte backbone is necessary to form binding sites for surfactant aggregates. Notably, Figure 24 decorated the hardcover of Volume 37 of the Surfactant Science Series [3].

Lundahl *et al.* proposed a model for the protein-SDS complex where an endcapped cylindrical SDS micelle is surrounded by a protein helix as seen in Figure 25 [118]. They named their model "the flexible helix." This model seems to overestimate the hydrogen-bonding formed between SDS head groups and peptide groups.

Chen and Teixeira proposed models of BSA-SDS complexes based on the results of a SANS study as seen in Figure 26, where models for two BSA/SDS ratios are sketched [137]. They used the fractal dimension as well as the number and size of the clusters to characterize the complex. Benkhira *et al.* elaborated on a model after examining the hydrodynamic and scattering data as seen in Figure 27 [100].

FIG. 23 Necklace model by Nagarajan, and Kalpakcir. From ref. 135, with permission from the American Chemical Society.

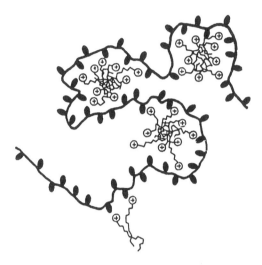

FIG. 24 Necklace model for polyion-oppositely charged surfactant ion interaction. From ref. 136 with permission from the American Chemical Society.

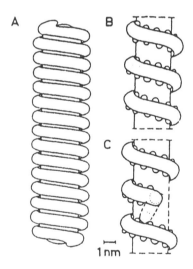

FIG. 25 The flexible helix model. The geometry is determined by the size of SDS rod (radius =1.6 nm) around which a polypeptide chain forms a helix with a radius of 2.4 nm. (A) A model for BSA-SDS complex, (B) Less densely pitched helix as estimated by Reynolds and Tanford, and (C) A hydrophobic polymer segment is traversing the SDS core, leading to a flexible complex. From ref. 118, with permission.

BSA/SDS ~ 1/1

Fractal Packing of Micelle-Like Cluster

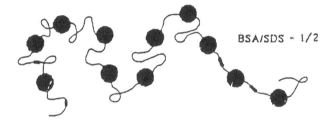

BSA/SDS ~ 1/2

FIG. 26 Fractal model for BSA-SDS complex. Basically a necklace model with of possible fractal structure. Small cylinders in the chain show a collapsed structure, e.g., α-helix. From ref. 137 with permission.

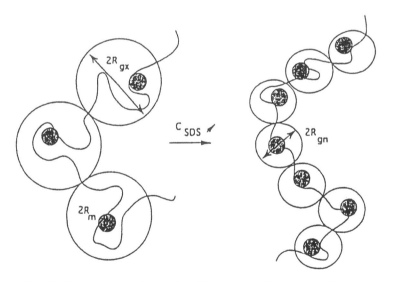

FIG. 27 Hydrodynamic detail of necklace model. R_m: radius of the micelles (clusters), R_{gx} (R_{gn}): radius of the coli of x (n) segments between two clusters. From ref. 100 with permission.

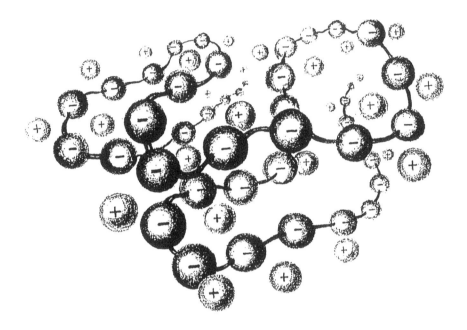

FIG. 28 Takagi free draining necklace model From ref. 128, with permission from Gendai Kagaku.

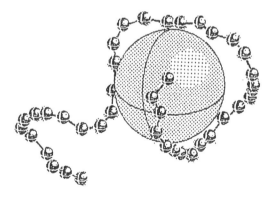

FIG. 29 Monte Carlo simulation of polyion-micelle aggregate. From ref. 139 with permission from the American Chemical Society.

The idea of a free draining necklace model was represented in the illustration accompanying a short comment by Takagi, shown in Figure 28 [138]. Recently, these schematic drawings have been joined by the graphical representation of simulation studies such as described in Chapter 5 of this book. A typical example based on Monte Carlo studies performed at the University of Lund is depicted in Figure 29. [139].

Indeed, we note that already a quarter of a century has elapsed since the first description of the necklace model of polymer-surfacatant complexes, establishing this model as the most feasible configuration on which to base theoretical descriptions and new experiments.

REFERENCES

1. K.Shinoda, T. Nakagawa, B. Tamamushi, and T. Isemura, *Colloidal Surfactants*, Academic Press, New York, 1963.
2. E. Lucassen-Reynders, (ed.) *Anionic surfactants: Physical Chemistry*, Marcel Dekker, New York, 1981.
3. D.N. Rubingh and P.M. Holland, (eds.), *Cationic surfactants: Physical Chemistry* Marcel Dekker, New York, 1991.
4. M.J. Schick, (ed.), *Nonionic surfactants: Physical chemistry* Marcel Dekker, New York, 1987.
5. Y. Moroi, *Micelles*, Prenum Press, New York, 1992.
6. P.J. Flory, *Principles of Polymer Chemistry*, Cornell University Press, Ithaca N.Y., 1953.
7. S.A.Rice and M. Nagasawa, *Polyelectrolyte Solutions*, Academic Press, New York, 1961.
8. H. Morawetz, *Macromolecules in solutions*, Interscience Pub., New York, 1965.
9. M.M. Breuer and I.D. Robb, *Chem. Ind.* 530 (1972).
10. I.D. Robb, in ref. 2, pp.109-142.
11. E.D. Goddard, *Colloid. Surf.*, *19*: 255, 301 (1986).
12. K. Hayakawa and J.C.T. Kwak, in ref. 3, pp. 189-248.
13. E.D. Goddard and K.P. Ananthapadmanabhan (eds.), *Interactions of Surfactants with Polymers and Proteins*, CRC Press, Boca Raton, 1992.
14. J. Steinhardt and J.A. Reynolds, *Multiple Equilibria in Proteins*, Academic Press, New York, 1969.
15. L. Piculell, F. Guillement, K. Thuresson, V. Shubin, and O. Ericsson, *Advan. Colloid Interface Sci.,63:* 1-21 (1996).
16. Y.C. Wei and S.M. Hudson, *Rev. Macromol. Chem. Phys.*, *C35*: 15 (1995).
17. J.C. Brackman and J.B.F.N. Engberts, *Chem. Soc. Rev.*, *22*: 85 (1993).
18. H. Arai, M. Murata, and K. Shinoda, *J. Colloid Interface Sci.*, *37*: 223 (1971).
19. K. Shirahama, *Colloid Polym. Sci.* *252*: 978 (1974).
20. K. Shirahama and N. Ide, *J. Colloid Interface Sci.*, *54*: 450 (1976).

21. T.L. Hill, *An Introduction to Statistical Thermodynamics*, Reading, Mass., 1960.
22. T.L. Hill, *Thermodynamics of Small Systems*, Vol. 1, Benjamin, New York, 1963.
23. *Proc. of Symposium on Associative Polymers*, Loen, Norway, 1995.
24. S. Harada, H. Sahara, and H. Oshikubo, *Repts. Progr. Polym. Phys. Jpn.,38*: 139(1995);   C. Maltesh and P. Somasundaran, *Colloids Surf., 69:*167 (1992).
25. Z. Zhen and C.-H. Tung, *Polymer, 33*: 812 (1992).
26. E.D. Goddard and P.S.Leung, *Langmuir, 8*: 1499 (1992).
27. S. Saito, in ref. 4, pp. 881-926, 1987.
28. F. van Voorst Vader, *Trans. Faraday Soc., 57*:110 (1961).
29. T. Gilanyi and E. Wolfram, *Colloids Surf., 3*: 181 (1981).
30. M. Dupeyrrat, *J. Chim. Phys., 61*: 306 (1964).
31. Y.C.Wei and S.H.Hudson, *Macromolecules, 26*: 4151 (1993).
32. K. Hayakawa, K. Fukuda, T. Maeda, M. Ikeda, and I. Satake, *Bull. Chem. Soc. Jpn., 66*: 2744 (1993).
33. D.M. Painter, D.M. Bloor, N. Takisawa, D.G. Hall, and E. Wyn-Jones, *J. Chem. Soc., Faraday Trans, 1, 84*: 2087 (1988).
34. I. Satake, T. Hatakenaka, T. Maeda, and K. Hayakawa, *Rep. Fac. Sci. Kagoshima Univ(Math., Phys., & Chem.), 27*: 35 (1994).
35. K. Shirahama, *Proc. 46th Symposium on Colloid and Interface Chemistry*, CSJ, Tokyo, 1993.
36. C.J. Davidsson, *PhD thesis*, University of Aberdeen, 1983.
37. C. Homberg, S. Nilsson, S.K. Singh, and L-O. Sundelöf, *J. Phys. Chem., 96*: 871 (1992).
38. K Anad, O.P. Yadav, and P.P. Singh, *Colloid Polym. Sci., 270*: 1201 (1992).
39. W. A. Wan-Badhi, W. M. Z. Wan-Yunus, D.M. Bloor, D.G. Hall, and E. Wyn-Jones, *J. Chem. Soc. Faraday Trans., 89*: 2723 (1993).
40. N. Kamenka, I. Burgaud, R. Zana, and B. Lindman, *J. Phys. Chem., 98*: 6785 (1994).
41. S. Saito and M. Yukawa, *J. Colloid Interface Sci., 30*: 211 (1969).
42. S. Saito and K. Kitamura, *J. Colloid Interface Sci., 35*: 346 (1971).
43. S. Saito, *J. Colloid Interface Sci., 24*: 227 (1967).
44. K. Shirahama, A. Himuro, and N. Takisawa, *Colloid Polym. Sci., 265*: 96 (1987).
45. K. Shirahama and S. Nagao, *Colloids Surf., 66*: 275 (1992).
46 K. Shirahama, M. Oh-ishi, and N. Takisawa, *Colloids Surf., 40*: 261 (1989).
47. ref. 2, p.119
48. K. Shirahama, K. Mukae, and H. Iseki, *Colloid Polym. Sci., 272*: 493 (1994).
49. J. Brackman and J.B.F.N. Engberts, *Langmuir, 8*: 424 (1992).
50. T. Seimiya, *Proc. 46th Symposium on Colloid and Interface Chemistry*, CSJ, Tokyo, 1993.
51. M.J. Blandamer, *Introduction to Chemical Ultrasonics*, Academic Press, New York, 1973, p. 11.
52. O. Anthony and R. Zana, *Langmuir, 10*: 4048 (1994).
53. A. Carlsson, B. Lindman, T. Watanabe, and K. Shirahama, *Langmuir, 5*: 1250 (1989).
54. K. Veggeland and S. Nilsson, *Langmuir, 11*: 1885 (1995).

55. W. Scholtan, *Makromol. Chem.*, *11*: 131 (1953).
56. T. Takagi and N. Kuroki, *J. Polym. Sci., Chem. Ed.,11*: 1889(1973).
57. I. Satake and J. T. Yang, *Biopolymers, 15*: 2263 (1976).
58. P. Hanssen and M. Almgren, *J. Phys. Chem., 100*: 9038 (1996).
59. T. Shimizu and J.C.T. Kwak, *Colloids Surf. A, 82*: 163 (1994).
60. T. Shimizu, *Colloids Surf. A, 84*: 239 (1994).
61. T. Shimizu , *Colloids Surf. A; 94*: 115 (1995).
62. S.M. Mel'nikov, V.G. Sergeyev, and K. Yoshikawa, *J. Am. Chem. Soc., 117*: 2401 (1995).
63. S.M. Mel'nikov, V.G. Sergeyev, and K. Yoshikawa, *J. Am. Chem. Soc., 117*: 9951 (1995).
64. Y.-C Wei and S.M. Hudson, *Macromolecules, 26*: 4151 (1993).
65. K. Shirahama and H. Nakamiya, *Proc. Yamada Conference on Ordering and Organisation in Ionic Solutions,*. (N. Ise and I. Sogami, eds.), World Scientific Publ., Singapore, p. 335-344 (1988).
66. T. Tanaka, *Phys. Rev. Lett., 40*: 820 (1978).
67. H. Inomata, S. Goto, and S. Saito, *Langmuir, 8*: 1030 (1992).
68. L. Piculell, D. Hourdet, and I. Iliopoulos, *Langmuir, 9*: 3324 (1993).
69. E. Kokufuta, Y-Q. Zhang, T. Tanaka, and A. Mamada, *Macromolecules, 26*: 1053 (1993).
70. M. Sakai, N. Satoh, K. Tsujii, Y-Q. Zhang, and T. Tanaka, *Langmuir, 11*: 2493 (1995).
71. Y. Murase, M. Sakai, T. Onda, K. Tsujii, and T. Tanaka, *Proc. 49th Symposium on Colloid and Interface Chem.*, CSJ, Tokyo, (1996).
72. H. Okuzaki and Y. Osada, *Macromolecules, 27*: 502 (1994).
73. H. Okuzaki and Y. Osada, *Macromolecules, 28*: 4554 (1995).
74. J.P. Gong and Y. Osada, *J. Phys. Chem., 99*; 10971 (1995).
75. K. Shirahama, S. Sato, M. Niino, and N. Takisawa, *Colloids Surf. A, 112*: 233 (1996).
76. T. Tanaka, in *Polyelectrolyte Gels*, (R.S. Harland and R.K. Prud'homme, eds.), ACS symposium Ser. 480, ACS, Washington, 1992, p.3.
77. C. Tanford, *Physical Chemistry of Macromolecules*, John Wiley, New York, 1961, pp.526-586.
78. R.F. Steiner and L. Garone, *The Physical Chemistry of Biopolymer Solutions*, World Scientific Publ., Singapore, 1991, pp.47-75.
79. M.L. Fishman and F.R. Eirich, *J. Phys. Chem.,75*: 3135 (1971).
80. F. Tokiwa and K. Tsujii, *Bull. Chem. Soc. Jpn., 46*: 2684 (1973).
81. M.L. Smith and N. Muller, *J. Colloid Interface Sci., 52*: 507 (1975).
82. R. Zana, P. Lianos, and J. Lang, *J. Phys. Chem., 89*: 41 (1985).
83. A.V. Hill, *J. Physiol.(London), 40*: 190 (1910).
84. R. Nagarajan, *Colloids Surf., 13*: 1 (1985).
85. E. Ruckenstein, G. Huber, and H. Hoffmann, *Langmuir, 3*: 382 (1987).
86. Y.J. Nikas and D. Blankschtein, *Langmuir, 10*: 3512 (1994).
88. G. Schwarz, *Eur. J. Biochem., 12*: 442 (1970).
89. D. Poland and H.A. Scheraga, *Theory of Helix-Coil Transitions in Biopolymers*, Academic Press, New York, 1970.

90. F.W. Schneider, C.L. Cronan, and S.K. Podder, *J. Phys. Chem.*, *72*: 4563 (1968).
91. G.S.Manning, *Accounts Chem. Res.*, *12*: 443 (1979).
92. A. Delville, *Chem. Phys. Lett.*, *118*: 617 (1985).
93. D. Poland, *Cooperative Equilibria in Physical Biochemistry*, Clarendon Press, 1978.
94. E.A. Guggenheim, *Mixtures*, Clarendon Press, London, 1952, pp.29-87.
95. T.H. Hill, *Statistical Mechanics*, McGraw-Hill, New York, 1956, pp.318-327.
96. K. Shirahama, T. Watanabe, and M. Harada, *The Structure, Dynamics, and Equilibrium Properties of Colloidal Systems*, (D.M. Bloor and E. Wyn-Jones, eds.), NATO ASI Series Vol. 324, Kluwer Academic Pub., London, 1990, pp. 161-172.
97. K. Shirahama, Y. Nishiyama, and N. Takisawa, *J. Phys. Chem.*, *91*: 5928 (1987).
98. K. Hayakawa, K. Fukuda, T. Maeda, and I. Satake, *Bull. Chem. Soc. Jpn, 66*; 2744 (1993).
99. B. Cabane and R. Duplessix, *Colloids Surf.*, *13*; 19 (1985).
100. A. Benkhira, E. Franta, and J. Francois, *J. Colloid Interface Sci.*, *164*: 428 (1994).
101. E. Rodens and M.L. Sierra, *Langmuir*, *12*: 1600 (1996).
102. E.A. Lissi and E. Abuin, *J. Colloid Interface Sci.*, *105*: 1 (1985).
103. R. Zana, W. Bianana-Limbele, N. Kamenka, and B. Lindman, *J. Phys.Chem.*, *96*: 5461 (1992).
104. K.Weber, J.R.Pringle, and M.Osborn, in *Methods in Enzymology, XXVI Enzyme Structure Part C*, (C.H.W.Hirs and S.N. Timasheff, eds.), Academic Press, New York, 1972, pp.3-27.
105. K. Weber and M. Osborn, *J. Biol. Chem.*, *244*: 4406 (1969).
106. J.A. Reynolds and C. Tanford, *Proc. Nat. Acad. Sci. U.S.*, *66*: 1002 (1970).
107. A. K. Wright, M.R. Thompson, and R.L. Miller, *Biochemistry*, *14*: 3224 (1975).
108. E.S. Rowe and J. Steinhardt, *Biochemistry*, *15*: 2579 (1976).
109. P.J. Rudd and B.R. Jennings, *J. Colloid Interface Sci.*, *48*: 302 (1974).
110. K. Shirahama, K. Tsujii, and T. Takagi, *J. Biochem.*, *75*: 309 (1974).
111. M.R. Karim, S. Shinagawa, and T. Takagi, *Electrophoresis*, *15*: 1141 (1994).
112. I. Noda, M. Nagasawa, and M. Ota, *J. Am. Chem. Soc.*, *86*: 5075 (1964).
113. D. Stigter and K.J. Mysels, *J. Phys. Chem.*, *59*: 45 (1955).
114. K. Aoki, *Bull. Chem. Soc. Jpn.*, *29*: 369 (1956).
115. J. Xia, P. L. Dubin, and Y. Kim, *J. Phys. Chem.*, *96*: 6805 (1992).
116. P. Bahadur, P. Dubin, and Y. K. Rao, *Langmuir*, *11*: 1951 (1995).
117. M. Samso, J-R. Daban, S. Hansen, and G.R. Jones, *Eur. J. Biochem.*, *232*: 818 (1995).
118. P. Lundahl, E. Greijer, M. Sandberg, S. Cardell, and K-O. Eriksson, *Biophys. Biochim. Acta, 873*: 20 (1986).
119. S-H. Chen and J. Teixeira, *Phys. Rev. Lett, 57*: 2583 (1986).
120. X.H. Guo, N.M. Zhao, S.H. Chen, and J. Teixeira, *Biopolymers*, *29*: 335 (1990).
121. N.J. Turro X-G. Lei, K.P. Ananthapadmanabhan, and M. Aronson, *Langmuir, 11*: 2525 (1995).

122. K. Shirahama, K. Kameyama, and T. Takagi, *J. Phys. Chem., 96*: 6817 (1992).
123. A.M. Malovikova, K. Hayakawa, and J.C.T. Kwak, *J. Phys. Chem., 88*: 1930 (1984).
124. S. Mori, H. Okamoto, T. Hara, and K. Aso, in *Particle Processing,*. (P. Somasundaran, ed.), American Institute of Mining Petroleum Engineering, New York, 1980, p.632.
125. T. Kato, *International Symposium on Polyelectrolytes and International Bunsen-Discussion-Meeting*, Berlin, 1995.
126. J. Skerjanc and K. Kogel, *J. Phys. Chem., 93*: 7913 (1989).
127. J. Reiter and I. R. Epstein, *J. Phys. Chem., 91*: 4813 (1987).
128. A.V. Verno, A.V. Andryushin, and V.I. Shimulis, *J. Phys. Chem., 95*: 4853 (1991).
129. H. Qiang and J. A. Schellman, *J. Phys. Chem., 96*: 3987 (1992).
130. T. Takagi, K. Shirahama, K. Tsujii, and K. Kubo, *Tanpakushitu, Kakusan, Koso (Protein, Nucleic Acid, Enzyme*, in Japanese*), 21*: 811 (1976).
131. K. Shirahama and T. Takagi, *Hyomen (Surface*, in Japanese*), 11:* 383 (1973).
132. E.D. Goddard and R.B. Hannan, *J. Colloid Interface Sci., 55*: 73 (1976).
133 B. Cabane and R. Duplessix, *J. Physique, 43*: 1529 (1982).
134. B. Cabane and R. Duplessix, *J. Physique, 48*: 651 (1987).
135. R. Nagarajan and B. Kalpakcir, *Polym. Repr. Am. Chem.Soc. Div. Polym. Chem., 23*: 41 (1981).
136. E.B. Abuin and J.C. Scaiano, *J. Am. Chem. Soc., 106*: 6274 (1984).
137. X.H. Guo, N.M. Zhao, S-H. Chen and J. Teixeira, *Biopolymers, 29*: 335 (1990).
138. T. Takagi, *Gendai Kagaku (Modern Chemistry*, in Japanese), 67 (1984).
139. T. Wallin and P. Linse, *J. Phys. Chem. 100*: 17873 (1996).

5
Models of Polymer-Surfactant Complexation

PER LINSE, LENNART PICULELL, and PER HANSSON   Physical Chemistry 1, Center for Chemistry and Chemical Engineering, Lund University, P.O. Box 124, S-221 00 Lund, Sweden

I.   INTRODUCTION

A striking feature of many (but not all) polymer/surfactant pairs is their complexation in dilute aqueous solution [1]. The standard evidence for such complexation is in terms of surfactant binding isotherms (Figure 1), obtained by methods such as potentiometry or surfactant self diffusion. The experiments show that when surfactant molecules are gradually added to a polymer solution, it forms a type of supramolecular aggregate at concentrations below the "normal" surfactant critical micellization concentration (cmc). The cmc is the concentration where the surfactant molecules would self-associate into micelles in the polymer-free solution under otherwise identical conditions.

To make a good physical picture of such a complexation phenomenon is, of course, essential for any attempts at modelling it. This is not trivial, however, and different pictures have been invoked over the years. Indeed, as we will argue, different pictures may be suited for different situations.

The very notion of a binding isotherm seems to suggest what one may call a "polymer-centered" viewpoint, where the polymer molecule is thought to contain certain sites that may bind surfactant molecules. In this picture, the polymer-surfactant interaction constitutes the strong perturbation to the bulk surfactant solution, which is regarded merely as a reservoir of binding ligands. The problem with the polymer-centered picture is, however, that it more or less ignores the fact that surfactant molecules are strongly hydrophobic. They basically wish to either join or form hydrophobic aggregates. In fact, it may generally be assumed that hydrophobic association is essential in aggregates involving surfactant molecules in an aqueous environment.

The recognition of the strong tendency for surfactant molecules to self-associate has inspired the now prevailing "surfactant-centered" picture, according to which polymer-surfactant complexation is essentially a surfactant micellization, albeit perturbed by the presence of the polymer [2-9]. This viewpoint has proven very fruitful, especially to explain the cooperative binding of surfactant molecules to certain polymers (Figure 1, solid curve). The critical association concentration (cac) may then simply be regarded as the surfactant cmc in the polymer solution. An inherent feature of the surfactant-centered picture is that the polymer-bound surfactant molecules should normally form finite aggregates even in the limit of saturation binding to an infinite polymer molecule. The latter feature, which is confirmed by recent experiments [10-17], does not emerge naturally from linear, polymer-centered binding models.

However, there are also important cases when no threshold concentration for surfactant binding is observed; there is no cac. One such case, illustrated in Figure 1 (dashed curve), is the binding of surfactant molecules to a hydrophobically associating water-soluble polymer (HAP). An HAP is a polymer with a hydrophilic main chain, onto which a small fraction of strongly hydrophobic groups ("hydrophobes") have been attached, either as end-groups or as side-groups. It is now realized that the reason for the absence of the cac is that the HAP hydrophobes form "hydrophobic domains," or HAP micelles, even in the absence of surfactant molecules [18-21]. These HAP micelles have the capacity to bind, or solubilize, individual surfactant molecules. This is quite different from the situation where surfactant self-association is a prerequisite for the complexation (as shown in Figure 1, solid curve). The relevant physical picture here seems be that of mixed micellization, where the surfactant

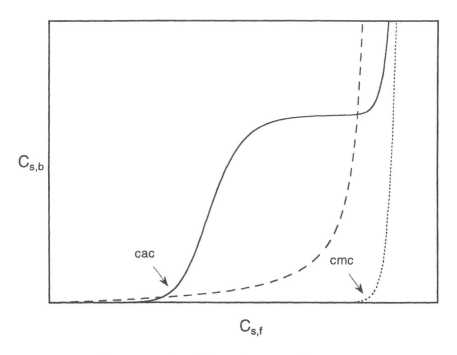

FIG. 1 Schematic representation of binding isotherms, showing the concentration of micellized surfactant ($C_{s,b}$) as a function of concentration of free surfactant ($C_{s,f}$) for solutions of surfactant alone (dotted curve), of surfactant and polymer with cooperative surfactant binding starting at a critical association concentration (solid curve), and of surfactant and a hydrophobically associating polymer (dashed curve). The critical micelle concentration (cmc) and critical aggregation concentration (cac) are indicated.

molecule and the HAP hydrophobes participate in the mixed micelles on more-or-less equal terms.

In the light of the above observations, and for the purpose of making appropriate models, it seems useful to distinguish three idealized categories of polymer-surfactant complexes, listed below in the order of increasing involvement of a hydrophobic attraction between the polymer and the surfactant molecules.

1. Pure micelles of ionic surfactant molecules formed in the presence of an oppositely charged, hydrophilic polyelectrolyte. Examples are the complexes

formed by sodium hyaluronate and cationic surfactants of the alkyl trimethylammonium type [12]. For these systems, the polymer-surfactant attraction is purely electrostatic, and the surfactant micelles may be assumed to be separated from the oppositely charged polyelectrolyte by a layer of water molecules. The binding isotherm always shows a cac.

2. Mixed micellar complexes involving the polymer main chain. This type includes some of the "classic" complexes of ionic surfactant molecules with nonionic homopolymers such as PEO, PVP, PVA, and PNIPAm. Here, there is an intimate molecular contact between the surfactant molecules and the polymer chain. Still, the polymer is not sufficiently hydrophobic to form hydrophobic aggregates on its own. Hence, it has no solubilizing capacity. The polymer-surfactant complex must therefore always involve self-associated surfactant molecules, and the binding isotherm always displays a cac. As for category 1, the concept of polymer-induced micellization is still useful.

3. Mixed micellar complexes involving the hydrophobes of an HAP. The important feature here, as pointed out above, is that the HAP may form micelles on its own, with the capacity to solubilize individual surfactant molecules. The situation is closely analogous to that of mixed surfactants [22].

The classification is clarified in the interaction schemes in Figure 2, where an arrow symbolizes a hydrophobic attraction, and the thickness of the arrow shows its strength. Note that categories 2 and 3 also include systems where the polymers and surfactant molecules are oppositely charged, giving important electrostatic contributions to the stability of the complexes. Cases in point are the complexes formed by cationic surfactant molecules with PSS [13, 23, 24] (category 2), or hydrophobically modified NaPA [20] (category 3), respectively. Finally, there are mixed cases of categories 2 and 3, where both the polymer main chain and attached hydrophobes form mixed micelles with added surfactant molecules.

The details of the micelles formed in the various categories should be different. For category 1 complexes, a useful approximation may be to model the surfactant aggregates as uniform spheres of a definite size, independent of the degree of binding. This assumption may still be useful, although less self-evident, for category 2 complexes, while it should break down altogether for category 3 complexes. In the latter case, we expect a transition from HAP-type aggregates to surfactant-type aggregates, depending on the stoichiometry of the mixed micelles. Similarly, the surfactant aggregation numbers of the complex, as obtained from suitable experimental techniques such as fluorescence, give no automatic information on the "micellar sizes" of category 3 complexes.

With the above physical pictures in mind, it is our purpose with this chapter to critically review various approaches as quantitative or qualitative

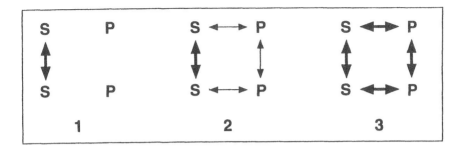

FIG. 2 Schematic representation of the role of hydrophobic interactions (arrows) in the three categories of complexes formed by polymer (P) and surfactant (S), as discussed in the text. A thicker arrow symbolizes a stronger hydrophobic attraction.

models of polymer/surfactant complexes. As should already be apparent, we will restrict ourselves to complexes formed in dilute or (for HAP) semi-dilute solutions of the polymer, and systems where the hydrophobic aggregates formed are closed and micelle-like. This means that we will exclude all precipitates, such as the lamellar or hexagonal structures formed in salt-free solutions of stoichiometric mixtures of oppositely charged polymer and surfactant molecules [25]. Moreover, the polymer is generally assumed to be linear, flexible, and solvent-swollen.

We have chosen to present the various theoretical approaches in the order of increasing complexity. The simplest approaches are given in section II. The usefulness of these models is primarily to provide physical insight into the nature of polymer-surfactant complexation. Thus, section II will illustrate differences between polymer-centered and surfactant-centered models applicable for category 1 and 2 complexes. Section II also contains a discussion of category 3 complexes on the basis of the mixed micelle concept. Models that provide more molecular detail, but which are also more computationally demanding, are discussed in section III. The chapter ends with a summary of the present state-of-the-art of models polymer-surfactant complexation, and their applicability to the various categories of complexes.

The selected set of theories and models are by no means complete. We have made a selection which reflects our experience in the area. Our objective has been to consider approaches which have, at least, some elements of molecularity, starting from simple ones and successively approaching more extended models and theories.

II. SIMPLE MODELS

A. Hydrophobic Polyelectrolyte Plus Oppositely Charged Surfactants

In this section, we will describe and compare one polymer-centered and one surfactant-centered approach to the category 1 complexation of oppositely charged polymer and surfactant. While both approaches succeed quite well in reproducing experimentally observed binding isotherms [17, 26], their implications for the aggregate structures are quite different. In the first two subsections we present the models, while in the third we compare their predictions with experimental data. We have chosen the C_{12}TAB-NaPA-NaBr system as a relevant reference system since the complexation is most likely of pure electrostatic nature and there exist a great body of experimental data from different sources. It is also evident from the extensive binding studies in the literature [26] that the binding behavior in this system is quite typical.

1. The Zimm-Bragg model

In 1976 Satake and Yang [27] used an analysis based on the Zimm-Bragg (ZB) theory [28] in an attempt to explain the coil-helix transition in polypeptide chains induced by the binding of surfactants. Ever since Shirahama et al. [29] in 1981 applied this type of analysis to the cooperative binding of an anionic surfactant to a cationic synthetic polyelectrolyte in solution, it has been used extensively to quantify the interactions between surfactants and polyelectrolytes of opposite charge. The model, including related and extended treatments, was reviewed recently by Hayakawa and Kwak [26] and by Wei and Hudson [30]. A detailed description is also presented in Chapter 4 of this volume.

In the ZB model, the polyelectrolyte possesses a linear array of binding sites. The binding of a surfactant molecule to one of these sites is characterized by the binding constant K_0 if both neighboring sites are empty of surfactant molecules, otherwise with the binding constant K_0u. The additional factor u, referred to as the "cooperativity parameter," is related to the interaction free energy ($-kT\ln u$) between neighboring surfactant molecules on the lattice. From the statistical analysis [27, 31] it follows that the fraction of sites, β, occupied by the surfactant is

$$\beta = \frac{1}{2} + \frac{K_0 u C_{s,f} - 1}{2\left\{\left(K_0 u C_{s,f} - 1\right)^2 + 4 K_0 C_{s,f}\right\}^{1/2}} \tag{1}$$

where $C_{s,f}$ is the concentration of free surfactant molecules. The average number, m, of surfactant molecules in an uninterrupted sequence is given by

$$m = \frac{2\beta(u-1)}{\left\{4\beta(1-\beta)(u-1)+1\right\}^{1/2} - 1} \tag{2}$$

The assumption of nearest neighbor interactions implies that all geometric constraints on the surfactant aggregates are removed, *i.e.*, there are no packing effects favoring a particular shape or aggregation number. The sizes of the clusters are limited only by the configurational entropy of the occupied and unoccupied sites.

2. Closed association model

A more recent approach to category 1 complexation adopts the surfactant-centered viewpoint, and considers the complexation as micelle formation in the field of the oppositely charged polyelectrolyte. For electrostatic reasons, the local concentration of oppositely charged surfactant molecules near the polyelectrolyte is higher than the global one. This suggests the simple, but fruitful picture that, when the global surfactant concentration is equal to the cac in the polymer system, the local surfactant concentration near the polyelectrolyte is close to the cmc for the surfactant in a polymer-free solution. Calculations using the Poisson-Boltzmann equation supporting this notion has been performed by Skerjanc *et al.* [32] and by Löfroth *et al.* [33].

A serious problem with the (polymer-centered) ZB model is the poor description of the surfactant properties. In conflict with the predictions of Eq. (2), recent results show that polyelectrolyte-surfactant complexation is characterized by the formation of globular aggregates with aggregation numbers largely independent of the degree of binding [12-16]. On the basis of the surfactant-centered view, one can construct thermodynamic binding models on the same level of theory as the ZB model. In the following model, referred to as the closed association (CA) model [17], the aggregation number of the micelles is limited to a single value. This is a simple, but conceptually important, improvement of the isodesmic description used in the ZB model.

The surfactant molecules can be in three different states: (i) free in bulk, (ii) singly bound to the polyelectrolyte, and (iii) in closed aggregates bound to the polyelectrolyte. The equilibrium between states (i) and (ii) is considered as an exchange reaction between surfactant ions S and simple counterions ions C of the same charge:

$$S_f + C_b \rightarrow S_b + C_f, \qquad\qquad K_{ex.}$$

Subscripts f and b denote free and bound states, respectively. The equilibrium constant K_{ex} can be related to a conditional binding constant K_0 for the surfactant binding by:

$$K_0 = \frac{K_{ex}}{C_{c,f}} \tag{3}$$

where $C_{c,f}$ is the concentration of free counterions, including counterions arising from additional salt. In excess of salt, i.e. when $C_{c,f}$ is constant, K_0 has the same interpretation as in the ZB model. The self-assembly of the bound surfactant molecules into aggregates S_N with the single aggregation number N, still bound to the polyelectrolyte, is described by the equilibrium constant K^N

$$NS_b \leftrightarrow S_N, \qquad\qquad K^N$$

From these assumptions, the following binding isotherm was obtained [17,34]

$$\beta = \frac{K_0 C_{s,f} + \left(K_0 K C_{s,f}\right)^N}{1 + K_0 C_{s,f} + \left(K_0 K C_{s,f}\right)^N} \tag{4}$$

The cooperative binding constant $K_0 u$ in the ZB model, describing the binding of a surfactant molecule next to an occupied site on the polyelectrolyte, is here replaced by the cooperative binding constant $K_0 K$ for the process of bringing one (out of N) free surfactant molecule into a polyelectrolyte-bound surfactant micelle. By allowing for the latter equilibrium only, thus neglecting the possibility of non-cooperative surfactant binding, the binding isotherm becomes

$$\beta = \frac{\left(K_0 K C_{s,f}\right)^N}{1 + \left(K_0 K C_{s,f}\right)^N} \tag{5}$$

which, in excess salt (where K_0 is a true constant), is identical to the Hill-Barcroft equation [35]. On the other hand, neglecting the cooperative binding gives the Langmuir isotherm

$$\beta \; = \; \frac{(K_0 C_{s,f})^N}{1 + K_0 C_{s,f}} \tag{6}$$

The effect of simple salt and polyelectrolyte concentration on the binding isotherm can be investigated by correcting for the dependence of K_0 on $C_{c,f}$ [17]. Within the model, the concentration of free simple counterions is

$$C_{c,f} \; = \; C_{salt,tot} - C_{c,b} \tag{7}$$

$C_{salt,tot}$ is the global concentration of counterions from the polyelectrolyte and from excess salt and $C_{c,b}$ is the concentration of bound counterions.

3. Binding isotherms: predictions and comparisons with experiment

The surfactant binding described by the CA model must always involve the replacement of polyelectrolyte bound counterions. Such a description is only reasonable for polyelectrolytes - tyically of high charge density - that are associated with a number of closely bound counterions. β is (by definition) equal to unity when the number of bound surfactant molecules is equal to the number of polyelectrolyte-bound counterions in the surfactant-free system. In principle, β may be larger than one but the additional binding is not described by the model since it would change the net charge of the complex (and therefore K_0).

In contrast, β values representing experimental data are calculated as the number of bound surfactant ions per polyelectrolyte monomer (or charged group). If the counterions in a polyelectrolyte solution can be approximated as either "bound" or "free," the two definitions differ by a factor (β') equal to the number of bound counterions per polyelectrolyte charge. (The maximum number of bound counterions is lower than the number of polyelectrolyte charges; full charge compensation by small ions is not achieved for entropic reasons.)

In the ZB model the nature of the binding sites is not specified, but when Eq. (1) is used to analyze experimental binding data, the number of binding sites is usually interpreted as the number of polyelectrolyte charges in the system. When applied to highly charged polyelectrolytes we prefer to define the number of "empty" sites as the number of bound counterions. This interpretation gives the same definition of β (and K_0) as in the CA model.

Although the factor β' may differ with external conditions, the following comparisons between theory and experiments were performed by rescaling the experimental degree of binding to agree with the model definition at large surfactant concentrations. Figure 3 shows the normalized binding isotherm for $C_{12}TAB$ in a 0.5 mM solution of NaPA containing 10 mM NaBr together with the isotherms from the ZB model and the CA model including its simplified limits (Eq. (5), Hill isotherm, and Eq. (6), Langmuir isotherm). The theoretical curves were calculated using $K_0 = 100$ M^{-1}, $K = u = 27$, and $N = 60$. The value of K_0 was selected to describe an identical distribution of surfactant ions and simple counterions below the cac ($K_{ex} = 1$), and the value of the aggregation number was taken from experiments [15]. The remaining parameter, K, was fitted so that the calculated isotherms coincide with the experimental isotherm at $\beta = 0.5$. We used the result from the Hill and the ZB models that $K = u = [K_0(C_{s,f})_{\beta=0.5}]^{-1}$, which is also a good approximation in the CA model when $K \gg 1$.

The CA model predicts a sharp increase of β at a critical concentration which is identified as the cac of this model. Below the cac, almost all polyelectrolyte-bound surfactant molecules are non-aggregated and above the cac almost all of them are self-assembled into polymer-bound micelles. The crossover occurs at $C_{s,f} \approx 1/K_0 K$, where aggregation becomes favorable and cooperative surfactant binding will dominate. It is evident that with the present set of parameters, the Hill and the CA isotherms are nearly indistinguishable. A small difference occurs at small $C_{s,f}$ where the absence of bound surfactant monomers in the Hill model leads to a lower β. As expected, the Langmuir isotherm follows the CA isotherm well at $C_{s,f} <$ cac, but remains low above the cac. In the ZB model, the possibility to form small (linear) aggregates is responsible for the deviation from the Langmuir isotherm already at very low surfactant concentrations and above the cac the cooperativity is not sufficiently strong.

It is obvious from Figure 3 that the isotherms predicted by the CA and Hill models are steeper than the experimental curve. A comparison with other systems confirms the generality of this observation as well as the fact that the binding is less favorable than predicted when $\beta > 0.5$ [17]. The derivative $dlnC_f/d\beta$ measures the change in chemical potential of the surfactant with β. This may be considered to have two contributions: (i) changes in the interaction free energy of the surfactant in the micelles and (ii) the law of mass action. In the present models, where K is a constant, only the second contribution is included. Thus, apart from the effect of a maximum degree of surfactant binding per polyelectrolyte, the cooperativity of the binding is determined only by N. In qualitative agreement with the latter prediction, the available experimental results indeed indicate that there is a correlation between large aggregation numbers and steep binding isotherms (*i.e.* high cooperativity) [17].

In the treatment above we have fixed some of the parameters by using data from other sources. Figure 3 shows that the ZB isotherm does not agree with

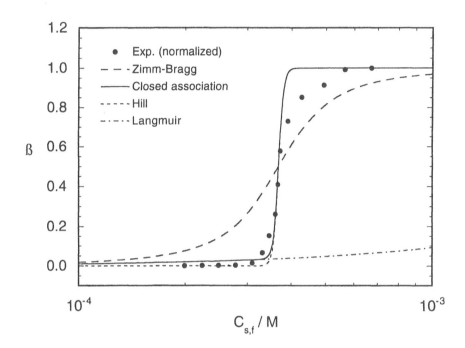

FIG. 3 Experimental (normalized) binding isotherms for $C_{12}TAB$ in 0.50 mM NaPA in the presence of 10.0 mM NaBr and model binding isotherms calculated from Eq (1) (Zimm-Bragg), Eq (4) (Closed Association), Eq (5) (Hill), and Eq (6) (Langmuir), using $K_0 = 100$ M^{-1}, $K = u = 27$, and $N = 60$, when applied. K or u was fitted, whereas K_0 and N were fixed, see text. Experimental data are adapted from Hansson and Almgren, ref 15, (original β values are divided by $\beta' = 0.66$).

the experimental data with the selected values of the parameters. In applications of the ZB model, however, usually both parameters are fitted and in many cases good fits are achieved. In Figure 4 we give the results from the ZB model where K_0 and u are treated as free fitting parameters. The fitted curve describes the binding isotherm quite well, but the predicted variation in the aggregation number (*i.e.*, the number of surfactant molecules in an uninterrupted sequence) is in conflict with experimental data. According to Eq. (2) the average aggregation number is predicted to increase from 8 at $\beta = 0.05$ to 155 at $\beta = 0.95$. As mentioned above, experiments show that the aggregation number is more or less independent of β. Clearly, the quality of the fit can not be used as a criterion for the reliability of the model. Figure 4 shows that a very similar fit can be obtained from the CA model by fitting all parameters (K_0, K, N). The values of K_0 and K are very close to K_0 and u from the ZB fit. It should be mentioned that, owing to the covariance of K_0 and K in the CA model, the fit is not unique. However, independently of the values of K_0 and K, the fitted value of N is ca. 20, which is considerably lower than the experimental value of 60.

Up to this point we have only discussed the binding in excess salt. Under such conditions, the shape of experimental binding isotherms (in a lin-log presentation) is relatively independent of the actual salt concentration. As is evident from Figure 5a, the apparent cooperativity does increase gradually with the addition of salt to an initially salt-free system, but the effect vanishes once the added salt is in excess. There is no change in the average aggregation number in the range studied [15]. The effect at low salt can be rationalized by the CA model (or the Hill model) simply by correcting for the accumulation of simple counterions in the bulk with increasing β [17]. Theoretical curves calculated using (i) Eq. (5) with the same values of N, K, and $K_{ex} = 1$ as in Figure 3 and (ii) Eqs. (3) and (7) to express the dependence of K_0 on $C_{c,f}$ are shown in Figure 5b for $C_{salt,tot} = 0.5 - 10.5$ mM. With $C_p = 0.5$ mM and $\beta' = 0.65$, this means that K_0 is reduced by 65% as β increases from 0 to 1 in the case of no salt, but only by 3% in the presence of 10 mM salt.

To conclude: the law of mass action can, at its best, provide us with a qualitative understanding of the interactions between highly charged hydrophilic polyelectrolytes and surfactants of opposite charge. While the polymer-centered ZB model is able to reproduce experimental data, it does so at the cost of microscopic predictions of micellar sizes that are in qualitative disagreement with experimental results. It appears that the concept of a finite micelle is necessary to capture even the most essential features of the surfactant binding.

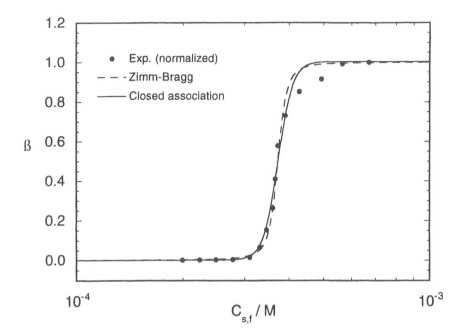

FIG. 4 Experimental (normalized) binding isotherms for $C_{12}TAB$ in 0.50 mM NaPA in the presence of 10.0 mM NaBr and model binding isotherms calculated from Eq. (1) (Zimm-Bragg) with fitted parameters $K_0 = 2.4$ M^{-1} and $u = 1100$, and using Eq. (4) (Closed Association) with fitted parameters $K_0 = 2.4$ M^{-1}, $K = 1100$, and $N = 21$.

B. The Mixed-Micelle Approach for Hydrophobically Associating Polymers

One class of polymers that interact particularly strongly with surfactants are the hydrophobically associating water-soluble polymers (HAP) mentioned in the introduction. (Other common names for HAP are "hydrophobically modified polymers," "associative polymers" or "associative thickeners"; in Chapter 10 of this volume they are called "water-soluble associative polymers" or WSAP.) Experiments generally show that added surfactants have large effects on the rheology [20,21,36-45] and phase behavior [20,21,41,44,46] of aqueous HAP solutions. It is now well established that the hydrophobic moieties of the HAP, the HAP hydrophobes, have the capacity to self-associate into finite

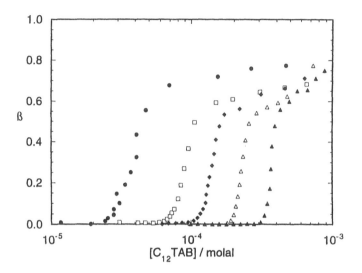

FIG. 5a Experimental binding isotherms for $C_{12}TAB$ in 0.50 mM NaPA in the presence of 0, 1.25, 2.50, 5.00, 10.0 mM NaBr (from left to right). Data taken from Hansson and Almgren (ref. 15) were obtained from potentiometric measurements using a surfactant sensitive membrane electrode.

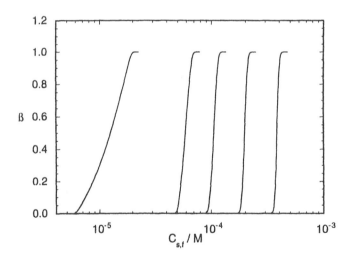

FIG. 5b Model binding isotherms calculated from Eq. (5) using $K_{ex} = 1$, $K = 27$, and N = 60 at $C_{salt,tot} = 0.5$, 1.75, 3, 5.5, and 10.5 mM (from left to right). Eqs (3) and (7) and $C_{c,b} = 0.66C_p(1-\beta)$ with $C_p = 0.5$ mM were used to calculate K_0. $0.66C_p$ is the total concentration of ions (surfactant + simple counterions) bound to the polyelectrolyte.

"hydrophobic domains," or HAP micelles, even in the absence of surfactant molecules [18-21]. The latter property may be considered as the characteristic feature of an HAP, distinguishing it from category 2 polymers which also form mixed micelles with surfactant molecules but which lack the ability to self-associate into micelle-like hydrophobic domains. We will here discuss only the simplest case, where only the HAP hydrophobes join into the mixed micelles, i.e., pure category 3 complexes. This simplifying assumption is not valid for some commonly studied HAP/surfactant mixtures, such as mixtures of SDS with hydrophobically end-capped PEO [42,47,48] The HAP may then be considered as a polymerized surfactant. Accordingly, the natural concentration unit for describing its mixtures with surfactant molecules is the concentration of HAP hydrophobes, C_h. No well-defined "sites" for surfactant binding occur on an HAP; rather, the binding isotherm is conveniently expressed in terms of the binding ratio $\beta_{sh} \equiv C_{s,b}/C_{h,b}$, where C_s stands for the surfactant concentration, and the subscript b denotes a species (HAP hydrophobe or surfactant molecule) that is "bound" in the mixed micelles.

It has recently been shown that various physical properties, such as rheology and phase behavior, may be understood in terms of the stoichiometries of the mixed micellar HAP/surfactant complexes [20,21,49-51]. We may immediately distinguish the following composition regimes, which should correspond to different properties of the mixture.

$\beta_{sh} \ll 1$: In this "surfactant solubilization" regime, the mixed micelles are essentially HAP micelles, with some incorporated surfactant. The structure of the system should be only marginally perturbed by the surfactant.

$\beta_{sh} \sim 1$: This is a "transition regime," where important changes in system properties (micellar aggregation number, functionality of mixed micellar crosslinks) are expected to result from changes in the micellar composition.

$\beta_{sh} \gg 1$: In this "cooperative binding" regime (cf. below) the mixed micelles should be similar to pure surfactant micelles.

$\beta_{sh} > N_s$: When the binding ratio exceeds the aggregation number N_s of the pure surfactant micelle in the (otherwise identical) polymer-free solution, pure surfactant micelles should occur. This might be called the "micelle excess" regime.

It is important to realize that it is not safe to assume that the average composition of the mixed micelles is given by the global ratio of surfactant molecules to hydrophobes. In general, $\beta_{sh} \neq C_{s,tot}/C_{h,tot}$, although the approximation $C_{h,b} \approx C_{h,tot}$ may often be valid, except when the HAP is a highly charged polyelectrolyte. The fact that two or more hydrophobes are located on the same polymer molecule in an HAP serves to bring the micellization concentration down below that of a comparable surfactant [21]. However, the concentration of free surfactant, $C_{s,f}$, is typically not negligible for an ionic surfactant with a high cmc. In fact, $C_{s,f}$ often dominates in those experimental situations where large effects of added surfactant are found [21, 49,50]. A good knowledge of the binding isotherm is therefore essential for a correct estimate of the mixed micellar composition. In addition, the concentration of free ionic surfactant may be an important factor in its own right, as it plays the role of an external salt, contributing to the salting-out of polymer-surfactant complexes [21,50].

We know of no model for the binding of surfactant to HAP which considers the details of the HAP architecture (hydrophobe lengths, substitution patterns *etc.*), although such factors should no doubt be important in a quantitative model. Useful semi-quantitative estimates have been however obtained [49,50], by using the well-known pseudo-phase separation model developed for the mixed surfactant micelles [22,52,53]. In this model, the micelles are treated as a separate phase in equilibrium with a solution phase containing solvent and free surfactant. The micellar phase is considered to be infinite, and is only described in terms of the micellar composition. Inherent in this approximation is the assumption that the compositions of the phases are independent of the phase volumes. Specifically, the total concentration of free surfactant is equal to the global cmc of the mixed micelle, independently of the concentration of micelles.

We will here only consider explicitly the simplest case, when there is ideal mixing within the micellar phase and the solution phase is also ideal. Above the cmc of a mixed surfactant solution, the concentration of free surfactant in equilibrium with the mixed micellar phase is then [22]

$$C_{s,f} = x_{s,b}\, cmc \tag{8}$$

where $x_{s,b}$ is the mole fraction of surfactant in the mixed micellar phase and cmc again is the critical micellization concentration for the pure surfactant in an otherwise identical solution. If Eq. (8) is obeyed for mixed micelles consisting of surfactant molecules and HAP hydrophobes, we immediately obtain the surfactant binding isotherm, describing the variation of the mixed micellar composition with the concentration of free surfactant, as

$$\beta_{sh} = x_{s,b}/(1 - x_{s,b}) = (C_{s,f}/cmc)/[1 - (C_{s,f}/cmc)] \tag{9}$$

The isotherm in Eq. (9) is given in terms of the "reduced monomer concentration" C_{sf}/cmc. A graph of this universal (for ideal mixtures) isotherm is shown in Figure 6. Note that the gradual binding at low $C_{s,f}$ changes over to a "cooperative binding" as $C_{s,f}$ approaches the cmc. The latter

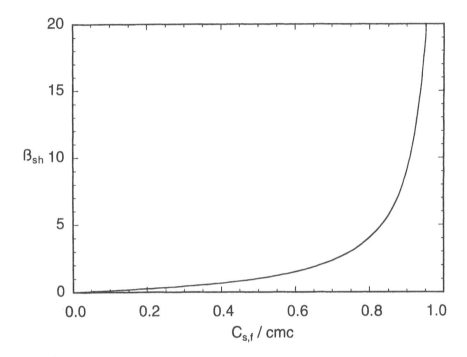

FIG. 6 Binding isotherm for the binding of a surfactant to a mixed micelle when there is ideal mixing in the micelle Eq. (9).

feature is simply a consequence of the fact that the free surfactant concentration can never exceed cmc. Crude as it may seem, this model is actually quite helpful in real systems.   Many of the most interesting features of HAP/surfactant mixtures appear in the cooperative binding regime, where the surfactant dominates in the mixed micelles [21,49,50], and the isotherm in Figure 6 is, of course, expected to be increasingly accurate as x_{sb} approaches unity.

More elaborate versions of the pseudo-phase separation model have been developed for surfactant mixtures [22]. Non-ideality in the mixed micelles may be handled by introducing activity coefficients, which can be given physical interpretations in terms of nearest neighbor interaction parameters in the regular solution approximation. It is not immediately apparent, however, that the latter approach would represent a real improvement for mixed HAP/surfactant micelles. Presumably, major non-ideality effects should here be due to the constraints imposed by the hydrophilic polymer chain, to which the hydrophobes are attached. It is possible, however, that the regular solution approximation could be useful to treat mixed micelles displaying a miscibility gap, *i.e.*, systems where, for one reason or another, the HAP hydrophobes and the surfactant molecules are not miscible in all proportions within the mixed micelles.

Special considerations are necessary for mixtures involving charged species. For an ionic surfactant binding to a non-ionic HAP, a small deviation from the isotherm in Figure 6 is expected in the form of a relatively larger initial slope at the lowest surfactant concentrations. At very low degrees of binding, the probability of finding two surfactant molecules in the same micelle is negligible. Hence, there is no repulsion, and the binding should be the same as for a nonionic surfactant. For $\beta_{sh} > 1/N_h$, (N_h being the aggregation number of the hydrophobes in the HAP micelles), however, most micelles will be charged, and a repulsion sets in. This transition from a non-cooperative binding at low binding ratios to an anti-cooperative binding at higher degrees of binding has been observed for the binding of ionic surfactants to nonionic micelles [54] and phospholipid vesicles [55].

Stronger charge effects are seen for oppositely charged species. For an HAP where the hydrophobes bear the charges, and each hydrophobe contains one charged group, the mixture should be analogous to one of oppositely charged surfactants. The binding is thus expected to be very strong (in practice, almost quantitative [21]) for $\beta_{sh} < 1$, where the net charge of the mixed micelles is opposite to that of the surfactant. Only for surfactant-dominated micelles is the binding isotherm expected to become similar to that in Figure 6.

A different case is the one where the charged groups are on the hydrophilic chain of the HAP. Then Eq. (8) may still be useful, if the cmc is taken to be the cmc of the surfactant in an identical system, but void of the HAP hydrophobes. This corresponds to the cac of the category 1 complex of the surfactant with a polyelectrolyte identical to the HAP, but without the hydrophobes. A further complication in a mixture involving a highly charged HAP is that the latter may have a high cmc in the absence of surfactant. A case in point is hydrophobically modified polyacrylate [20]. In this case, the initial surfactant solubilization regime of the binding isotherm is absent; no binding should occur until that surfactant concentration is reached where the global cmc for the mixed micelle is exceeded.

The isotherms obtained for the binding of charged surfactants to oppositely charged globular proteins often display striking similarities with those for oppositely charged HAP and surfactant [56]: an early, high-affinity binding is observed, followed by an anti-cooperative binding as the equivalent surfactant concentration exceeds that of the protein, and, finally, a cooperative binding as the monomeric surfactant concentration approaches the cmc [57,

58]. Nevertheless, it is important to appreciate that there are important differences in the early part of the binding isotherm, where a phase separation occurs. The binding to the protein typically results in the formation of a stoichiometric precipitate, in equilibrium with a solution of either pure protein or, above the equivalent total surfactant concentration, a soluble non-stoichiometric complex [57,59]. Thus, the protein may be involved in two different complexes in the separating phases, where the composition of at least the precipitated complex seems fixed, regardless of the global composition of the system. In contrast, in any given HAP/surfactant mixture (at mixing ratios below the micellar excess regime) there is usually only one type of mixed complex, but the stoichiometry of this complex varies with the composition of the mixture [21].

Finally, the structure of single aggregates formed by end-capped water soluble polymers and neutral surfactants have been investigated by solving simple and lattice based models by Monte Carlo simulations [60,61]. The results showed that the length of the solvophobic part of the surfactant and the polymer-to-surfactant concentration ratio affected the aggregates formed and hence also should affect the viscosity of the solution.

III. MORE ADVANCED MODELS

In this section we will examine several more advanced models used to describe polymer-surfactant complexation belonging to categories 1 and 2. All those given here start from the surfactant-centered picture, *i.e.*, the polymer-surfactant complexation is essentially a surfactant micellization, however perturbed by the presence of the polymer.

A. Thermodynamic Models

By breaking down the surfactant self-association process in the presence of the polymer into several simpler steps and by assigning free energy contributions to each step, it is possible to gain further insight into the polymer-surfactant complexation. Such a division of the self-assembly process is an important part of thermodynamic models which are used to predict, *e.g.*, the size distribution of aggregation numbers and the cmc. These types of models have successively grown from initial work by, *e.g.*, Tanford [62,63], and Ruckenstein and Nagarajan [64] and have been applied to the self-assembly of surfactant molecules into micelles, mixed micelles, and solubilization in micelles [64-66], to diblock copolymers in solution [67], and also to the complexation between surfactant micelles and neutral polymers [65,68-71].

Common to all the theories describing the polymer-surfactant complexation is the treatment of the surfactant self-assembly in a polymer-free solution. From the thermodynamic equilibrium among free (singly dispersed) surfactant molecules and surfactant molecules residing in micelles with different aggregation numbers, the size distribution of micelles in a polymer-free solution is given by

$$X_N = X_1{}^N \exp(-g \, \Delta\mu^{\circ}{}_N / kT) \tag{10}$$

where X_1 is the mole fraction of free surfactant, X_N the mole fraction of the aggregates containing N surfactant molecules and $\Delta\mu^{\circ}{}_N$ is a central quantity denoting the difference of the standard chemical potential per surfactant molecule in the aggregate of N surfactant molecules and the standard chemical potential of the free surfactant in the solution. Given an explicit expression of $\Delta\mu^{\circ}{}_N$, the size distribution of the aggregates can be obtained from Eq. (10), and if some criterion of the cmc is supplied, the cmc can also be evaluated. An often used simplification is to consider the aggregates as a pseudophase in equilibrium with free surfactant molecules. The critical mole fraction of surfactants, X_{cmc}, is in this approximation given by [69]

$$kT \ln X_{cmc} \approx \Delta\mu^{\circ}{}_N \big|_{N=N'} \tag{11}$$

where N' is obtained from the solution of

$$d(\Delta\mu^{\circ}{}_N)/dN = 0 \tag{12}$$

In the polymer-free case, $\Delta\mu^{\circ}{}_N$ is assumed to be composed of several terms and is often expressed as

$$\Delta\mu^{\circ}{}_N = \Delta\mu_{w/hc} + \Delta\mu_{hc/mic} + \Delta\mu_{\sigma} + \Delta\mu_{steric} + \Delta\mu_{elec} \tag{13}$$

The first term denotes the energy change associated with the transfer of a surfactant tail from water to a liquid hydrocarbon while the second one represents the difference between a liquid hydrocarbon and the somewhat different micellar interior. The third term accounts for the free energy of formation of an interface separating the micellar core and the surrounding water. The remaining two terms refer to the steric repulsion among the surfactant head groups at the core-water interface and the electrostatic interaction among the head groups at the micellar surface (for ionic surfactants), respectively. The explicit form of these terms varies in the literature.

In the models for polymer-surfactant solutions, it is explicitly assumed that the surfactant molecules form aggregates in a similar fashion as in the polymer-free solution. However, the free energy change for this process is now modified by the wrapping of polymer molecules around the aggregates. The various theories in this area differ in how the polymer enters the models and how it changes the free energy of the micellization.

A simple approach to describe the effect of the polymer merely involves modifications in Eq. (13). The use of Eqs. (12) and (11) together with the modified Eq. (13) then gives the aggregation number and the cmc of the polymer-bound complexes in the pseudophase approximation [69]. The modifications in Eq. (13) are: (i) an enhanced shielding of the micellar core from water provided by the polymer as described by an additional parameter in the expression of $\Delta\mu_\sigma$, (ii) an increase in the steric repulsion in the head group region due to the presence of the polymer described by changes in $\Delta\mu_{steric}$ by using the same additional parameter as in (i), and (iii) an addition of a new term in Eq. (13) representing the reduction of unfavorable water-polymer contacts by bringing polymer segments from water into contact with the hydrophobic surface of the micellar core. In (i) and (iii), the modification leads to a lowering of the free energy, whereas in (ii) the free energy increases due to a reduction of the entropy. The same approach has also been applied to calculate the cmc and the aggregation number using the full size distribution with the modifications (i) and (ii) applied [65].

In a more extended description, polymer-free and polymer-bound surfactant micelles appear in the model (on an equal level) and the two types of aggregates may have different aggregation numbers [68,70]. The total surfactant concentration X_{tot} is now given by

$$X_{tot} = X_1 + MX_M + NnX_p[X_N/(1 + X_N)] \tag{14}$$

where X_1 is the mole fraction of free surfactant, X_M the mole fraction of free micelles with M surfactant molecules and described by Eq. (10), whereas X_p denotes the mole fraction of polymer, n the maximum number of micelles that can complex with one polymer molecule, and X_N the mole fraction of polymer-bound aggregates with aggregation number N. Thus, the polymer enters the model through its mole fraction, but the number of micelles complexed with one polymer molecule is predetermined to n and it is the same for all polymer molecules. The quantity X_N in Eq. (14) is given by

$$X_N = X_1^N \exp[-N(\Delta\mu^\circ_N + \Delta\mu^\circ_{ads})/kT] \tag{15}$$

where $\Delta\mu^\circ_{ads}$ denotes the contribution from the polymer complexation to the standard chemical potential of the self-assembly as discussed above. Here, the view of Ruckenstein et al. is that the polymer changes the environment outside a micelle and hence affects the surface tension in the $\Delta\mu_\sigma$ term [70]. The new term $\Delta\mu^\circ_{ads}$ accounts for (i) the decrease of the interfacial free energy between the hydrophobic core and the more hydrophobic environment and (ii) the increase in the interfacial free energy between the headgroups and the new environment. The view taken by Nagarajan is slightly different: $\Delta\mu^\circ_{ads}$ contains a polymer-specific shielding area describing (i) a reduction of the surface energy of the micelle core and (ii) an increase of the steric repulsion in the head-group region caused by the presence of the polymer. Hence, the approach of Ruckenstein et al. would correspond to category 1 and since Nagarajan's view requires polymer-micelle contact it corresponds to the category 2 case. However, the outcome of the model is hardly affected by this minor difference.

The effect of the presence of polymer on the free surfactant concentration is illustrated in Figure 7 using the thermodynamic model for model parameters corresponding to SDS in aqueous solution. First, we see that for the polymer-free solution the free surfactant concentration levels off at ca. 8 mM, which is the characteristic cmc of SDS in aqueous solution at room temperature. Secondly, Figure 7 shows that upon an addition of a polymer (with parameters to mimic the interaction of PEO with SDS) the cac becomes smaller than the cmc (reduces to ca. 3.5 mM) and the cac is essentially independent of the polymer concentration. At increasing surfactant concentration, the polymer molecules may become saturated with surfactant micelles and the free surfactant concentration rises again to ca. 8 mM (as illustrated by the curve for $nX_p = 0.2$ mM in Figure 7). The appearance of free micelles in the solution occurs at a global surfactant concentration (denoted as cmc') that of course is larger than the cmc in a polymer-free solution and cmc' depends on the polymer concentration.

For ionic surfactants, $\Delta\mu^\circ_N$ in Eq. (13) contains an electrostatic term $\Delta\mu_{elec}$ which describes the interaction among the charged headgroups. At increased salt concentration this repulsion is screened and $\Delta\mu_{elec}$ is reduced. Figure 7 also shows the reduction of the cmc in the presence of 50 mM 1:1 electrolyte. Such a reduction of the cmc is also well known from the experimental literature. The addition of salt to the polymer-containing SDS solution also reduces the cac, now down to ca. 1 mM. In this simple model, the effects on the cmc of the polymer and the salt enter independently. Of course, the salt reduces the headgroup repulsion, and in the view of Ruckenstein, the polymer reduces the surface tension between the micellar core and the nearby polymer solution. In reality, cross effects exist, but still the influences on the

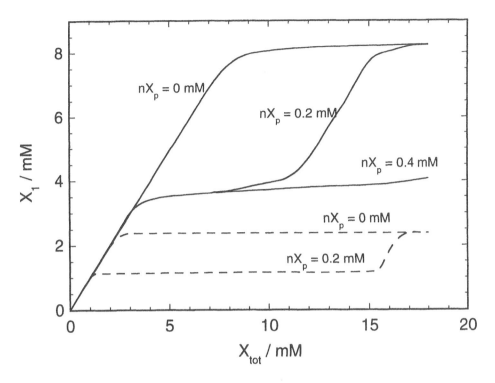

FIG. 7   Calculated equilibrium concentration of free surfactant (X_1) as a function of the total surfactant concentration (X_{tot}) for parameters corresponding to SDS in salt-free solution (solid curves) and in 50 mM 1:1 electrolyte solution (dashed curves) at indicated polymer concentration. Data taken from Ruckenstein et al., ref. 70.

cmc of the presence of polymer and salt are operating in the same direction. This is in clear contrast to polyelectrolytes in systems with oppositely charged surfactants, where an addition of salt increases the cac (cf. Figure 5a), whereas an addition of salt to polyelectrolyte-free surfactant solution decreases the cmc.

In the approach by Nikas and Blankschtein, the surfactant and the polymer are described on an equal level [71]. Besides solvent, surfactant molecules, and free micelles with different aggregation numbers, the solution also contains polymer molecules explicitly. The polymer molecules may form complexes with surfactant molecules involving different number of micelles of the same aggregation number which again is described by thermodynamic equilibrium constants. The mole fraction of polymer molecules with m(N) micelles of aggregation number N, $X_{p,m(N)}$, is given by

$$X_{p,m(N)} = X_{p,0}\{X_1^{Nm(N)} \exp[-Nm(N)(\Delta\mu°_N + \Delta\mu°_{ads} + \Delta\mu°_{conn})/kT]\} \qquad (16)$$

where $\Delta\mu°_N$ is given by an equation similar to Eq. (13), $\Delta\mu°_{ads}$ describes the change in $\Delta\mu°_N$ due to the adsorption of polymer on the surfactant micelles (again the explicit nature of $\Delta\mu°_{ads}$ is different as compared to above). The last and new term, $\Delta\mu°_{conn}$, describes the free energy change of binding more than one micelle to a single polymer molecule. The contribution arises from (i) the electrostatic repulsion between the charged micelles and (ii) the entropic cost of stretching the polymer chain between two wrapped micelles. The last term makes it possible to describe successive binding of micelles to a given polymer molecule, and for a given surfactant concentration, a distribution of the number of micelles complexed to a polymer molecule is obtained. Beside the effect of the polymer on the cac through the shielding of unfavorable hydrophobic core-water contacts, the solvency of the polymer as such should influence the suppression of the cmc. The explicit treatment of the polymer makes it possible to examine such influence of the solvency.

Figure 8 displays the cac/cmc ratio as function of the parameter for hydrophobic core-polymer interaction (χ_p) at two different polymer solvencies (χ) as obtained by Nikas and Blankschtein. First, we see that the cac/cmc ratio reduces as the shielding becomes more effective (lower χ_p). Such effects have also been demonstrated by Nagarajan [65] and by Ruckenstein et al. [70]. Moreover, Figure 8 also shows that a worsening of the solvency condition for the polymer with otherwise identical conditions (higher χ), reduces the cac and reduces the critical magnitude of χ_p at which cac starts to deviate from the cmc. These model calculations also show that the aggregation number of the polymer-bound micelles decreases as χ_p becomes more negative and as χ increases. The changes of the aggregation numbers, also given in Figure 8 (broken lines), follow the trends of the cac, and the change in cac and N are both related to the reduction of the interfacial micellar core-water interaction.

As mentioned, the explicit modelling of the polymer chains makes it possible to examine the stepwise buildup of polymer-bound micelles. Figure 9 shows the predicted number of surfactant molecules per polymer chain as function of the free surfactant concentration at different polymer chain lengths. For the shortest polymer with 100 segments, the chain can only complex at most one micelle with ca. 13 surfactant molecules. The increase to 200 segments leads to slight reduction of the cac and an increase to 14 surfactant molecules, but still only at most one micelle per chain. The further increase to 400 segments now displays a stepwise complexation of up to two micelles per

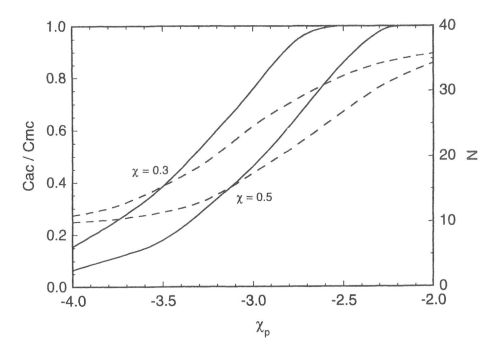

FIG. 8 Calculated cac/cmc ratio (solid curves, left axis) and polymer-bound micellar aggregation number of the first micelle (m(N) = 1) (dashed curves, right axis) as a function of the free energy change associated with the adsorption of one polymer segment from water to the micellar core surface (χ_P) at indicated polymer solvency given by interaction parameter (χ). $r_p = 400$, $C_{p,seg} = r_pC_p = 0.1$ mM (concentration of polymer segments), $b = 4$ Å (segment-segment separation), and $t = 25°C$. Data taken from Nikas and Blankschtein, ref. 71.

chain and a chain with 800 segments can bind three micelles. The successive binding of micelles to the polymer molecules reflects the intermicellar repulsive electrostatic energy arising among micelles bound to the same polymer chain. Obviously, the model predicts that the shortest chains ($r_p = 100$ and 200) are too short to wrap two micelles and simultaneously keep them sufficiently separated. Nikas and Blankstein reported that a stepwise binding has not yet been observed experimentally and that such a behavior probably requires nearly monodisperse polymers [71].

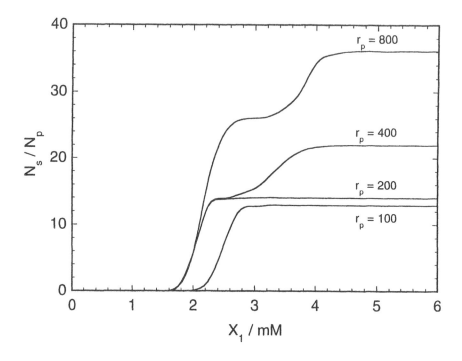

FIG. 9 Calculated average number of surfactants per polymer chain (N_s/N_p) as a function of free surfactant concentration (X_1) at indicated number of polymer segments (r_p). $C_{p,seg}$ = 0.1 mM, b = 4 Å, and t = 25°C. Data taken from Nikas and Blankschtein, ref. 71.

B. Mean-Field Lattice Models

The self-consistent mean-field lattice theory, as initially developed by Scheutjens and Fleer [72], is a versatile theory for describing (i) the adsorption of chain molecules such as surfactants or polymers on solid or fluid interfaces and (ii) for modelling the self-assembly of chain molecules in bulk solution. This theory has also been extended to describe polyelectrolytes by Böhmer et al. [73] and to deal with the micellization of ionic surfactants [74]. Recently the micellization of charged surfactants in the presence of polyelectrolytes has been modelled using the self-consistent mean-field lattice theory [75], and the cmc, the aggregation number, and structural information were obtained as a result of a molecular model with specified interaction between the different components.

In the application of the self-consistent mean-field theory to surfactant-polyelectrolyte complexation [75], the interaction between one surfactant micelle and its surrounding polymer solution is considered using a spherically symmetric lattice. In more detail, the space is divided into concentric shells, and each shell is further divided into lattice cells, which contain one surfactant segment, one polyelectrolyte segment, one salt species, or solvent molecule in each. The conformations of the chain molecules are described as random walks on the spherical lattice. Interactions are regarded as extending only to segments or solvent molecules in adjacent lattice cells. The random-mixing (mean-field) approximation is applied within each layer separately, and thus radial concentration profiles are obtained with the micelle in the center of the lattice. The model is in principle applicable to all three categories of complexation, although the inital work is restricted to categories 1 and 2 [75].

There are two different types of interactions in the model: electrostatic (charge-charge) and nonelectrostatic (the rest). The nonelectrostatic interaction between species in adjacent lattice sites is described by Flory-Huggins χ-parameters [76]. In line with the random-mixing approximation, the charged species (charged polyelectrolyte segments and salt species) are assumed to interact with a potential of mean force, Ψ_i, which depends only on the radial distance (layer number i). The potential of mean force is related to the charge density through Poisson's equation

$$\varepsilon_0 \varepsilon_r \nabla^2 \Psi_i = -\rho_i \tag{17}$$

where $\varepsilon_0 \varepsilon_r$ is the dielectric permittivity of the medium, ∇^2 the Laplacian, and ρ_i the total charge density in layer i. The charges of the species are located on surfaces in the middle of each lattice layer, the space between the charged surfaces is free of charge. A uniform dielectric permittivity is used.

The computation involves a self-consistent determination of the segment distribution (i.e., the radial distribution of surfactant and polyelectrolyte segments, salt species, and water molecules). The species volume fraction, ϕ_{Ai} (volume fraction of species A in layer i), is simply related to n_{xsi}, the number of sites in layer i occupied by segments of rank s (the s:th segment in a chain) belonging to component x according to

$$\phi_{Ai} = \frac{1}{L_i} \sum_x \sum_{s=1}^{r_x} \delta_{A,t(x,s)} n_{xsi} \tag{18}$$

where L_i is the number of sites in a layer i and r_x is the number of segments in component x ($r_x = 1$ for salt species and water, $r_s > 1$ for surfactant and $r_p > 1$ for polyelectrolyte). The Kronecker δ selects only segments of rank s of component x if they are of type A.

The expression for the segment distribution is more complex, since the correct weight of all conformations, as well as the connectivity of the chains, has to be taken into account. Starting with the partition function, n_{xsi} is obtained by using a matrix method and is given by [77]

$$n_{xsi} = C_x \{\Delta_i^T \cdot [\prod_{s'=r_x}^{s+1} (W^{t(x,s')})^T] \cdot s\} \cdot \{\Delta_i^T \cdot [\prod_{s'=2}^{s} (W^{t(x,s')})] \cdot p(x,1)\} \quad (19)$$

where C_x is a normalization factor related to the bulk volume fraction of component x, $W^{t(x,s)}$ is a tridiagonal matrix comprising elements which contain factors describing the lattice topology as well as weighting factors for each segment of rank s belonging to component x, and p(x,1) is a vector describing the distribution of the first segment of component x among the layers, Δ and s being elementary column vectors. The weighting factor G_{Ai} for species A in layer i entering W is given by

$$G_{Ai} = \exp(-u_{Ai}/kT) \quad (20)$$

The species potential u_{Ai} entering in Eq. (20) can be expressed as

$$u_{Ai} = u'_i + kT \sum_{A'} \chi_{AA'} (\langle \phi_{A'i} \rangle - \phi_{A'}^b) + q_A \psi_i \quad (21)$$

if the species potentials are defined with respect to the bulk solution with zero electrostatic potential, i.e., if $u_A^b = 0$. The angular brackets indicate an averaging over three adjacent layers. The species independent term u'_i in Eq. (21) ensures that the space is completely filled in layer i; u'_i being related in a continuous model to the lateral pressure. In bulk, u' becomes zero. The remaining two species dependent terms describes the mixing energy for species A in layer i being diminished by the mixing energy for species A in bulk and the electrostatic energy of species A carrying charge q_A in the electric field ψ_i. At distances far from the micelle, ϕ_{Ai} approaches ϕ_A^b, ψ_i approaches zero, and hence u_{Ai} becomes zero. Since u_{Ai} is needed for obtaining ϕ_{Ai} using Eqs. (18-20), and since u_{Ai} depends in turn on ϕ_{Ai} according to Eq. (21), Eqs. (18-21) need to be solved self-consistently. In addition, the electrostatic potential, which enters in Eq. (21) and depends on ϕ_{Ai}, has to fulfill Poisson's equation (Eq. (17) with $\rho_i = \Sigma_A q_A \phi_{Ai}$.

The solution of Eqs. (17-21) has two branches, one with homogeneous concentration profiles throughout the lattice (corresponding to the case below the cac) and one with a single micelle formed at the center of the lattice and in equilibrium with specified bulk concentrations of the components. The total volume fraction of component x, ϕ_x^{tot}, is obtained by dividing the solution into subsystems containing one micelle and its accompanying solution and ϕ_x^{tot} is the sum of the excess and bulk volume fractions according to

$$\phi_x^{tot} = \Gamma_x/V_s + \phi_x^b \qquad (22)$$

where $\Gamma_x = \Sigma_i L_i(\phi_{x\,i} - \phi_x^b)$ is the excess of component x in one subsystem and V_s the volume of the subsystem. Finally, the aggregation number of a micelle is given by $N = \Gamma_s/r_s$. Further details on the evaluation of the micellization is given elsewhere [75].

We will below give some selected results obtained for a model representing the self-assembly of charged surfactant molecules in the presence of an oppositely charged polyelectrolyte [75]. The surfactant is represented as $A_{14}B_3$, where A is a tail segment and B a head-group segment carrying $-1/3$ elementary charge. The polyelectrolyte contains 1000 segments and all of them carry the fraction τ of an elementary charge of positive sign. Furthermore, T = 298 K, $\varepsilon_r = 80$, and a hexagonal lattice with a lattice spacing d = 3.1 Å is employed. If nothing else is stated, all Flory-Huggins interaction parameters are zero, except for those involving the hydrophobic tail of the surfactant where $\chi = 2$ is used. For technical reasons, the precise cmc and cac are difficult to locate due to lattice effects, see, e.g., ref. 75. We therefore provide the free surfactant concentration $C_{s,f}$ at given total surfactant concentrations C_s, recalling that $C_{s,f}$ is a reasonable approximation to the cmc, or the cac, when C_s exceeds the cmc, or the cac, respectively.

Figure 10 shows the free surfactant concentration and the aggregation number in a 10 mM 1:1 salt solution free of polyelectrolyte (symbols on the abscissa) at $C_s = 311$ mM. The model system predicts a cmc ≈ 0.8 mM for the $A_{14}B_3$ surfactant (assuming that cmc $\approx C_{s,f}$) and an aggregation number N = 56 in a 10 mM 1:1 salt solution. Moreover, Figure 10 also shows the free surfactant volume fraction and the aggregation number in the salt solution with polyelectrolytes at different linear charge densities ($\tau > 0$) present at $C_s = 0.52$ mM. It is seen that $C_{s,f}$ decreases as the linear charge density is increased. At $\tau = 0.5$, corresponding to a charge-charge separation of *ca.* 6 Å, $C_{s,f}$ is reduced by a factor of 2 as compared to $\tau = 0.2$ (*ca.* 15 Å between consecutive charges). The aggregation numbers of the micelles formed are also given in Figure 10 and it is seen that N increases with τ at constant C_s.

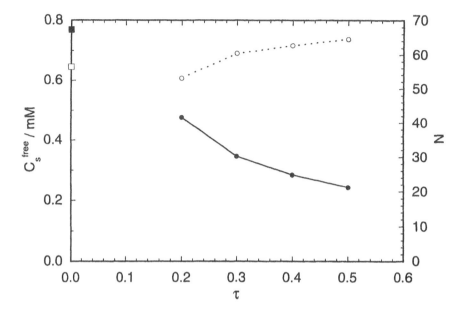

FIG. 10 Calculated free surfactant concentration ($C_{s,f}$, filled circles) and aggregation number (N, open circles) as a function of the linear charge density of the polyelectrolyte (τ) from the mean-field lattice theory. $C_{p,seg} = 2.0$ mM and $C_{salt} = 10$ mM. Other conditions are given in the text. The free surfactant concentration (filled square on the left axis) and the aggregation number in polyelectrolyte-free solution (open square) are also given.

The theory makes it possible to obtain radial distribution functions of all the components from the center of the micelle and outward. The main part of Figure 11 shows that the center of the micelle (layers 1-5) is nearly entirely composed of tail segments of the surfactant, whereas the head group is preferentially located further out (in layers 6-9). These distributions are not very much affected by the presence of the polyelectrolyte (cf. ref. 75 for the polyelectrolyte-free case). In the presence of a polyelectrolyte in the solution, there is a strong accumulation of polyelectrolyte segments outside the hydrophobic core of the micelle and this polyelectrolyte layer nearly superimposes on the volume fraction profile of the charged head segments. However, the inset shows that the head distribution decays rapidly to its intermicellar value whereas the polyelectrolyte volume fraction decays more

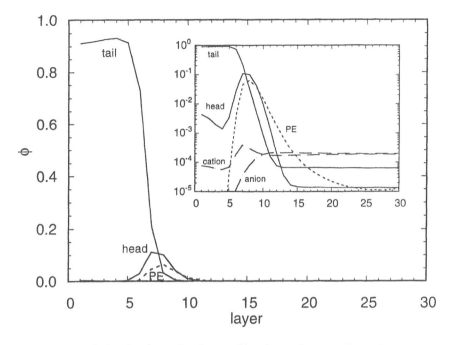

FIG. 11   Calculated volume fraction profiles for surfactant tail, surfactant head, polyelectrolyte segments, cations, and anions of a polyelectrolyte complexing a charged surfactant micelle from the mean-field theory. The insert shows the same data on a lin-log scale. $\tau = 0.5$, $C_s = 0.52$ mM, $C_{p,seg} = 2.0$ mM, and $C_{salt} = 10$ mM. Other conditions are given in the text.

gradually, showing that the micelle has a rather defined surface and that there is an enhanced concentration of polyelectrolyte outside the micelle. The inset also shows the volume fraction profiles of the salt species. The small ions are nearly evenly distributed outside the micelle with the cations (counterions to the surfactants) somewhat accumulated in the headgroup region and the anions (counterions to the polyelectrolyte) avoiding the micelle being slightly enhanced outside the micelle. We believe that the qualitative picture is correct although the neglect of charge and density correlations within the layers limits the quantitative prediction of the model.

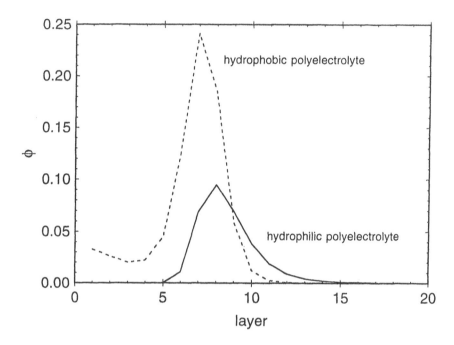

FIG. 12 Calculated volume fraction profiles of polyelectrolyte segments for a hydrophilic polelectrolyte at $C_s = 0.52$ mN and a hydrophobic polyelectrolyte at $C_s = 0.02$ mN of polyelectrolyte complexing a charged surfactant micelle from the mean-field theory. $\tau = 0.2$, $C_{p,seg} = 2.0$ mM, and $C_{salt} = 10$ mM. Other conditions are given in the text.

The lattice model can also be used to investigate the effects of a polyelectrolyte that is more hydrophobic, but still water soluble (a category 2 complex), on the complexation. Such a system can be realized by increasing the polyelectrolyte-water interaction parameter while at the same time making the polyelectrolyte-surfactant tail parameter less repulsive. In the following, we will consider the complexation between surfactant and a hydrophobic polyelectrolyte characterized by $\tau = 0.2$, $\chi_{polymer,tail} = 0.4$, (2.0) and $\chi = 1.6$ (0.0) for all the other polyelectrolyte interactions (the values in parentheses denote the interaction parameters for the previous hydrophilic polyelectrolyte). Figure 12 shows that the increased hydrophobicity of the polyelectrolyte leads to a stronger accumulation of the polyelectrolyte segments in the surfactant headgroup region and a slight shift toward the center of the micelle as well as a non-negligible penetration of the polyelectrolyte into the micellar core. Moreover, $C_{s,f}$ is reduced by more than one order of magnitude and N is increased of from 53 to 72. The effect of the hydrophobicity of the polyelectrolyte becomes less accentuated for the higher linear charge density τ = 0.5. Experimentally it is known that the cac and the aggregation number tends to decrease with increasing hydrophobicity of the polymer [13,17].

C. Monte Carlo Simulations

Monte Carlo and molecular dynamics simulations are two statistical techniques for solving a given model system [78,79]. The advantage of these methods is that in principle it is possible to study systems of any complexity without introducing any additional approximations. Since these methods are statistical in nature, the results suffer from statistical uncertainties, although the magnitudes of the uncertainties are normally controlled by the simulation length.   Monte Carlo simulations have been applied to investigate polyelectrolytes confined to charged planar and curved surfaces [80,81]. Recently, similar simulations have been made for investigating the complexation between one micelle and one hydrophilic polyelectrolyte [82-84] (category 1 complexation).   The scope was to study and explain how the cac/cmc ratio depends on different system parameters as the flexibility and linear charge density of the polyelectrolyte and the surfactant tail length, and for that purpose a simple model was found to be sufficient.

In these investigations [82-84], it was assumed that the cac/cmc ratio is related to the change in the free energy of the complexation between one micelle with accompanying counterions and one polyelectrolyte with its counterions, ΔA_{pe}, according to [82,85]

$$\frac{cac}{cmc} = \exp\left[\frac{\Delta A_{pe}}{NkT}\right] \tag{23}$$

where N is the micellar aggregation number. The assumptions inherent in Eq. (23) are that the aggregation number is not affected by the presence of the polyelectrolyte, and that the volume-pressure work caused by the addition of polyelectrolyte may be neglected, making it possible to identify the Gibbs free energy with the Helmholtz free energy.

The idea of the approach is to select a simple model system, but a system that still contains all essential aspects of a real system.   The main approximations involved are as follows: (i) The so-called primitive model is used: water is treated as a dielectric medium and enters the model only through its dielectric permittivity, and all other constituents are described in terms of hard spheres with point charges in the centers of the spheres. (ii) The cell model is applied, *i.e.*, one micelle and/or one short polyelectrolyte plus counterions are enclosed in a spherical cell. (iii) The presence of free surfactant is neglected at this stage, but is included on a simplified level after the simulations. (iv) The polyelectrolyte is modelled as a chain of charged hard spheres (beads) joined by harmonic bonds with the chain flexibility controlled by harmonic angular energy terms. (v) The micelle is described as a hard sphere with fixed charge and radius, and hence the aggregation number is assumed not to change upon addition of the polyelectrolyte.

Figure 13 illustrates the cell model of the initially separated micellar (M) and polyelectrolyte (P) solutions and the final mixed micelle-polyelectrolyte (MP) solution as well as two decoupled intermediate states (further described below). The radii of the cells are selected so that the volume is conserved upon mixing and that they can accommodate the fully stretched polyelectrolyte. The M and MP systems contain one micelle with fixed position in the center of the cell and freely moving monovalent counterions, and the P and MP systems contain one polyelectrolyte with n_{bead} charged beads and monovalent counterions of the polyelectrolyte. The micelle has z_{mic} unit charges and a radius R_{mic}, whereas the number of polyelectrolyte charges is normally twice as large and opposite in sign.

The total energy of a system (M, P, or MP) can be written as

$$U = U_{nonbond} + U_{bond} + U_{angle} \tag{24}$$

where the nonbonded energy is given by the Coulomb interactions and the hard sphere overlap repulsion among the charged hard spheres. The dielectric permittivity is constant throughout the system and hence surface polarization is neglected [86]. The total bond energy between adjacent beads in the chain is given by

$$U_{bond} = \sum_{i=1}^{n_{bead}-1} k_{bond} f(r_{i,i+1}, r_0)\left(r_{i,i+1} - r_0\right)^2 \tag{25}$$

with $r_{i,j}$ denoting the distance between bead i and j, r_0 the bead-bead separation constant, and k_{bond} the force constant. The factor $f(r_{i,i+1}, r_0)$ entering in Eq. (25) facilitates the thermodynamic integration, but has no other consequences (see ref 82 for further details). The bare flexibility of the polyelectrolyte is regulated by the angular energy given by

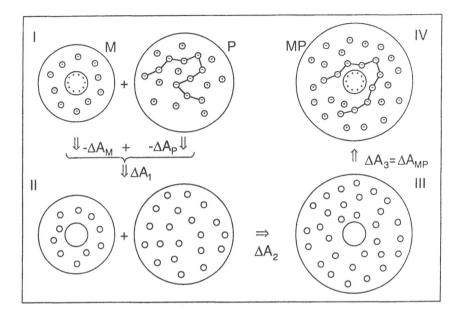

FIG. 13 Illustration of the initial micellar (M) and polyelectrolyte (P) systems (state I), the final mixed micelle-polyelectrolyte (MP) system (state IV), and the two intermediate and decoupled systems (states II and III). The pathway for the calculation of the Helmholtz free energy of mixing, ΔA_{pe}, is also indicated (see text for details).

$$U_{angle} = \sum_{i=2}^{n_{bead}-1} k_{ang}(\alpha_i - \alpha_0)^2 \qquad (26)$$

where α_i is the angle formed by the vectors $r_{i+1} - r_i$ and $r_{i-1} - r_i$ (α_i is referred to as the angle between consecutive beads) with an equilibrium angle $\alpha_0 = 180°$ and a force constant k_{ang}. Calculations have been performed where the chain flexibility (controlled by k_{ang}), the linear charge density of the polyelectrolyte (controlled by r_0), and the size and charge of the micelle (described by R_{mic} and z_{mic}) were varied independently.

In addition to the initial and final systems described above (states I and IV), Figure 13 shows two intermediate states. These are related to state I and IV, respectively, but the interactions are decoupled except for the hard sphere repulsion, *i.e.*, there are no electrostatic, no bond, and no angular energy terms. By using the four states, ΔA_{pe} for the mixing of the M and P systems to obtain the MP system can be expressed as a sum of three terms according to $\Delta A_{pe} = \Delta A_1 + \Delta A_2 + \Delta A_3$, where $\Delta A_1 = -\Delta A_M - \Delta A_p$ and $\Delta A_3 = \Delta A_{MP}$. Here, ΔA_M, ΔA_p, and ΔA_{MP} denote the free energy differences between the coupled and decoupled M, P, and MP systems, respectively.

The free energy contributions ΔA_M, ΔA_p, and ΔA_{MP} are obtained by thermodynamic integration [78]

$$\Delta A \equiv A(1) - A(0) \equiv \int_0^1 \frac{\partial}{\partial \lambda} A(\lambda) d\lambda = \int_0^1 \left\langle \frac{\partial}{\partial \lambda} U(\lambda) \right\rangle d\lambda \qquad (27)$$

where λ is a coupling parameter which continuously brings the decoupled systems into the interacting ones as λ varies from 0 to 1 and ‹ ... › denotes a canonical ensemble average obtained by Monte Carlo simulations. The precise nature of $U(\lambda)$ and further details are found in ref. 82. Monte-Carlo simulations according to the Metropolis algorithm [78] were performed on the M, P, and MP systems in order to evaluate the ensemble averages as described by Eq. (27), and to obtain structural information of the fully coupled systems. The free energy of mixing the two hard sphere systems to a single one, ΔA_2, is obtained by integrating the Carnahan-Starling compressibility along an isotherm [87,88]. Finally, a correction term is added to ΔA to obtain the free energy of the complexation at infinite dilution and with the effects of free surfactants taken into account [82].

The effect of the electrostatic interaction on the micellization of charged surfactants in the presence of an oppositely charged, hydrophilic polyelectrolyte has been investigated in a series of MC simulations [82-84] using the model described. In particular, we note that in the model there are no nonelectostatic attractive surfactant-polyelectrolyte interactions which promote a complexation (pure category 1 complexation). Figure 14 shows that the model gives a cac/cmc ratio of the order 0.01 to 0.1. Moreover, the suppression of the cmc increases with decreasing separation between the charges of the polyelectrolyte and with increasing flexibility, here measured by the persistence length of the corresponding neutral polymer.

Figure 15 displays experimental and model results for the cac/cmc ratio for different surfactant chain lengths with constant micellar surface charge density. The model parameters used to describe the polyelectrolyte in the calculations (see the caption of Figure 15) come close to those relevant for alginate [84]. Figure 15 indeed shows a satisfactory agreement between simulation data and experimental data with respect to (i) the magnitude of the depression of the cmc and (ii) the cac/cmc dependence on the surfactant chain length.

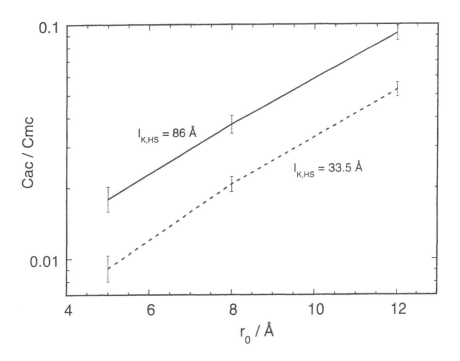

FIG. 14   Calculated cac/cmc ratios from MC simulations as a function of the bead-bead separation constant (r_0) for a more flexible, $l_{K,HS} = 33.5$ Å, (dashed curve) and a stiffer polyelectrolyte, $l_{K,HS} = 86$ Å, (solid curve). The symbol $l_{K,HS}$ denotes the statistic Kuhn segment length of the corresponding noncharged polyelectrolyte. $R_{mic} = 15.0$ Å and $z_{mic} = 20$. The error bars denote a 95% confidence limit. Data taken from Wallin and Linse, ref. 83.

The difficulties of mapping a real polyelectrolyte onto a model one and the differences in the experimental conditions (polyelectrolyte and salt concentration) make a more detailed comparison difficult. Nevertheless, the agreement in Figure 15 supports the notion that the electrostatic interaction is the main driving force for the reduction of the cmc in systems containing charged surfactants and oppositely charged and hydrophilic polyelectrolytes.

The MC simulation combined with the thermodynamic integration provides us with information of the energy and entropy changes for all the steps in Figure 13. Here, we will only consider the division of the free energy for the *full* mixing process of the micellar and the polyelectrolyte solution.

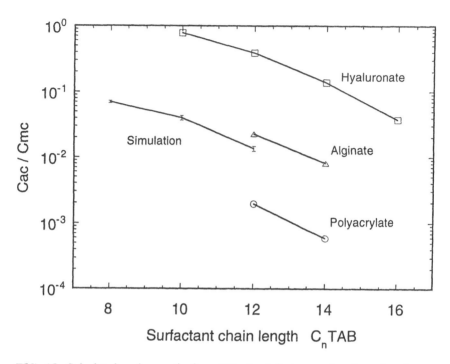

FIG. 15 Calculated cac/cmc ratio from MC simulations as a function of surfactant chain length (n) for $l_{K,HS}$ = 86 Å and r_0 = 8 Å and experimental ones for different polyelectrolytes. The polyelectrolyte concentrations and amount of added salt differ among the experimental systems, 0.5 - 1.0 mM and 0 - 10 mM, respectivley. In the model, R_{mic} and z_{mic} were calculated from n to describe micelles formed by alkyltrimethylammonium cations, and the calculated results are for an infinitely dilute system without added electrolyte but the effects of the free surfactans are taken into account. The error bars denote a 95% confidence limit. Data taken from Wallin and Linse, ref. 84 (simulation); from Thalberg and Lindman, ref. 89 (hyaluronate); and from Hayakawa *et al.*, ref. 90 (alginate and polyacrylate).

Figure 16 (next page) shows the free energy and its energy and entropy components as a function of the charge-charge separation along the polyelectrolyte chain. It is obvious that the mixing is energetically as well as entropically favored making ΔA_{pe} strongly negative. The energetic contribution dominates over the entropic one, but the difference between them decreases as ΔA_{pe} becomes less negative either by a reduction of the linear

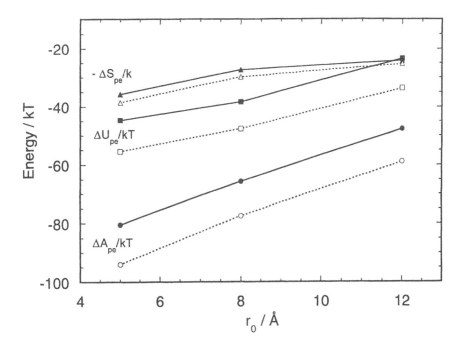

FIG. 16 Calculated change in energy (ΔU_{pe}), entropy (ΔS_{pe}), and free energy (ΔA_{pe}) for the mixing process given in Figure 13 as a function of the bead-bead separation constant (r_0) for $l_{K,HS} = 33.5$ Å (open symbols) and $l_{K,HS} = 86$ Å (filled symbols). The ordinate is in reduced unit per system. Parameters as in Figure 14. The largest estimated uncertainties are $\sigma(\Delta U/kT) = 0.6$, $\sigma(\Delta S/k) = 0.6$, and $\sigma (\Delta A/kT) = 0.3$. Data taken from Wallin and Linse, ref. 83.

charge density or by an increase in the chain stiffness. The energetic contribution arises from a large reduction of the electrostatic energy by the complexation process, despite the already large attractive electrostatic energy between the micelle and its counterions and similarly between the polyelectrolyte and its counterions. The entropic part stems mainly from the release of the small ions accumulated close to the micelle and the polyelectrolyte. The contribution to the mixing free energy from the internal degrees of freedom of the polyelectrolyte counteracts the complexation, but are much smaller in magnitude as compared to the contribution from the charges.

Finally, snapshots of the micelle-polyelectrolyte complex at two different flexibilities are shown in Figure 17. The images illustrate graphically how the

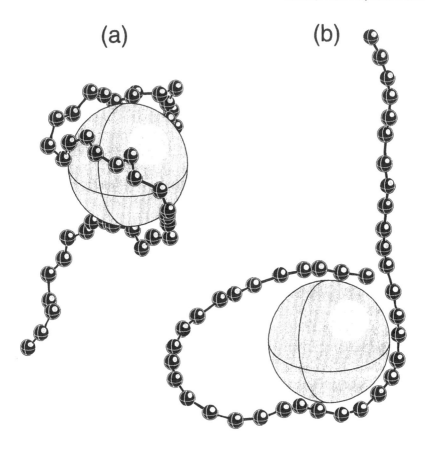

FIG. 17   Images of the micelle-polyelectrolyte complexation taken from MC simulations for (a) a flexible polyelectrolyte and (b) a fairly stiff polyelectrolyte. The small ions are omitted for clarity. Parameters: $r_0 = 5$ Å, $R_{mic} = 15.0$ Å, and $z_{mic} = 20$. Data taken from Wallin and Linse, ref. 82.

bare flexibility of the polyelectrolyte affects its structural behavior close to the oppositely charged micelle. For the flexible polyelectrolyte (a), the chain can easily wrap around the micelle and the micellar charge is neutralized by polyelectrolyte charges located close to the micelle. In the other case where the bare persistence length is of the order of the micellar dimension (b), the polyelectrolyte is wrapped less tightly around the micelle and obviously the possibility for the polymer to replace small counterions of the micelle is diminished.

IV. SUMMARY AND OUTLOOK

In this chapter we have presented several theories for polymer-surfactant complexation, covering the range from simple equilibrium models to detailed mean-field lattice approaches and computer-demanding simulations of molecular models. It is still not realistic to make *ab initio* predictions for the polymer-surfactant complex formed, and all results presented here therefore have to rely on some assumptions regarding the nature of the polymer-surfactant complexation, such as the surfactant-centered viewpoint for the complexation of surfactants with oppositely charged and/or slightly hydrophobic polymers. However, this viewpoint is strongly supported by experiments, and by well-established *a priori* knowledge of surfactant self-assembly. The hydrophobically modified polymers constitute a special case where we believe that the mixed micelle picture is the most appropriate one.

Given this surfactant-centered view, a central aspect in all approaches is the possibility of additional stabilization of micelles formed in the neighborhood of polymer molecules. In those cases where we have oppositely charged surfactants and hydrophilic polyelectrolytes (category 1 complexes), cac << cmc have been found with purely hydrophilic polymers which do not penetrate into the micellar core, using a simple Poisson-Boltzmann model, a lattice model, and computer simulations. Hence, in these cases the electrostatic interactions are solely responsible for the additional stabilization of the micelles. The computer simulations are able to confirm experimental cac/cmc ratios for different surfactant tail lengths, and further analyses of the results showed that the complexation was energetically as well as entropically favored. The energetic stabilization is associated with the fact that the surfactant and polyelectrolyte charges are spatially localized closer to each other as compared the more extended double layers formed in the separated systems. The entropic stabilization was an effect of the release of the small ions from the micelles and polyelectrolytes.   The model used in the mean-field approach is more fundamental than the model used in the simulation studies, since the surfactant molecules enter the lattice model individually and the actual micellization process in the presence of the polyelectrolyte is considered. However, the price of the increased level of detail is that a statistical mechanical approximation (the mean-field approach) is needed to make headway.

For category 2 complexation, the hydrophobic attraction between surfactant molecules and polymers facilitates the surfactant micellization and hence the complexation. These types of effects are more subtle and more difficult to model in a satisfactory manner as compared to the charge interactions. In the mean-field lattice model, these effects are introduced via the effective Flory-Huggins interaction parameters and similar or related terms are used in the advanced thermodynamic models. These terms are sometimes evaluated from more profound theories, but often they are used as free fitting parameters or determined from independent experiments analyzed with similar theory.

In the advanced thermodynamic models, the values of the free energy contributions that enter in Eq. (13) or similar expressions can in principle be obtained from other sources. However, in practice, they have to be (slighly) adjusted to reproduce experimental cmc, cac, and aggregation numbes. The two first terms in Eq 13, describing the transfer of a surfactant tail to the micellar interior, dominate $\Delta\mu°_N$. The cmc is essentially controlled by this contribution, whereas the aggregation number is determined by the balance of the derivatives of $\Delta\mu_{w/hc} + \Delta\mu_{hc/mic}$ and $\Delta\mu_{steric} + \Delta\mu_{elec}$ with respect to N. It should be clear that these parameters have to be assigned values describing the micellization in polymer-free solutions before the effect of a polymer can be taken into account (cf. Eq. (16)).

In view of the more refined descriptions, the closed association model only provides a simple phenomenological description of the self-assembly and the complexation and its predictive power is thus expected to be limited. As shown in Figure 3, the model can not predict the correct slope of the binding isotherms. We believe that one weakness is the fact that long-range electrostatic interactions are not well described by equilibrium constants.

It should be clear that the polymer-surfactant complexation is a non-trivial process involving many degrees of freedom, controlled by different types of interactions. A few of these are quite strong (electrostatic interactions), whereas others are weaker but numerous (nonpolar interactions). Meanwhile, both a polymer molecule and a micelle have large possibilities of internal responses. This complexity contributes to the diversity of existing approaches. It also explains the fact that most attempts to describe the complexation on a molecular, or nearly on a molecular, level is restricted to the binary complexation of one micelle with one polymer molecule. This is still far from the experimental situation, especially for more concentrated (phase-separated) systems and physical gels.

We believe that developments in this subject will continue to proceed along different directions. Simple approaches are of great value for the examination of experimental data. The mean-field approach has the potential of describing the complexation of all categories in a uniform way, since it is straightforward to introduce branched polymers with different interaction parameters for the main chain and the side chains. This would also give us a more detailed description of category 3 complexation which is still lacking. Finally, with the seemingly never ending increase in computer resources, the possibilities of simulating more detailed models are obvious. The possibility of network formation can also be explored. Still, good knowledge of the experimental systems is required to make balanced and useful models.

ACKNOWLEDGEMENTS

PL wishes to thank the Swedish National Science Research Council (NFR), LP the Swedish Reseach Coucil for Engineering Science (TFR), and PH the Center for Amphiphilic Polymers from Renewable Resouces for financial support.

ABBREVIATIONS

EHEC	ethyl hydroxyethyl cellulose
HAP	hydrophobically associating polymer
HEC	hydroxyethyl cellulose
NaHy	sodium hyaluronate
NaPA	sodium polyacrylate
PEO	poly (ethylene oxide)
PVP	poly (vinyl pyrrolidone)
PVA	poly (vinyl alcohol)
PNIPAm	poly (N-isopropylacrylamide)
$C_n E_m$	$CH_3(CH_2)_{n-1}(OCH_2CH_2)_m OH$
SDS	sodium dodecyl sulfate

REFERENCES

1. E.D. Goddard, in *Interactions of Surfactants with Polymers and Proteins* (E.D. Goddard and K.P. Ananthapadmanabhan, eds.), CRC Press, Boca Raton, 1993, Ch. 1.

2. H. Arai, M. Murata, and K. Shinoda, *J. Colloid Interface Sci. 37*: 223 (1971).

3. B. Cabane, *J. Phys. Chem. 81*: 1639 (1977).

4. J.P. Santerre, K. Hayakawa, and J.C.T. Kwak, *Colloids Surf. 13*: 35 (1985).

5. E.D. Goddard, *Coll. Surf. 19*: 255 (1986).

6. E.D. Goddard, *Coll. Surf. 19*: 301 (1986).

7. L. Piculell and B. Lindman, *Adv. Coll. Int. Sci. 41*: 149 (1992).

8. B. Lindman and K. Thalberg, in *Interactions of Surfactants with Polymers and Proteins* (E.D. Goddard and K.P. Ananthapadmanabhan, eds.), CRC Press, Boca Raton, 1993, Ch. 5.

9.   P. Hansson and B. Lindman, *Curr. Opinion Colloid Interface Sci. 1*: 604 (1996).
10. F.M. Witte and J.B.F.N. Engberts, *Colloids Surf. 36*: 417 (1989).
11. J. van Stam, M. Almgren, and C. Lindblad, *Prog. Colloid Polym. Sci. 84*: 13 (1991).
12. K. Thalberg, J. van Stam, C. Lindblad, M. Almgren, and B. Lindman, *J. Phys. Chem. 95*: 8975 (1991).
13. M. Almgren, P. Hansson, E. Mukhtar, and J. van Stam, *Langmuir 8*: 2405 (1992).
14. J.J. Kiefer, P. Somasundaran, and K.P. Ananthapadmanabhan *in Polymer Solutions, Blends, and Interfaces* (I. Noda and D.N. Rubingh, eds.), Elsevier, Amsterdam, 1992, Vol. 11, pp. 423-444.
15. P. Hansson and M. Almgren, *J. Phys. Chem. 99*: 16684 (1995).
16. O. Anthony and R. Zana, *Langmuir 12*: 1967 (1996).
17. P. Hansson and M. Almgren, *J. Phys. Chem. 100*: 9038 (1996).
18. R. Tanaka, J. Meadows, G.O. Phillips, and P.A.Williams, *Carbohydr. Pol. 12*: 443 (1990).
19. C. Senan, J. Meadows, P.T. Shone, P.A. Williams, *Langmuir 10*: 2471 (1994).
20. B. Magny, I. Iliopoulos, R. Zana, and R. Audebert, *Langmuir 10*: 3180 (1994).
21. F. Guillemet and L. Piculell, *J. Phys. Chem. 99*: 9201 (1995).
22. *Mixed Surfactant Systems* (K. Ogino, and M. Abe, eds.), Marcel Dekker, New York, 1993.
23. E. Abuin, J.B. Scaiano, *J. Am. Chem. Soc. 106*: 6274 (1984).
24. Z. Gao, J.C.T. Kwak, and R.E. Wasylishen, *J. Colloid Interface Sci. 126*: 371 (1988).
25. M. Antonietti, C. Burger, and J.J. Effing, *Adv. Matter 7*:751 (1995).
26. K. Hayakawa and J.C.T. Kwak in *Cationic Surfactants: Physical Chemistry* (D.N. Rubingh and P.M. Holland, eds.), Marcel Dekker, New York, 1991, Ch. 5.
27. I. Satake and J.T. Yang, *Biopolymers 15*: 2263 (1976).
28. B.H. Zimm and J.K. Bragg, *J. Phys. Chem. 31*: 526 (1959).
29. K. Shirahama, H. Yuasa, and S. Sugimoto, *Bull. Chem. Soc. Jpn. 54*: 375 (1981).
30. Y.-C. Wei, S.M. Hudson, *Rev. Macromol. Chem. Phys. C35(1)*: 15 (1995).
31. G. Schwarz, *Eur. J. Biochem. 12*: 442 (1970).
32. J. Skerjanc, K. Kogej, and G. Vesnaver, *J. Phys. Chem. 92*: 6382 (1988).
33. J.-E. Löfroth, L. Johansson, A.-C. Norman, and K. Wettström, *Prog. Colloid Polym. Sci. 84*: 8 (1991).
34   The expression in ref. 17 corresponding to Eq. (4) reads (with the present notation) $\beta = (K_0C_{s,f}+N(K_0KC_{s,f})N)/(1+K_0C_{s,f}+N(K_0KC_{s,f})N)$. The differences arise from different concentration units of the bound micelles.
35. D. Poland, *Cooperative Equilibria in Physical Biochemistry*, Oxford University Press, 1978.
36. R. Gelman, in *TAPPI Proceedings of the International Dissolving Pulps Conference*, Geneva, 1987, p.159.
37. A.J. Dualeh and C.A. Steiner, *Macromolecules 23*: 251 (1990).
38   R. Tanaka, J. Meadows, P.A. Williams, and G.O. Phillips, *Macromolecules 25*:1304 (1992).
39. I. Iliopoulos, T.K. Wang, and R. Audebert, *Langmuir 7*: 617 (1991).

40. S. Biggs, J. Selb, and F. Candau, *Langmuir 8*: 838 (1992).
41. E.D. Goddard and P.S.Leung, *Colloids Surf., 65*: 211 (1992).
42. T. Annable, R. Buscall, R. Ettelaie, P. Shepherd, and D. Whittlestone, *Langmuir 10*: 1060 (1994).
43. A. Sarrazin-Cartalas, I. Iliopoulos, R. Audebert, and U. Olsson, *Langmuir 10*: 1421 (1994).
44. U. Kästner, H. Hoffman, R. Dönges, and R. Ehrler, *Colloids Surf. A, 82*: 279 (1994).
45. B. Nyström, K. Thuresson, and B. Lindman, *Langmuir 11*: 1994 (1995).
46. E.D. Goddard, *J. Colloid Interface Sci. 152*: 578 (1992).
47. K. Persson, G. Wang and G. Olofsson, *J. Chem. Soc., Faraday Trans. 90*: 3555 (1994).
48. K. Persson and B.L. Bales, *J. Chem. Soc., Faraday Trans. 91*: 2863 (1995).
49. L. Piculell, K. Thuresson, and O. Ericsson, *Faraday Discuss. 101*: 307 (1995).
50. L. Piculell, F. Guillemet, K. Thuresson, V. Shubin, and O. Ericsson, *Adv. Colloid Interface Sci. 63*: 1 (1996).
51. K. Thuresson, O. Söderman, P. Hansson, and G. Wang, *J. Phys. Chem. 100*: 4909 (1996).
52. J.H. Clint, *J. Chem. Soc. Faraday. Trans 1*. 71:1327 (1975).
53. P.M. Holland and D.N. Rubingh, *J. Phys. Chem. 87*: 1984 (1983).
54. K. Shirahama, Y. Nishiyama, and N. Takisawa, *J. Phys. Chem. 91*: 5928 (1987).
55. M. Takasaki, N. Takisawa, and K. Shirahama, *Bull. Chem. Soc. Jpn. 60*: 3849 (1987).
56. K.P. Ananthapadmanabhan in *Interactions of Surfactants with Polymers and Proteins* (E.D. Goddard and K.P. Ananthapadmanabhan, ed.), CRC Press, Boca Raton, Fla., 1993, Chapter 8.
57. K. Fukushima, Y. Murata, N. Nishikido, G.Sugihara, and M. Tanaka, *Bull. Chem. Soc. Jpn. 54*: 3122 (1981).
58. K. Fukushima, Y. Murata, G. Sugihara, and M. Tanaka, *Bull. Chem. Soc. Jpn. 55*: 1376 (1982).
59. A.K. Morén and A. Khan, *Langmuir 11*: 3636 (1995).
60. A. Balazs and J.Y. Hu, *Langmuir 5*: 1253 (1989).
61. A. Balazs and J.Y. Hu, *Langmuir 5*: 1230 (1989).
62. C.J. Tanford, *Phys. Chem. 24*: 2469 (1974).
63. C.J. Tanford, *Proc. Nat. Acad. Sci. 71*: 1811 (1974).
64. E. Ruckenstein and R. Nagarajan, *J. Phys. Chem. 79*: 2622 (1975).
65. R. Nagarajan, *Adv. Colloid Interface Sci. 26*: 205 (1986).
66. S. Puvvada and D. Blankschtein, *J. Chem. Phys. 92*: 3710 (1990).
67. R. Nagarajan and K. Ganesh, *J. Phys. Chem. 90*: 5843 (1989).
68. R. Nagarajan, *Colloid. Surf. 13*: 1 (1985).
69. R. Nagarajan, *J. Chem. Phys. 90*: 980 (1989).
70. E. Ruckenstein, G. Huber, and H. Hoffmann, *Langmuir 3*: 382 (1987).
71. Y. J. Nikas and D. Blankschtein, *Langmuir 10*: 3512 (1994).
72. J.M.H.M. Scheutjens and G. Fleer, *J. Phys. Chem. 83*:1619 (1979), *ibid. 84*:178 (1980).

73. M.R. Böhmer, O.A. Evers, and J.M.H.M. Scheutjens, *Macromolecules 23*:2288 (1990).
74. M.R. Böhmer, L.K. Koopal, and J. Lyklema, *J. Phys. Chem. 95*: 9569 (1991).
75. T. Wallin and P. Linse, *Langmuir*, to be published.
76. P.J. Flory, *Principles of Polymer Chemistry*, Cornell University Press, Ithaca, NY, 1953.
77. P. Linse, and M. Björling, *Macromolecules 24*: 6700 (1991).
78. M.P. Allen and D.J. Tildesley, *Computer Simulations of Liquids*, Claredon Press, Oxford, 1987.
79. *Monte Carlo and Molecular Dynamics Simulations in Polymer Science* (K. Binder, ed.), Oxford University Press, Oxford, 1995.
80. M.K. Granfeldt, S.J. Miklavic, S. Marcelja, and C.E. Woodward, *Macromolecules, 23*: 4760 (1990).
81. M. K. Granfeldt, B. Jönsson, and C. E. Woodward, J. Phys. Chem. 95:4819 (1991).
82. T. Wallin and P. Linse, *Langmuir 12*: 305 (1995).
83. T. Wallin and P. Linse, *J. Phys. Chem. 100*: 17873 (1996).
84. T. Wallin and P. Linse, *J. Phys. Chem. B 101*: 5506 (1997).
85. D.F. Evans and H.Wennerström, *The Colloidal Domain*, VCH Publishers, NY, 1994.
86. P. Linse, *J. Phys. Chem. 90*: 6821 (1986).
87. N.F. Carnahan and K.E.Starling, *J. Chem. Phys. 51*: 635 (1969).
88. J.P. Hansen and I.R. McDonald, *Theory of Simple Liquids*, Academic Press, London, 1976.
89. K. Thalberg and B. Lindman, *J. Phys. Chem. 93*: 1478 (1989).
90. K. Hayakawa, J.P. Santere, and J.C.T. Kwak, *Macromolecules, 16*: 1642 (1983).

6

NMR Studies of Polymer-Surfactant Systems

PETER STILBS Physical Chemistry, Royal Institute of Technology, S-100 44 Stockholm, Sweden

I.   INTRODUCTION

As a technique, nuclear magnetic resonance (NMR) recently celebrated its 50-year anniversary. It is safe to claim that for the last 40 years no other physico-chemical method has come close to rivalling it with regard to versatility as a general, quantitative and detailed source of information at the molecular level. The gap just seems to widen.

In the continued development of NMR during the last decades the majority of new methods are in the field of multidimensional NMR, for the purpose of structural determination of organic compounds or biological macromolecules. We have also witnessed extensive new use of NMR in biomedicine, both as a spectroscopic method and as an imaging tool that complements X-ray and positron emission tomograph (PET) methods. To a large extent, these new NMR-based procedures rely on the general developments in the field of electronics and, in particular, on that of digital computing.

With regard to studies of polymer and surfactant systems in solution, the main NMR "tools" are quite different from those used in the determination of molecular structure. Of course, NMR spectroscopy can be applied in a similar manner as other types of spectroscopy or physico-chemical methods, *e.g.* studies based on solution conductivity, density, viscosity etc. To monitor chemical shift changes as a function of solution composition would be the simplest NMR-based approach.   However, NMR offers many other experimental dimensions. The multi-nuclear detectability (protons, deuterium, carbon-13, oxygen-17, fluorine-19, sodium-23 etc.) is of course one unique characteristic of "Nuclear Magnetic Resonance." However, dynamic information that is not in any way accessible from other techniques is nowadays conveniently available from either NMR spin relaxation data or from NMR-based, multi-component studies of molecular self-diffusion. These are the most valuable NMR-tools in the present context.

In polymer-surfactant systems the basic spin relaxation approach [1] may provide detailed information on (picosecond to microsecond) local molecular dynamics and order of alkyl chains, water or counterions in the system. In principle, it can be based on data for any NMR-accessible type of nuclei in the system, at any location in a molecule.

The multi-component self-diffusion approach [2] provides information on overall molecular displacements on a much longer timescale (typically on the order of 100 milliseconds, depending on experimental conditions and the instrumental set-up). Generally, multi-component self-diffusion data provide easily interpretable data that are particularly sensitive to changes in aggregation or binding of the system.   The two families of methods are complementary, as well as highly selective and applicable to very complex systems. Simultaneous multicomponent self-diffusion measurements can now be regarded as standard experiments.

If necessary, isotopic labelling can sometimes also be applied to further enhance the selectivity and sensitivity of either of the methods in question. Particularly common in the present context is selective deuterium labelling for the purpose of studying deuterium spin relaxation at a particular location, at a high sensitivity and spectral selectivity. This strategy provides quantitative information on the reorientation and order of a particular C-H bond vector, in a very unambiguous way.

II. SCOPE OF THE PRESENT REVIEW

It does not appear meaningful to provide a new, detailed account of the underlying NMR-spectroscopic theory and basic methodology in this context, and at this point in time. This subject is already well covered in a review by Söderman and Stilbs on "NMR Studies of Complex Surfactant Systems" only a few years back [1]. Also, very few new NMR-methodological methods and concepts have been introduced in any recent studies of polymer-surfactant systems - the overwhelming majority of papers are based on strategies already established for simple surfactant or polymer systems in solution.

A few reviews and methodological papers dealing with NMR-based field gradient methods for measuring self-diffusion and molecular transport in solution have recently appeared, however. They will be discussed together with some final remarks in the concluding section this review, since they do contain elements of significant future value related to the study of polymer-surfactant systems. An attempt to make a reasonably complete literature survey has been made. While all papers will be listed, many share a great deal of similarity, both with regard to choice of system and from a methodological standpoint, only key issues and results of a few selected papers will be discussed. The main aim of the review is primarily to convey to the reader a reasonably balanced idea with regard to what can/cannot be done by current NMR methods, and what would be the most appropriate NMR-based approach. It is furthermore assumed that the reader has some previous basic knowledge of NMR.

III. INTRINSIC PROBLEMS IN NMR INVESTIGATIONS OF POLYMER-SURFACTANT SYSTEMS

Necessarily, a three-component system polymer-surfactant-water is much more complex than single aqueous solutions of surfactant or polymer alone, not only from an aggregation/interaction point of view, but also from basic considerations about the detection of a useful NMR spectrum. Polymers or surfactants in solution do not really have well-resolved NMR-spectra to begin with, a problem that becomes amplified in mixed systems. This is not only because of added signal overlap, but also because of line broadening and faster spin relaxation rates that can occur as a direct result of enhanced aggregation and specific interactions. In general, line broadening effects reduce sensitivity quite extensively in all types of NMR experiments, and they are normally the limiting experimental parameter.

Going to higher magnetic field might seem like an obvious cure for the basic "spectral resolution of peaks"-problem, since one would assume that NMR-spectral dispersion should be linear with the applied magnetic field. However, for macromolecular systems this is only partly true, since the overall reorientation dynamics is normally at such a spectral range that linewidths increase with magnetic field, and thereby do change "unfavorably" from a mere resolution point of view. On the other hand, the line broadening and spin relaxation effects described actually do provide a rich source of detailed quantitative information about characteristics of the aggregates. Data on size, shape and local chain dynamics become accessible this way provided a proper interpretation of the magnetic field-dependence and other experimental parameters is made. Only a few studies along the lines of the full experiment and analysis have been achieved to date, however.

IV. EARLY NMR-STUDIES OF POLYMER-SURFACTANT SYSTEMS IN SOLUTION

It appears that the subject of polymer-surfactant interaction, as studied by NMR methods has only been reviewed twice: in the introductory section of a Conference Proceedings paper "NMR Studies of Interactions between Neutral Polymers and Anionic Surfactants in Aqueous Solution" by Gao and Kwak in 1991 [3] and in Section 6 of the already mentioned review by Söderman and Stilbs of 1994 [1]. The latter contains a rather complete summary of NMR-based methodology applicable to studies of polymer-surfactant interaction, as previously pointed out.

Gao and Kwak's 1991 review introduces well the fundamental questions of the subject, and summarizes fully the existing experimental work at the time. The dominant early mode of approach was based on a monitoring of proton, carbon-13 or fluorine-19 NMR chemical shifts of selected parts of the polymer or surfactant molecules. Pioneering investigations by Oakes (on Bovine Serum Albumin/SDS systems, cf. Fig. 1) [4], by Muller et al. on fluorine labelled

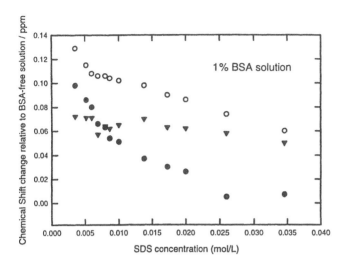

FIG. 1 The chemical shift data on which Oakes' pioneering NMR study of SDS binding to Bovine Serum Albumin at pH 5.6 was based (data taken from Table I of ref. 4). Filled circles represent proton chemical shift values for the α-protons of SDS, open circles the unresolved -(CH$_2$)$_n$- bandshape, and triangles the methyl group signal. Using several assumptions with regard to the association model and the chemical shifts of SDS protons at different levels of binding, these data were subsequently fitted to a model of the type given in Eq. (2), to produce a binding isotherm.

surfactant/protein [5] and surfactant/polymer [6] systems, and by Cabane [7] on the PEO/SDS system are summarized in this review, together with a few other and similar investigations. Generally, as a result of altered aggregation conditions with system composition or temperature, chemical shifts of various parts of the molecules involved will also change. However, the effects are generally quite small, and their origin is difficult to pinpoint. Nonetheless, with proper measurement procedures with regard to shift referencing and other relevant experimental parameters even very minor chemical shift changes do become experimentally significant. As such, they can simply be utilized as "phenomenological parameters" for obtaining binding isotherms or to indicate interaction. Any further interpretation in terms of actual changes in the environment of molecules or details of specific interactions between molecules will normally be at best speculative.

A case where one can actually (at least semi-quantitatively) interpret chemical shift effects is in situations where NMR-active nuclei are in the vicinity of an "aromatic-ring" resulting in so-called aromatic ring-current induced shifts. These reflect a proximity between a chemically shifted group of peaks, and some (anisotropically shielding) aromatic ring present in the system. This moiety can either be naturally present in the polymer (and will then shift the surfactant signals), or it can be artificially added by basing the experiment on a selected surfactant that has an aromatic ring (which will instead shield some of the polymer signals). Early papers along these lines were presented by Gao and Kwak and co-workers [8,9]. One should note in this context that the origin of such chemical shift effects is a time-averaged "dipolar-like interaction" (*i.e.* proportional to $< (3\cos^2 \theta - 1)/r^3 >$, where r represents the distance between the shifted nucleus in question and the center of the aromatic ring, θ is the angle between the normal of the aromatic plane and the nucleus, and <...> represents a time-average). For a non-preferential relative orientation between the nucleus and the aromatic ring, this quantity becomes zero to first order, regardless of any close proximity in space, however [10]. Also, there have been suggestions in the literature of a specific interaction between Br⁻ (counter)ions and aromatic rings. In the case of surfactants like DTAB or CTAB this would be likely to cause secondary effects with regard to the ordering of other moieties in an aggregate, in particular for the head-group. One should therefore exercise caution with regard to any generalizing of structural conclusions about micellar ordering and similarity to other families of surfactants [10].

Some semi-quantitative conclusions based on spin relaxation and linewidth observations were also presented in the early papers by Oakes and by Cabane just mentioned. Information about aggregation and local mobility can be obtained in this way: an increased linewidth (or reduced spin relaxation time) is assumed to correspond to a reduced degree of reorientational mobility. For the same purpose, Gao, Wasylishen and Kwak also made the first attempts to make a full analysis of carbon-13 spin relaxation data on a polymer-surfactant system (DOTAB/PSS) [11]. The same authors applied a newly developed paramagnetic spin relaxation method [12,13] in an investigation of PEO/SDS interaction [3,14]. This approach is based on electron-nuclear spin relaxation effects of added paramagnetic ions in the aqueous solution phase. It provides a pathway to quantitative information on the time-averaged overall contact between a part of a molecule (surfactant or polymer) and the aqueous pseudophase. Molecules in the aggregated state are assumed to be unaffected, while those in the aqueous phase are fully exposed to the spin-relaxing effects of the paramagnetic ions. By a straightforward analysis, the "free/bound" ratio (*i.e.* binding isotherm-type information) of different species can be quantified. Of course, the method is not applicable in a situation where the charges of functional groups on the aggregated molecules and the paramagnetic ions or species have opposite sign.

Another proven method for quantifying "free/bound" situations and "binding isotherms" is NMR-based multicomponent self-diffusion monitoring by the FT-PGSE (Fourier Transform Pulsed-Gradient Spin-Echo) method [2]. This approach relies on the often vastly different self-diffusion rates for the same molecule in "free" or 'bound/aggregated' state. It was first applied to polymer-surfactant systems by Carlsson *et al.* 15]. The basic PGSE experiment [16] relies on the use of pulsed linear magnetic field gradients (of amplitude g, duration δ and separation Δ) that are applied during a so-called spin-echo experiment [17], involving two or more radiofrequency pulses (in the simplest case separated in time by τ. Under these conditions the amplitude of the spin-echo (which occurs at a time 2τ after the initial radiofrequency pulse) attenuates from its full value A(0) according to the so-called Stejskal-Tanner relation [16]:

$$A(2\tau) = A(0) \exp(-2\tau/T_2) \exp(- D (\gamma g \delta)^2 (\Delta - \delta/3)) \qquad (1)$$

Here T_2 represents the (disturbing and experimentally limiting) transverse spin-spin relaxation time of the nuclei in question, and γ their gyromagnetic ratio. Subsequent Fourier transformation of the composite echo in the time-domain separates the contributions at each frequency, just like in the normal basic pulsed FT-NMR experiment [2]. Figure 2 illustrates a typical application of FT-PGSE NMR to a polymer-surfactant system.

V. SUMMARY OF NMR-BASED INVESTIGATIONS OF POLYMER-SURFACTANT INTERACTION

To date at least 60 papers in this general category have appeared. Some of these use one of the NMR-approaches as the main or only tool. Others just refer to complementary NMR spectra in an investigation that is mainly based on some other experimental technique(s). Multicomponent FT-PGSE self-diffusion monitoring is the currently most popular NMR-experimental approach. The summary given below should be almost complete from 1988 and onwards: It also includes the early papers that can be regarded as pioneering in some respect. The reader is referred to the review by Gao and Kwak [3] for some additional, early (chemical shift-based) work in this field. Unless otherwise indicated, all studies were made on aqueous solutions of polymer-surfactant systems.

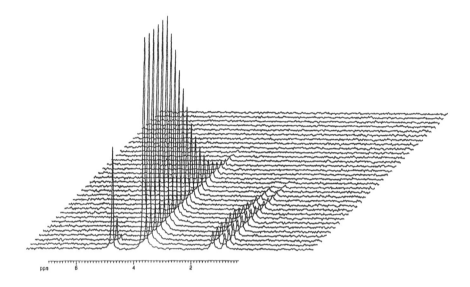

FIG. 2 A 300 MHz proton stimulated echo FT-PGSE data set on a solution of 0.1% PEO polymer (M=6·10^5) and 0.5% C12E5 surfactant in heavy water (P. Stilbs and W. Brown, unpublished work). The magnetic field gradient strength (0.22-3.52 T/m·A) as well as the duration of the magnetic field gradient pulses (1-7 ms) was varied in the experiment, in a fashion that was sorted with respect to their product, which increases upwards in the series. The respective frequency-domain echo decays follow the Gaussian-like decay behaviour implied in Eq. (1). From left to right the signals correspond to (rapidly diffusing) residual water and protons (HDO) at 4.8 ppm, a composite PEO and C12E5 band at 3.6 ppm and and two pure C12E5 bands near 1 ppm. The latter assignments are based on an analysis that leads to polymer and surfactant D-values of 2.7·10^{-12} m^{2}s^{-1} and 3.9·10^{-12} m^{2}s^{-1} respectively, as further discussed below (*cf.* Figures 5a and b).

A. Studies Based on Chemical Shift Changes and Semi-Quantitative Linewidth Information

This type of approach normally aims at an evaluation of binding isotherms, and to indicate specific interaction patterns at the molecular level. A general assumption is made that a molecule has a characteristic chemical shift value (δ), depending on its state of aggregation, and that chemical exchange between all states is rapid on the pertinent NMR time-scale (here of the order of milliseconds). With these assumptions, the observed time-averaged shift value thus becomes quantifiable in terms of a "free/bound-ratio" and possibly also in terms of more complex aggregation models through computer modelling of experimental data, to fit an equation of the following type:

$$\delta_{obs} = p_{free}\,\delta_{free} + p_{bound(1)}\,\delta_{bound(1)} + p_{bound(2)}\,\delta_{bound(2)} + ... \qquad (2)$$

Here the sum of the p-values are presumed to add up to 1. Including more than one bound "state" is sometimes justifiable in order to mimic experimental data properly. It is also required in case a more complex interaction model is implied by complementary information.

The work cited here is based on proton chemical shifts, unless otherwise indicated. Systems studied include aqueous solutions of: a) Bovine Serum Albumin (BSA) and SDS [4] or ω-trifluoro dodecyl- or tridecylsulfate [5]. The BSA/SDS system has later been re-investigated, as based on combined proton NMR chemical shift measurements and FT-PGSE NMR self-diffusion data [18]; b) PEO and ω-trifluoro dodecylsulfate, as monitored in fluorine-19 spectra [6]; c) PEO and SDS [7]; d) PEO and Sodium ω-phenyldecanoate [19]; e) comparative studies of polyelectrolytes like poly(maleic acid-co-styrene), poly(maleic acid-co-ethylene), poly(styrene sulfonate), poly(vinyl sulfonate) and DTAB [8]; f) poly(maleic acid-co-butyl vinyl ether) and DOTAB [9]; g) a review of e-f, and further comparative investigations along similar lines of PEO, PVA and PVP and ω-phenyldecanoate [3]; h) poly(L-lysine) and SDS, as monitored in carbon-13 spectra [20]; i) styrene-ethylene oxide block copolymers and a range of anionic and cationic surfactants [21]; j) poly(vinyl pyrrolidone) (PVP) and SDS or AOT, as monitored in carbon-13 spectra [22,23]; k) PEO-POP-PEO copolymers and SDS, as monitored in carbon-13 spectra [24]; l) PEO and Carbon-13 enriched SDS, monitored in carbon-13 spectra down to very low concentrations of SDS [25]; m) Hydroxypropyl methyl cellulose and SDS [26]; n) hydrophobically modified acrylamides and alkylbenzene sulfonates [27,28] and sodium 4-(p-butylphenyl)butane-1-sulfonate and reduced and carboxyamido-methylated BSA, as based on ring-current induced proton chemical shifts [29]; o) solubilization of aromatic compounds into poly(maleic acid-co-butyl vinyl ether)-DOTAB systems [9]; p) hydrophobically modified poly(acrylamide)s and alkylbenzenesulfonates, as based on proton chemical shift and qualitative linewidth considerations [28]; q) Gelatin and SDS, as based on carbon-13 chemical shift data [30] and linewidth changes.

B.   Studies Based on Proton Electron-Nuclear Spin Relaxation from Paramagnetic Ions in the Aqueous Phase

Such studies lead to a quantitative assessment of the fraction of non-aggregated species, *i.e.* binding isotherm-type information. Only two groups have so far exploited the possibilities of this approach and the only polymer-surfactant systems studied so far are PEO with SDS or sodium ω–phenyldecanoate [3,13,14] and surfactant-modified ethoxylated urethane dispersions with solubilized alcohols [31].

C.   Studies Based on Nuclear Spin Relaxation Data

This type of experimental information is closely related to the reorientation behaviour of individual alkyl chain C-H bond vectors or that of inorganic counterions and is therefore a source of information for a quantitative assessment of local mobility and order of different parts of polymer and surfactant molecules.   One should note that in early work based on simple linewidth considerations or single-frequency spin relaxation data, one was not generally aware of the need to use the nowadays established experimental and evaluation procedures required for a full analysis of the experiment [1].   The origin of these complications is that the motional spectrum of a polymer or large aggregate in solution will partly be in the so-called non-extreme narrowing NMR condition, *i.e.* part of the characteristic correlation time distribution will be above the inverse of the angular NMR frequency.   On the other hand, spin relaxation data on such systems contain more information than in the extreme narrowing condition experienced by small molecules in solution.

Also, one should note that proton spin relaxation or linewidth data are normally useless for any further quantification attempts, since inter- and intramolecular proton spin relaxation contributions are difficult or impossible to separate.   The phenomenon of "spin diffusion" is a further complicating factor for protons in particular.   Strategies based on carbon-13 or deuterium spin relaxation are much recommended in this context [1], since spin relaxation here is exclusively intramolecular, and thus translatable into molecular reorientation rates. Figures 3a and 3b illustrate results of an analysis of multi-field deuterium spin relaxation in a system of SDS and a C12 hydrocarbon end-capped, 200 unit PEO model associative thickener in aqueous solution [32].

It should be pointed out that a major confusion with regard to nomenclature exists in the literature of polymer-surfactant interaction and NMR. Jones in his pioneering work on polymer-surfactant interaction [33] regrettably introduced the notations T_1 and T_2 for the critical surfactant concentrations at which "interaction between polymer and surfactant first occurs" and "the concentration at which the polymer is saturated with surfactant," respectively. That use has unfortunately continued to this date. Ever since the late 1940s, however, an altogether different use of these symbols had been established in the NMR literature; *i.e.* to denote the very central parameters "longitudinal spin-lattice relaxation time" and "transverse spin-spin relaxation time," respectively. In this chapter this NMR notation will be adhered to.

Systems studied include aqueous solutions of: a) PEO and SDS [7] (by carbon-13 and sodium-23 spin relaxation, and a semi-quantitative interpretation); b) poly(styrene sulfonate) and DOTAB, by carbon-13 spin relaxation, applying a full quantitative analysis [11]; c) sodium hyaluronate/TTAB, as studied by multi-field carbon-13 spin relaxation data, including a full analysis of the spin relaxation behavior (through differential line broadening observation and analysis of the spin coupled carbon-13 quartet signal of the TTAB methyl groups) [34]; d) the viscoelastic poly(vinyl methyl ether)/CTAB/sodium salicylate system, studied by the same approach [35]; e) sodium poly(styrene sulfonate) and deuterium methyl labelled (zwitterionic) hexadecyl phosphocholine and CTAB [36] by using deuterium bandshape and signal intensity information; f) a model associative thickener based on a poly(ethylene oxide) backbone with ether-linked terminating alkyl groups and α–deuterium labelled SDS, using multi-field deuterium spin relaxation, applying a full analysis to the data [32]; g) PEO and SDS, as studied by carbon-13 spin relaxation [37,38]; h) poly(acrylamides) and alkylbenzenesulfonates, using semi-quantitative proton linewidth observations [28]; i) hydrophobically modified poly(acrylamides) and SDS, using the same strategy [27]; j) EHEC or hydrophobically modified EHEC and (α–position) deuterium-labelled SDS, as based on deuterium spin-spin relaxation rates; k) BSA and SDS, as based on semiquantitative linewidth observations in deuterium spectra of per-deuterated SDS [39]; l) gelatin and SDS, as based on semiquantitative carbon-13 linewidth considerations [30].

D.   Studies Based on FT-PGSE NMR Multicomponent Self-Diffusion Measurements

Self-diffusion data are a very direct source of information about aggregation processes. For unrestricted diffusion during a selected time span, the characteristic self-diffusion coefficient, D, simply translates into a mean square

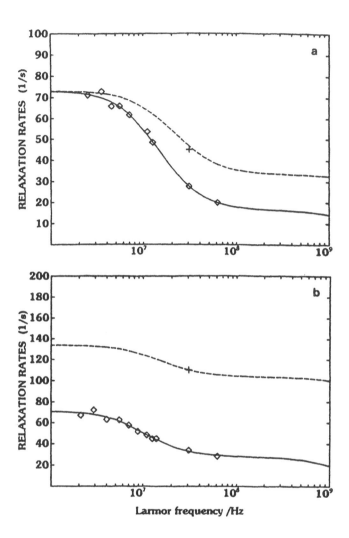

FIG. 3 Deuterium spin relaxation rates for a) α-deuterium labelled SDS and b) a similarly α-C12 alkyl deuterium labelled, end-capped, 200 unit PEO "model associative polymer" (MAP) In a mixed system of 4% MAP and 4% SDS. These are results of two separate studies - the studied samples contained only one deuterium-labelled component at a time, of course. The solid lines represent fits of the experimental data to a theoretical (a) two-step (with correlation times 5 ns for overall and 32 ps for local motions of SDS, respectively, and an order parameter of 0.135) and (b) three-step spin MAP relaxation model (with the first two correlation times in a similar range as in (a), and the third in the microsecond range). From ref. 32, with permission from the American Chemical Society).

displacement in space ($<\Delta r^2>$) during the observation time Δt, through the Einstein relation:

$$<\Delta r^2> = 6D\Delta t \tag{3}$$

One should note that the mean square displacement of even large macromolecules during the selected (method-dependent) time span of the experiment Δt is generally very much larger than the average macromolecular diameter. Therefore the quantity ($<\Delta r^2>$) normally requires no further interpretation and becomes easy to visualize. In polymer-surfactant systems very dramatic changes of self-diffusion behavior may also occur with variations in system composition; a reduction of a component's self-diffusion coefficient by three orders of magnitude is not at all uncommon. It is evident that such data very directly mimic a formation of large aggregates or 3-dimensional networks in solution.

In binding studies, the self-diffusion approach relies on a relative comparison of time-averaged self-diffusion rates between a "bound" and "free" state, meaning that in the simplest two-state situation the effective self-diffusion coefficient D(obs) will be given by:

$$D(obs) = p\, D(bound) + (1\text{-}p)\, D(free) \tag{4}$$

where the degree of binding p may assume values in the range $0 < p < 1$. This equation can be rewritten into the form:

$$p = (D(free)\text{-}D(obs))/(D(free)\text{-}D(bound)) \tag{5}$$

Hence, such experiments do provide the information needed for obtaining binding isotherm information for example, just by a simple comparison of experimentally available self-diffusion coefficients. It is evident that the method is most accurate around p-values of the order of 0.5, and becomes uncertain near p-values of 0 or 1. It is also easy to see that the experimental detail and precision needed for situations more complex than a simple "free/bound" case may be much more difficult to achieve. However, this would be true for any other experimental approach as well.

Before the development of the FT-PGSE NMR technique [2], the acquisition of multi-component self-diffusion data was a very cumbersome task indeed, requiring synthetic work to provide radioactive isotope labelling of the compounds to be studied, as well as tedious and time-consuming measurement procedures. Today, a simple NMR measurement of multi-component self-diffusion can be completed in a matter of minutes (*cf.* Figure 2 above).

The macroscopic viscosity of polymer systems in solution varies considerably with solution composition, which is a quite central problem in the present type of experimental approach. There is still controversy and confusion in the literature on whether one should account for the macroscopic solution viscosity to correct experimental self-diffusion (or spin relaxation) data. While such a "viscosity correction" is highly justified for molecules in normal solution (*cf.* the friction coefficient f for a molecule with radius r in a continuous medium with a viscosity η as given by $f = 6\pi r\eta$), it is clearly irrelevant for a small molecule in a solution of macromolecules. The reason is easy to see: experimental observations show unambiguously that the self-diffusion rate of the solvent or any small molecule is only very marginally correlated with the typically rapid and huge increase of solution viscosity with polymer concentration. Studies on gel systems point to the same conclusion, showing that small molecules (like water) in a rigid gel framework diffuse almost as rapidly as they do in a normal aqueous solution (see *e.g.* refs. 40 and 41). The presence of macromolecules in solution or in a gel framework primarily serves as an obstacle for small molecules in the system, forcing them to use a longer path in the normal process of diffusing between two points in space. The time-scale of the diffusion experiment is normally very much longer (milliseconds to days, depending on the technique used) than it takes to diffuse a macromolecular diameter. For this reason the apparent experimental self-diffusion coefficient of *small* molecules in the FT-PGSE NMR experiment and almost all other techniques simply becomes reduced in proportion to this obstruction effect. An exception will be the neutron spin-echo experiment, which has very much shorter characteristic time-scale in the nanosecond range (see *e.g.* refs. 42-44 for some applications to polymer systems in solution).

The factor to account for in describing the resulting apparent lowering of the effective time-averaged self-diffusion coefficient of small molecules here is therefore the obstruction effect, rather than the solution viscosity. Several attempts to quantify such obstruction effects on self diffusion have been made to date. In a first approximation it is linearly related to the volume fraction of obstructing objects in solution, and to their general shape. A useful reference that corrects the theory behind previous quantification attempts was published some years ago [45]. It provides a general treatment of a variety of obstructing object shapes and the concentration dependence of their obstruction effect. A more elaborate treatment of obstruction effects on self-diffusion, together with a literature update on the subject can be found in a recent paper by Jóhannesson and Halle [46].

It would appear that the "obstruction-effect approach" should be an interesting pathway for the investigation of aggregate shapes in solution. However, in practice this rarely works, unless the system chemistry allows self-diffusion monitoring of relatively high concentration spans without any change in aggregate shape. Intrinsically, a high experimental precision also is required to allow a distinction between various aggregate shapes.

Clearly, the underlying effect behind the typically very strong increase in solution viscosity with concentration of a high polymer solution is some form of an entanglement effect. Quantification and theoretical treatment of polymer and polymer aggregate diffusion necessarily is quite complex and difficult to generalize, in particular if strong specific interaction patterns operate together with the basic entanglement effects. A number of theoretical [47-53] and experimental studies [54-62] on polymer diffusion in solution have appeared. The references given are by no means complete, but should serve as a good starting point for further study.

One should note in this context that Eq. (3) only holds in the limit of large Δt. Large, in this context, has a different meaning for small molecules and macromolecules. Uhlenbeck and Ornstein derived a generalized equation, in their classical treatment of the subject [63]:

$$\langle \Delta r^2 \rangle = 6D\left[\Delta t - \tau_c(1 - \exp(1 - \Delta t/\tau_c))\right] \qquad (6)$$

where the correlation time $\tau_c = m/f$, f being the coefficient of friction and m the mass of the Brownian particle. For small molecules in liquids the correlation time is of the order of a nanosecond, while the shortest imaginable NMR spin-echo experiments will exceed a fraction of a millisecond, and are more typically tens or hundreds of milliseconds. For macromolecules, however, this correlation time may be such that it actually affects (and thus becomes measurable by) the typical NMR PGSE experiment, and therefore may significantly affect any results and conclusions or introduce serious experimental artifacts, if proper conditions for the experiment are overlooked. A very lucid discussion on the appropriate interpretation of the NMR PGSE experiment under these conditions was recently given by Callaghan and Stepišnik, in an excellent review on "Generalized Analysis of Motion Using Magnetic Field Gradients" [64]. Notably, only four previous experimental investigations of such non-Gaussian diffusion seem to have been made at the time of writing of that review, none of which concern polymer-surfactant systems.

In a case of gradual self-association of polymers or polymer-surfactant systems into different families of 3-dimensional networks and other structures, the situation becomes even more complex indeed with regard to self-diffusion behaviour. Quantification attempts of experimental data would really be futile, were it not that quite dramatic self-diffusion coefficient changes of three orders of magnitude or so often result with changes in solution concentration and composition. Under such conditions the experimental data will again become interpretable in terms of general association patterns or solution structure at a reasonable level of confidence. Figures 4a, b and c illustrate a typical binding/aggregation study [65] of SDS and a C12 end-capped, 200 unit PEO in heavy water. Spin relaxation in the same system was illustrated in Figures 3a and 3b.

Returning to actual experimental studies of polymer-surfactant systems in the past, we find that polymer self-diffusion data have not always been accessible with the experimental set-up used, primarily due to fast spin-spin relaxation of the polymer protons. Many of the investigations cited below are for that reason based on self-diffusion data for the surfactant only (in particular the cellulose-based polymer type studies). Of course, any quantitative conclusions may become considerably less reliable under such conditions. However, modern equipment has a much better experimental performance compared to what was available only a few years back. Primarily, the improvements are the introduction of i) self-shielded magnetic field gradient coils and ii) very much more powerful and stable current-regulated magnetic field gradient drivers (see *e.g.* references cited in ref. 1).

In the work cited here, all studies are based on proton FT-PGSE measurements. Systems studied include: a) ethyl hydroxyethyl cellulose (EHEC) and DOTAB [15]; b) sodium hyaluronate and DTAB or DOTAB [66]; c) hydroxypropyl methyl cellulose (HPMC) and SDS d) agarose gels, containing $C12E6$ and $C12E8$ [67]; e) a model "associative thickener" based on a 200 EO unit poly(ethylene oxide) backbone with ether-linked terminating alkyl groups and SDS [32,65]; f) PEO-POP-PEO triblock polymers in AOT-type isooctane/water-type microemulsions [68]; g) poly(vinyl methyl ether) in (viscoelastic) CTAB/sodium salicylate/water systems [69]; h) EHEC and $C12E8$ [70]; i) PEO and SDS [38,71]; j) PEO and custom-made anionic ethoxylated surfactants, probably designed for enhanced oil recovery research [72]; k) more than 6 additional papers along the lines of the pioneering FT-PGSE application to polymer-surfactant systems of a) above to EHEC and CTAB or SDS systems [73-78]; l) HEUR-type thickeners and DTAB or SDS [79]; m) a commercial, non-ionic, random copolymer (named UCON) of ethylene oxide and propylene oxide and $C12E4$ [80]; n) BSA and SDS [18]; o) sodium hyaluronate and TTAB [81]; p) gelatin and anionic surfactants, like SDS ˙ [30,82] and C8 to C14 SDS analogues [83]; q) a comparative study of PEO and hydrophobically modified PEO (alkyl end-capped) and SDS [84]; r) a mostly methodological paper on new experimental and evaluation procedures with regard to the proper extraction of multi-component self-diffusion coefficients in FT-PGSE, as illustrated on data from the EHEC/SDS system [85].

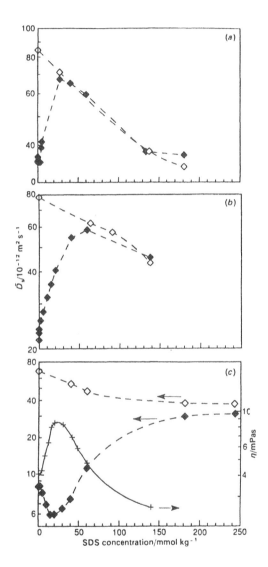

FIG. 4 FT-PGSE NMR-based self-diffusion "binding and aggregation studies" of SDS and a 200 unit PEO (open symbols), or a C12 end-capped 200 unit PEO (filled symbols) at three different polymer concentrations (a) 0.5%, b) 1.0% and c) 2.5% polymer). The solution viscosity is included as well in c). These data are interpretable in terms of structure-breaking effects of SDS on 3-dimensional networks originally formed by the hydrophobically modified, end-capped PEO. From ref. 65, with permission of The Royal Society of Chemistry.

E.   Studies on Polymer-Surfactant Systems in Non-Aqueous Solution

In aqueous solution, some polymer-surfactant systems will form insoluble precipitates that cannot be studied by any of the methods described so far. If a non-aqueous solvent can be found, one can then search for specific interaction patterns in the "dissolved complex" by standard high-resolution techniques. However, hydrophobic interaction is of course no longer operative, and ionic interactions in particular will similarly differ substantially between aqueous and non-aqueous systems.

Pools of water in inverse micelles of AOT/iso-octane/water-type systems and similar constitute a borderline case. Considerable interest has focused on protein systems in such pools, as it is known that enzymatic activity is often preserved. Systems studied include: a) amphiphilic polysulfonates, containing 1-adamantyl or 1-naphtyl methyl groups and the double-chain surfactant didodecyl dimethylammonium bromide (DDAB) in common organic solvents. These studies were based on standard proton NMR, NOESY and spin relaxation measurements [86]; b) POP-PEO-POP-type copolymers in Water/iso-octane/AOT-type microemulsions, using the FT-PGSE self-diffusion approach. [68]; c) albumin and α-chymotrypsin in inverted micelles of AOT or Brij-96 type, using 13-C T_1-data along the surfactant chains [87-89]; and d) two-dimensional proton NOESY an 13C T_1-studies on prolyl-tyrosyl-glycinamide in reverse AOT micelles [90].

F.   Studies on Polymer-Surfactant Systems in Liquid Crystals or in the Solid State

This family of studies relies on so-called "solids NMR" techniques, that can probe molecular order and dynamics in a lattice, and may be sensitive to the proximity between magnetically active nuclei. Application of normal high-resolution NMR techniques will not normally result in a detectable spectrum of a solid; nor is solids capability a standard feature of any common NMR spectrometer configuration. Solids-NMR is also often unresolved in the normal sense, due to line-broadening effects of different origin, some of which can be suppressed by various experimental procedures. Line-broadening and unfavorable spin-relaxation characteristics, as compared to the situation for liquid samples most often lead to a low experimental solids-NMR sensitivity, despite an intrinsically very high concentration of material.

Proton-based solids NMR work is also quite difficult for a number of fundamental reasons. Deuterium labelling at a specific location of a component molecule is a fairly common practice, however. This allows selective experiments at a quite high level of sensitivity and reasonable ease of interpretation [1]. Another popular experimental approach is based on carbon-13 CP/MAS NMR (cross-polarization/magic angle spinning), where different carbon-13 signals can be monitored at a reasonably high level of resolution.

Studies include: a) drug release-related investigations of PEO/Griseofulvin and SDS, DOTAB or poly(oxyethylene) dodecyl ether (Brij) in the solid state, using carbon-13 CP/MAS NMR to probe the local interactions and the structure of the solid solutions [91,92]; b) the liquid crystalline system PEO/cesium pentadecafluoro-octanoate (CsPFO)/water, using deuterium NMR bandshape information to probe phase structure changes with composition [93]; c) poly(styrene sulfonate) with the counterions octadecyltrimethylammonium or tetradecyltrimethylammonium, using carbon-13 CP/MAS, and two other more advanced solid-state NMR techniques, for the purpose of probing the structure and local dynamics of such compounds [94]; d) sodium poly(styrene sulfonate) in aqueous solution that precipitates upon addition and binding to the single or mixed systems of deuterium-labelled surfactants hexadecyl phosphocholine, and CTAB [36,95]. Formation of the complex was monitored through a gradual disappearance of the solution deuterium NMR signal. The structure of the precipitated material was subsequently target to investigations by deuterium solids-NMR methods.

VI. WHAT CAN WE LEARN FROM NMR-STUDIES OF POLYMER-SURFACTANT SYSTEMS?

NMR is well-resolved, in the notion that individual groups in a molecule normally provide separate spectra that can be assigned to that particular moiety. In principle, this spectrum thus mimics the local conditions of that particular group with regard to intramolecular and intermolecular interaction. It may also be possible to monitor the same group by several separate experiments. If needed, one can for example utilize data on more than one nucleus (*e.g.* protons, deuterium and carbon-13, in the case of a -CH_2-group, for example), and/or select one or more "NMR experiments" that are designed to be sensitive to some target molecular properties. Therefore, specific interaction patterns between polymers and surfactants on the molecular level normally become easily amenable for study at a good level of confidence and detail. Such information is not readily available from other experimental sources than NMR.

As stated in the Introduction section, some NMR experiments also do provide detailed and quantitative dynamic information on the molecular level, which is not available from any other experimental technique. The detail with regard to spin relaxation-based molecular dynamics information becomes very much enhanced by a multi-nuclear approach, and even more so by using more than one magnetic field in the experimental spin relaxation investigation. The experimental spin relaxation tools are readily available in most NMR-equipped laboratories. A good understanding of NMR-experimental design and proper data evaluation procedures are essential for any meaningful efforts in this area, however [1].

NMR-based self-diffusion measurements by the FT-PGSE technique [1,2] form the most popular experimental approach today. For the mere determination of polymer-surfactant binding isotherms there are "traditional" methods that often work much better and offer higher precision. Examples are equilibrium dialysis, conductivity studies and electrochemical measurements, as based on ion selective electrodes that are sensitive to the free surfactant activity in solution. A large number of such investigations are summarized in various chapters of a recent comprehensive review [96] and also in other chapters of the present monograph. It should be pointed out, however, that using the NMR self-diffusion approach may be far less cumbersome and time-consuming than the techniques mentioned. Being component-selective, it can in general also be applied to more complex systems. Also, one should note that the self-diffusion approach provides concentration-related information rather than the activity-related one that will result from the macroscopic "thermodynamic" methods mentioned.

Provided that the self-diffusion of the polymer part can be experimentally monitored, the FT-PGSE NMR-based approach can provide complementary dynamic and structural information that will not be accessible from conventional binding-isotherm determinations. Upon formation of large aggregates or 3-dimensional networks in solution, polymer inter-aggregate exchange rates will normally become slow on the pertinent NMR time-scale (milliseconds). Originally, experimental manifestations of this unusual exchange condition were misinterpreted as being due to polydispersity of the polymer material itself [97]. Under these slow-exchange conditions one must instead account for the *aggregate polydispersity* in solution, as well as for the concentration of free surfactant (*i.e.* the actual binding isotherm data) when interpreting the underlying FT-PGSE NMR experiment. Of course, this has been overlooked in the past and has so added systematic errors to a large number of FT-PGSE-based studies of polymer-surfactant diffusion. For a mere free/bound binding isotherm determination, the errors introduced may be marginal. On the other hand, completely erroneous conclusions may have been drawn when inferring more complex surfactant binding schemes while using a aggregate diffusion model which is too simple.

It is quite evident that a completely new type of combined structural/dynamic detail is inferable from such "polydisperse" self-diffusion data (see *e.g.* refs. 32, 65, 75, 78, 82, 84, 98-100), at the expense (and need) of increased experimental performance and evaluational complexity. Unfortunately, an error was introduced in the first paper in the above series [98], which has since propagated in the literature, and also been cited in a previous review [1]. Polydisperse self-diffusion in most of these PGSE-based experiment has conveniently been modelled using a Kohlrausch-Williams-Watts (KWW-type) stretched exponential distribution of diffusion coefficients [101], to provide a mean self-diffusion coefficient and a measure of the width of the distribution of self-diffusion coefficients (the β-value). While the basic analysis procedure should be correct, the inverse nature of an application to diffusion coefficients in relation to the original use of the KWW-distribution (for an analysis of correlation times) was overlooked. The resulting mean self-diffusion coefficient value from such an analysis is therefore not based on an average over D-values, but rather on an average over 1/D-values, which is clearly not the same quantity (we are grateful to Prof. P.T. Callaghan for pointing this out to us).

VII. FUTURE PROSPECTS

The molecular details of polymer-surfactant interaction are now open for investigation in greater detail, using a large family of modern pulsed NMR techniques that were originally designed for structural determination of biological macromolecules. In contrast to conditions for very dynamic objects like normal surfactant micelles where intermolecular interactions are very non-localized on the average, prospects should be considerably better for more rigidly clutched polymer-surfactant and protein-surfactant structural moieties in solution. Very few studies of this kind have so far appeared, however. A continued and extended application of FT-PGSE NMR-based studies of multi-component self-diffusion is foreseeable. This is primarily prompted by the interesting new possibility of studying aggregate self-diffusion under conditions of slow inter-aggregate polymer exchange, as discussed in the preceding section, and under any potentially significant non-Gaussian diffusion conditions, as discussed in the context of Eq. (6) above. Instrumental developments make measurements much easier than previously, and widen the experimental range quite extensively. This is essential in order to provide the necessary experimental information needed to allow conclusions on more complex interaction conditions than previously considered. Such studies could

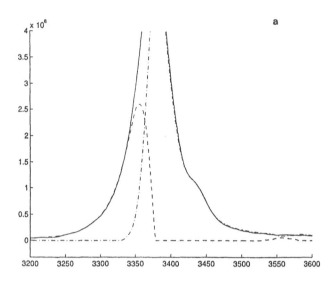

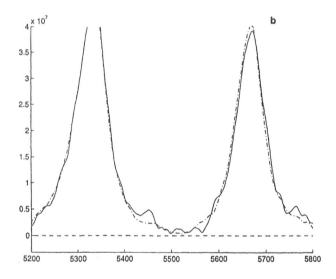

FIG. 5   The component bandshapes that result from a CORE analysis of the experimental data in Figure 2. a) Composite PEO-C12E5 band at 3.6 ppm; b) C12E5 band at 1 ppm. The x-axis corresponds to the frequency channel number from left to right (out of 8192) rather than the chemical shift value given in Figure 2. The y-axis gives the signal intensity, in arbitrary spectrometer units. Solid lines: spectral data of trace #1 in the data set; dashed line: PEO bandshape; dash-dotted line: C12E5 bandshape. Note the total impossibility to unravel the PEO-wing to the composite bandshape in a) by traditional peak intensity or peak integral monitoring.

prove very rewarding, with regard to new information about the aggregate microdynamics.

Recent advances with regard to data processing of FT-PGSE NMR data sets will definitely also be of great service here [85,102,103]. The so-called COmponent-REsolved (CORE) data analysis approach [85,102] in particular, allows a straightforward and confident application of FT-PGSE to very complex systems. Results of a CORE processing of the FT-PGSE data in Figure 2 is illustrated in Figures 5a and 5b.

CORE-processing also provides an effective Signal/Noise increase of a factor of 10 or so in comparison with previous methods of data evaluation, at no extra cost with regard to experimental time. The data processing also results in a complete unravelling of highly overlapping component bandshapes, without any assumptions whatsoever with regard to their actual form, as seen in Figures 5a and 5b. These valuable properties of the CORE analysis do seem to make it completely essential in future studies of polymer-surfactant and protein-surfactant systems by the FT-PGSE technique. Applying traditional evaluation methods to the data of Figure 2, for example, would not have lead to any meaningful results for the PEO diffusion rate. Figure 6 illustrates the power of the method even more clearly - the system studied contains gelatin, SDS and a non-ionic surfactant in heavy water, and the composite spectrum is very complex and overlapping. By pure "brute-force" 3-component data processing of the FT-PGSE data set, and no assumptions about component NMR bandshapes whatsoever, an almost non-random residuals map results. The very obvious non-random feature is an HDO spike in the first trace, where the water signal was not yet suppressed by the magnetic field gradient attenuation effect. Had the processing been based on 4 components instead of 3, that signal would have been accounted for as well, of course.

Another new development, named GRAM (the Generalized Rank Annihilation Method) [103] can be used for similar purposes. It is based an elegant simplification of a previously suggested multivariate FT-PGSE data processing approach [104]. GRAM-processing is very much less computer-intensive than CORE-processing, but is only applicable for a separation of purely exponential decay patterns. In the present form it is therefore inapplicable for FT-PGSE studies of polymer-surfactant aggregate diffusion under the prevailing conditions of slow polymer inter-aggregate exchange.

ACKNOWLEDGEMENTS

The author wishes thank all former collaborators in this field for stimulating discussions, and also the Swedish Natural Science Research Council (NFR) and the Carl Trygger Foundation for generous financial support.

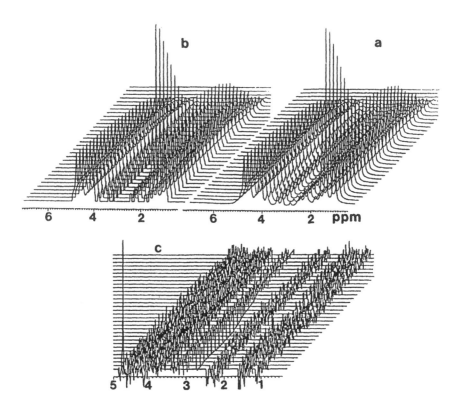

FIG. 6 a) Experimental and b) CORE-processed FT-PGSE data and c) residuals map on a gelatin/SDS/non-ionic surfactant system in heavy water, as fitted to a 3 components model without assumptions about the underlying component bandshapes. The residuals map reveals a slight trace of the HDO signal, which was not completely suppressed in trace #1. The component bandshapes and even closely similar diffusion coefficients are confidently found through such a CORE analysis (c.f. Figures 2 and 5), but will not be reproduced here for this system (P.C. Griffiths and P. Stilbs, unpublished results).

REFERENCES

1. O. Söderman and P. Stilbs, *Prog. Nucl. Magn. Reson. Spectrosc.* *26*: 445 (1994).
2. P. Stilbs, *Prog. Nucl. Magn. Reson. Spectrosc.* *19*: 1 (1987).
3. Z. Gao and J.C.T. Kwak, in *Surfactants in Solution*, (K.L. Mittal and D.O. Shah, eds.), Volume 11, Plenum Press, New York. 1991;p. 261-75.
4. J. Oakes, *J. Chem. Soc. Faraday Trans.1* *70*: 2200 (1974).
5. N. Muller and R.J. Mead, *Biochemistry 12*: 3831 (1973).
6. M.L. Smith and N. Muller, *J. Colloid Interface Sci.* *52*: 507 (1975).
7. B. Cabane, *J. Phys. Chem.* *81*: 1639 (1977).
8. Z. Gao, J.C.T. Kwak, and R.E. Wasylishen, *J. Colloid Interface Sci.* *126*: 371 (1988).
9. Z. Gao, R.E. Wasylishen, and J.C.T. Kwak, *Macromolecules 22*: 2544 (1989).
10. P. Stilbs, *J. Colloid Interface Sci.* *87*: 385 (1982).
11. Z. Gao, R.E. Wasylishen, and J.C.T. Kwak, *J. Phys. Chem.* *94*: 773 (1990).
12. Z. Gao, R.E. Wasylishen, and J.C.T. Kwak, *J. Phys. Chem.* *93*: 2190 (1989).
13. Z. Gao, J.C.T. Kwak, R. Labonte, D.G. Marangoni, and R.E. Wasylishen, *Colloids Surf.* *45*: 269 (1990).
14. Z. Gao, R.E. Wasylishen, and J.C.T. Kwak, *J. Phys. Chem.* *95*: 462 (1991).
15. A. Carlsson, G. Karlström, and B. Lindman, *J. Phys. Chem.* *93*: 3673 (1989).
16. E.O. Stejskal and J.E. Tanner, *J. Chem. Phys.* *42*: 288 (1965).
17. E.L. Hahn, *Phys. Rev. 80*: 580 (1950).
18. A.D. Chen, D.H. Wu, and C.S. Johnson, Jr. *J. Phys. Chem.* *99*: 828 (1995).
19. Z. Gao, R.E. Wasylishen, and J.C.T. Kwak, *J. Colloid Interface Sci.* *137*: 137 (1990).
20. Y. Inoue, N. Teraoka, Y. Suzuki, and R. Chujo, *Makromol. Chem.* *182*: 1819 (1981).
21. P. Bahadur, N.V. Sastry, Y.K. Rao, and G. Riess, *Colloids Surf.* *29*: 343 (1988).
22. K. Chari and W.C. Lenhart, *J. Colloid Interface Sci.* *137*: 204 (1990).
23. K. Chari, *J. Colloid Interface Sci.* *151*: 294 (1992).
24. M. Almgren, J. van Stam, C. Lindblad, P. Li, P. Stilbs, and P. Bahadur, *J. Phys. Chem. 95*: 5677 (1991).
25. R. Ramachandran and G.J. Kennedy, *Colloids Surf.* *54*: 261 (1991).
26. A. Hammarström and L.O. Sundelöf, *Colloid Polymer Sci. 271*: 1129 (1993).
27. J.J. Effing, I.J. McLennan, and J.C.T. Kwak, *J. Phys. Chem. 98*: 2499 (1994).
28. J.J. Effing, I.J. McLennan, N.M. van Os, and J.C.T. Kwak, *J. Phys. Chem. 98*: 12397 (1994).
29. K. Tsujii and T. Takagi, *J. Biochem. (Tokyo) 77*: 511 (1975).
30. D.D. Miller, W. Lenhart, B.J. Antalek, A.J. Williams, and J.M. Hewitt, *Langmuir. 10*: 68 (1994).
31. D.J. Lundberg, E. Fossum, and J.E. Glass, *Polym. Mater. Sci. Eng. 45*: 269 (1990).
32. S. Abrahmsén-Alami and P. Stilbs, *J. Phys. Chem. 98*: 6359 (1994).
33. M.N. Jones, *J. Colloid Interface Sci. 23*: 36 (1967).

34. T.C. Wong, K. Thalberg, B. Lindman, and H. Gracz, *J. Phys. Chem. 95*: 8850 (1991).
35. T.C. Wong, C.S. Liu, C.D. Poon, and D. Kwoh, *Langmuir 8*: 460 (1992).
36. P.M. Macdonald, D. Staring, and Y. Yue, *Langmuir 9*: 381 (1993).
37. K. Chari, B. Antalek, M.Y. Lin, and S.K. Sinha, *J. Chem. Phys. 100*: 5294 (1994).
38. B. Antalek and K. Chari, *Mod. Phys. Lett. B 9*: 1555 (1995).
39. N.J. Turro, X.-G. Lei, K.P. Ananthapadmanabhan, and M. Aronson, *Langmuir 11*: 2525 (1995).
40. W. Brown and P. Stilbs, *Chem. Scr. 19*: 161 (1982).
41. W. Brown, P. Stilbs, and T. Lindström, *J. Appl. Polym. Sci. 29*: 823 (1984).
42. D. Richter, L.J. Fetters, J.S. Huang, B. Farago, and B. Ewen, *J. Non-Cryst. Solids 131*: 604 (1991).
43. D. Richter, L. Willner, A. Zirkel, B. Farago, L.J. Fetters, and J.S. Huang, *Macromolecules. 27*: 7437 (1994).
44. M. Monkenbusch, D. Schneiders, D. Richter, B. Farago, L. Fetters, and J. Huang, *Physica B (Amsterdam) 213&214*: 707 (1995).
45. B. Jönsson, H. Wennerström, P.G. Nilsson, and P. Linse, *Colloid Polymer Sci. 264*: 77 (1986).
46. H. Johannesson and B. Halle, *J. Chem. Phys. 104*: 6807 (1996).
47. J. Klein, *Contemp. Phys. 20*: 611 (1979).
48. J. Klein, *Macromolecules 19*: 105 (1995).
49. J.S. Vrentas, J.L. Duda, and L.W. Ni, *Macromolecules 16*: 261 (1983).
50. M. Tirrell, *Rubber Chem. Technol. 57*: 523 (1984).
51. J.M. Petit, B. Roux, X.X. Zhu, and P.M. Macdonald, *Macromolecules 29*: 6031 (1996).
52. J.S. Vrentas, C.M. Vrentas, and N. Faridi, *Macromolecules 29*: 3272 (1996).
53. J.M. Zielinski, *Macromolecules 29*: 6044 (1996).
54. E.D. Von Meerwall, J. Grigsby, D. Tomich, and R. van Antwerp, *J. Polym. Sci., Polym. Phys. Ed. 20*: 1037 (1982).
55. V.A. Sevreugin, V.D. Skirda, and A.I. Maklakov, *Polymer 27*: 290 (1986).
56. M.C. Piton, R.G. Gilbert, B.E. Chapman, and P.W. Kuchel, *Macromolecules. 26*: 4472 (1993).
57. R.A. Waggoner, F.D. Blum, and J.M.D. Macelroy, *Macromolecules. 26*: 6841 (1993).
58. A. Bandis, P.T. Inglefield, A.A. Jones, and W.-Y. Wen, *J. Polym. Sci. 33*: 1495 (1995).
59. A. Bandis, P.T. Inglefield, A.A. Jones, and W.-Y. Wen, *J. Polym. Sci. 33*: 1505 (1995).
60. A. Bandis, P.T. Inglefield, A.A. Jones, and W.-Y. Wen, *J. Polym. Sci. 33*: 1515 (1995).
61. T. Cosgrove and P.C. Griffiths, *Polymer 36*: 3335 (1995).
62. T. Cosgrove and P.C. Griffiths, *Polymer 36*: 3343 (1995).
63. G.E. Uhlenbeck and L.C. Ornstein, *Phys. Rev. 36*: 832 (1930).
64. P.T. Callaghan and J. Stepišnik, *Adv. Magn. Opt. Reson. 19*: 325 (1996).

65. K. Persson, G. Wang, and G. Olofsson, *J. Chem. Soc. Faraday. Trans. 90*: 3555 (1994).
66. K. Thalberg, J. Van Stam, C. Lindblad, M. Almgren, and B. Lindman, *J. Phys. Chem. 95*: 8975 (1991).
67. M.H.G.M. Penders, S. Nilsson, L. Piculell, and B. Lindman, *J. Phys. Chem. 97*: 11332 (1993).
68. G. Fleischer, F. Stieber, U. Hofmeier, and H.F. Eicke, *Langmuir 10*: 1780 (1994).
69. K.F. Morris, C.S. Johnson, Jr., and T.C. Wong, *J. Phys. Chem. 98*: 603 (1994).
70. K. Zhang, M. Jonströmer, and B. Lindman, *J. Phys. Chem. 98*: 2459 (1994).
71. K. Chari, B. Antalek, and J. Minter, *Phys. Rev. Lett. 74*: 3624 (1995).
72. K. Veggeland and S. Nilsson, *Langmuir 11*: 1885 (1995).
73. R. Zana, W. Binana-Limbele, N. Kamenka, and B. Lindman, *J. Phys. Chem. 96*: 5461 (1992).
74. N. Kamenka, I. Burgaud, R. Zana, and B. Lindman, *J. Phys. Chem. 98*: 6785 (1994).
75. H. Walderhaug, B. Nyström, F.K. Hansen, and B. Lindman, *J. Phys. Chem. 99*: 4672 (1995).
76. H. Walderhaug, B. Nyström, F.K. Hansen, and B. Lindman, *Prog. Colloid Polym. Sci. 98*: 51 (1995).
77. K. Thuresson, O. Söderman, P. Hansson, and G. Wang, *J. Phys. Chem. 100*: 4909 (1996).
78. P. Hansson and B. Lindman, *Curr. Opin. Colloid Interface Sci. 1*: 604 (1996).
79. K. Zhang, B. Xu, M.A. Winnik, and P.M. Macdonald, *J. Phys. Chem. 100*: 9834 (1996).
80. K.W. Zhang, M. Jonströmer, and B. Lindman, *Colloids Surf. A 87*: 133 (1994).
81. M. Björling, Å. Herslöf-Björling, and P. Stilbs, *Macromolecules 28*: 6970 (1995).
82. P.C. Griffiths, P. Stilbs, A.M. Howe, and T. Cosgrove, *Langmuir 12*: 2884 (1996).
83. P.C. Griffiths, P. Stilbs, A.M. Howe, and T.H. Whitesides, *Langmuir 12*: 5302 (1996).
84. K. Persson, P.C. Griffiths, and P. Stilbs, *Polymer 37*: 253 (1996).
85. P. Stilbs, K. Paulsen, and P.C. Griffiths, *J. Phys. Chem. 100*: 8180 (1996).
86. Y. Morishima, M. Seki, S. Nomura, and M. Kamachi, *Proc. Jpn. Acad. Ser. B 69*: 83 (1993).
87. Y.E. Shapiro, N.A. Budanov, A.V. Levashov, N.L. Klyachko, Y.L. Khmelnitskii, and K. Martinek, *Collect. Czech. Chem. Commun. 54*: 1126 (1989).
88. Y.E. Shapiro, F.G. Pykhteeva, A.V. Levashov, and N.L. Klyachko, *Biol. Mem. 7*: 379 (1994).
89. Y.E. Shapiro, V.Y. Gorbatyuk, A.V. Levashov, and N.L. Klyachko, *Biol. Mem. 7*: 277 (1994).
90. Y.E. Shapiro, V.Y. Gorbatyuk, A.A. Mazurov, and S.A. Andronati, *Analyst 119*: 647 (1994).
91. M. Aldén, J. Tegenfeldt, and E. Sjökvist, *Int. J. Pharm. 83*: 7 (1992).

92. M. Aldén, J. Tegenfeldt, and E.S. Saers, *Int. J. Pharm.* *94*: 31 (1993).
93. A.M. Smith and M.C. Holmes, *Mol. Cryst. Liq. Cryst. Sci. Technol., Sect. A 260*: 605 (1995).
94. M. Antonietti, D. Radloff, U. Wiesner, and H.W. Spiess, *Macromol. Chem. Phys.* *197*: 2713 (1996).
95. D.J. Semchyschyn, M.A. Carbone, and P.M. Macdonald, *Langmuir 12*: 253 (1996).
96. Goddard ED and Anathapadmanabhan KP, (Eds.). *Interactions of Surfactants with Polymers and Proteins*. CRC Press: Boca Raton, Fla.: 1993.
97. K. Persson, S. Abrahmsén, P. Stilbs, F.K. Hansen, and H. Walderhaug, *Colloid Polymer Sci. 270*: 465 (1992).
98. H. Walderhaug, F.K. Hansen, S. Abrahmsén, K. Persson, and P. Stilbs, *J. Phys. Chem. 97*: 8336 (1993).
99. K. Persson, *Associative Polymers. Their Self-Association and Interaction with Surfactants*. Thesis, Royal Institute of Technology, Stockholm; 1995.
100. S. Abrahmsén-Alami, *Associative Polymers. Hydrophobically End-capped Poly(ethylene oxide). Aggregation, Solution Structure and Transport Dynamics*. Thesis, Royal Institute of Technology; 1996.
101. G. Williams and D.C. Watts, *Trans. Faraday Soc. 66*: 80 (1970).
102. P. Stilbs and K. Paulsen, *Rev. Sci. Instrum. 67*: 4380 (1996).
103. B. Antalek and W. Windig, *J. Am. Chem. Soc. 118*: 10331 (1996).
104. D. Schulze and P. Stilbs, *J. Magn. Reson. Ser. A. 105*: 54 (1993).

7

Fluorescence Methods in the Study of Polymer-Surfactant Systems

FRANÇOISE M. WINNIK and SUDARSHI T.A. REGISMOND
Department of Chemistry, McMaster University, 1280 Main St. W.,
Hamilton ON, L8S 4M1, Canada

I.   INTRODUCTION AND SCOPE

In this review we present the application of fluorescence techniques to study polymer/surfactant systems, discussing the various methods and materials in use. In the first section we consider the use of low molecular weight fluorescence probes to measure the critical aggregation concentration (cac) and the surfactant aggregation number (N_{agg}) of polymer-surfactant complexes and to characterize the microenvironment within the complexes. The next section deals exclusively with the association of fluorescently labelled polymers and surfactants. The final part of the review is dedicated to a description of selected recent investigations, with particular emphasis on the study of the complexes formed between surfactants and hydrophobically-modified polymers.

   The use of fluorescent probes and labels to obtain information about micellar environments is rather new [1]. The technique was applied first to gain a measure of the effective polarity of the probe microenvironment. The most common probe, pyrene, an extremely hydrophobic molecule of limited solubility in water, tends to partition into hydrophobic microdomains in water, such as surfactant micelles or polymer/surfactant aggregates. In 1984 Turro and coworkers reported a series of studies of polyvinylpyrrolidone/sodium dodecylsulfate (PVP/SDS) and polyethylene oxide/SDS (PEO/SDS) systems with pyrene as a fluorescence probe [2]. This work received the attention of physical chemists studying polymer/surfactant interactions since it outlined the basic principles of a simple yet powerful tool. At present, pyrene fluorescence is a well accepted method for determining critical aggregation concentration and aggregation numbers in polymer/surfactant systems. Abuin and Scaiano made an important early contribution to the study of charged polymer/surfactant systems by fluorescence [3]. Using the fluorescent probe 8-anilino-1-naphthalene sulfonic acid (ANS) as a sensor of "free" cationics in solution, they obtained direct information on the extent of binding of cationic surfactants to polystyrenesulfonate (PSS). In their extensive study of the cationic cellulosic Polymer JR/SDS system, Goddard and coworkers applied the fluorescence technique to determine the micropolarity of the micellar aggregates in the immediate pre- and post-precipitation zones [4]. Pyrene aldehyde was chosen as a gauge of the medium polarity. Overall, through these three studies the fluorescence technique gained credibility, inciting researchers to test the scope of the method in a wide range of systems.

Photochemical studies of polymer-surfactant interactions require that the system under scrutiny contains a luminescent dye which may be free to diffuse in solution (termed "probe" in this review) or attached via a covalent bond to the polymer or to the surfactant (called "label"). Luminescent probes can include fluorescent and phosphorescent dyes. The latter have not been used widely to study polymer/surfactant systems and are not included in this review. Fluorescent probes are used to determine 1) the cac of polymer/surfactant systems; 2) the aggregation number (N_{agg}) of polymer/surfactant complexes and 3) the microenvironment (polarity and viscosity) within the complexes.

A substantial body of research has been devoted to the photophysics and photochemistry in micellar environments. It has been the subject of several excellent reviews published over the last decade [5-10]. Hence, this review will be limited to technical and theoretical points of direct relevance to the study of polymer/surfactant systems. In Section 2, we review briefly the probe methods used to detect the occurrence of interactions between polymers and surfactants. Theories describing the behavior of fluorescent dyes and the main assumptions involved will be recalled succinctly as necessary. Section 3 is dedicated to the fluorescence label technique with emphasis on the preparation of labelled polymers and on the spectroscopic parameters commonly employed. Section 4 presents a review of the application of fluorescence in studies of polymer/surfactant interactions. Systems studied prior to 1991 have been compiled in a previous review [11] and hence will be omitted here, except if pertinent to the discussion. Also excluded are systems consisting of polyacids and cationic surfactants which are described by Zana in Chapter 10 of this volume. The photophysical data are evaluated in the context of models of the interactions and their conclusions are evaluated briefly against those originating from other experimental techniques.

II. THE STUDY OF POLYMER-SURFACTANT INTERACTIONS WITH FLUORESCENT PROBES

The environment immediately surrounding a molecule often determines its basic properties, such as solubility or optical response. As a corollary, it is possible to use the spectroscopy of "well-behaved" probes to detect the formation of microenvironments within a macroscopically-homogeneous solution and to correlate changes in photophysical parameters as a function of various stimuli to structural modifications within the self-assemblies. The majority of fluorescence probes are aromatic dyes which are preferentially solubilized within hydrophobic microdomains. While the precise location of a dye can never be ascertained unambiguously, excellent guidelines exist for a very limited number of probes.

A.   Characterization of the Physicochemical Properties of Polymer-Surfactant Assemblies

Fluorescence probe techniques have been used to detect the occurrence of polymer/surfactant interactions and to characterize the polymer/surfactant aggregates in terms of apparent microviscosity (η_{app}), and micropolarity, often expressed in terms of effective dielectric constant (ε_{eff}). Extracting absolute values for η_{app} or ε_{eff} is usually fraught with difficulties and great care should be taken when trying to quantify these values, and in comparing them with those of solvents. However, valuable information can be extracted from the changes of a given photophysical property upon progressive modification of a solution by various additives. The measured changes can then be used to infer changes in the immediate probe environment. Structures of common fluorescence probes are shown in Figure 1.

1.   Micropolarity

Changes in micropolarity are the basis of most photophysical techniques applied to detect the onset of polymer/surfactant interactions. The probes must exhibit in their emission spectrum some feature strongly dependent on solvent polarity. This property may be a medium-induced variation in the relative intensity of emission bands, in the wavelength of maximum emission, or in the fluorescence quantum yield. Illustrations of each effect are presented in the following section which reviews the photophysics of the three most common fluorescence probes, i.e. pyrene (Py), pyrene-1-carboxaldehyde (PCA), and 1-anilinonaphthalenesulfonate (ANS).

Pyrene (Py).   The key feature in the photophysical properties of Py is a solvent-induced variation in the ratio of emission band intensities [12]. These variations have been correlated with the polarity of the probe's immediate environment, yielding the empirical "py scale" of solvent polarity. This scale has been catalogued for a wide variety of experimental conditions, ranging from bulk liquids [13] to surfactant micelles and interfacial monolayers [8,9]. Despite the wide use of Py as a probe, there is still no general mechanistic understanding of the molecular processes responsible for the correlation between emission band intensity ratios and solvent polarity. A recent study [14] relating the polarity-dependent fluorescence response of Py to vibronic coupling has brought about a deeper understanding of the origin of the effect. Readers are referred to it for excellent background information on the photophysical processes involved, a subject beyond the scope of this review.

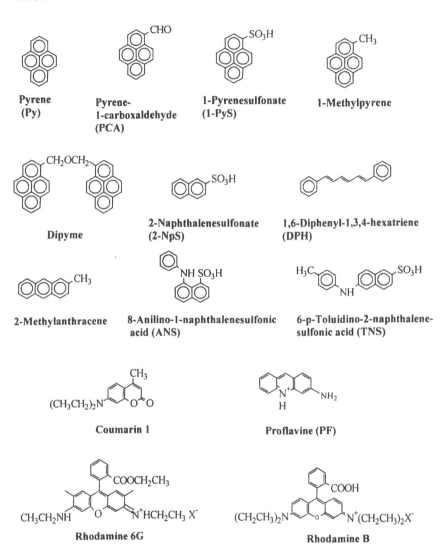

FIG. 1 Chemical structures of fluorescence probes discussed in this review.

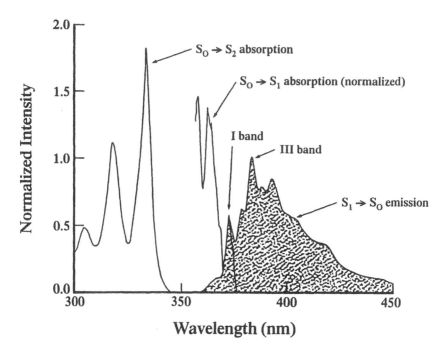

FIG. 2 Absorption and emission spectra of pyrene in cyclohexane. The (0-0) band in the absorption spectrum is normalized to the (0-0) emission transition. From ref. 14 with permission.

In practice, the solvent polarity dependence of the Py emission is expressed in terms of the ratio, I_1/I_3 (or its inverse I_3/I_1) of the intensities, I_1 and I_3, of the bands I and III corresponding to the $S_1^{v=0} \rightarrow S_0^{v=0}$ (0-0) and the $S_1^{v=0} \rightarrow S_0^{v=1}$ transitions, respectively, where S_1 and S_0 are the first singlet excited state and ground state of pyrene, respectively (Figure 2). The values typically range from about 1.9 in polar solvents (1.87 in water) to about 0.6 in hydrocarbons. The cmc and cac of surfactants can be obtained from measurements of the changes in I_1/I_3 as a function of surfactant concentration. The ratio I_1/I_3 decreases sharply at the onset of micelle formation, reflecting the preferential solubilization of pyrene in a hydrophobic environment. The effect is illustrated for polymer/surfactant systems in Figure 3 which presents the changes of the ratio I_1/I_3 in the emission of Py (*ca.* 3×10^{-7} M) in an aqueous solution of PNIPAM-C10 (1.87 g L^{-1}), a function of SDS and CTAC concentrations [15]. The ratio has a value of 1.78 in the absence of surfactant. The pyrene, in this case, indicates a polarity slightly lower than that of water, implying that the probe is partly solubilized in hydrophobic regions of the polymer. Addition

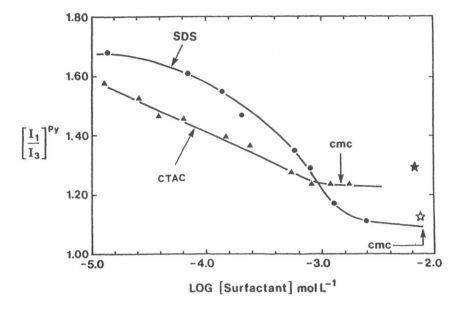

FIG. 3 Changes in the I_1/I_3 ratio of pyrene ($\sim 3 \times 10^{-7}$ M) emission in aqueous solutions of PNIPAM-C10 (1.87 gL^{-1}) as a function of surfactant concentration (logarithmic scale) for CTAC and SDS. Values of I_1/I_3 for pyrene in SDS and CTAC micelles are indicated by an open and full star, respectively. From ref. 48 with permission.

of surfactant begins to affect I_1/I_3 at [SDS] $\sim 1 \times 10^{-4}$ M. The cac of HPC/SDS (5.6 $\times 10^{-4}$ M) is about 15 times smaller than the cmc of SDS (indicated by an arrow in Figure 3). The value of I_1/I_3 at high [SDS] (1.05) is similar to that measured for Py in SDS micelles. Other polycyclic aromatic compounds which have been investigated as probes of micellar systems include phenanthrene and triphenylene [16]. To date they have not been used to investigate polymer/surfactant aggregates.

Pyrene-1-carboxaldehyde (PCA). The useful feature in the spectroscopy of PCA in solution is the dependence of its fluorescence on the dielectric permittivity (ε) of the solvent [17]. It has been quantified by an empirical linear relationship between the wavelength of maximum emission (λ_{max}^{em}) and ε. A study of the emission properties of PCA in a wide range of solvents suggests, however, that the dependence is much more complex [18]. Micelles formed in aqueous solutions of PCA (the concentration of the saturated solution is $\sim 10^{-6}$ M) readily incorporate the probe. The change in PCA microenvironment is detected by a blue-shift of the emission maximum, from 473 nm in water to 448 nm in the micellar core. Due to the presence of the polar carboxaldehyde group, PCA resides near the micelle-water interface and hence the fluorescence shifts can be interpreted in terms of the micropolarity of the micelle-water interfaces, in contrast to the information obtained from the spectroscopy of Py, incorporated within the palisade layer of micelles.

1-Anilinonaphthalene-8-sulfonate (ANS).   The quantum yield of fluorescence of ANS, Φ_f, defined as the ratio of the number of emitted photons to the number of absorbed photons, exhibits a marked dependence on the polarity of the solvating medium [19]. Many other factors may influence Φ_f as well. In the most general terms it can be written as:

$$\Phi_f = \frac{k_f}{k_f + k_q[Q] + k_{IC} + k_{ISC} + k_p + k_{ET}}$$

where [Q] is the concentration of any quencher present, k_f, k_q, k_{IC}, k_{ISC}, k_p and k_{ET} are the rate constants for fluorescence, quenching, internal conversion, intersystem crossing, photochemical reaction in the excited state, and energy transfer, respectively [20]. The increased fluorescence quantum yield of ANS in a micellar environment is a very sensitive way of detecting surfactant aggregation [21].

2.   Microviscosity

Fluorescence techniques for microviscosity determinations involve the analysis of some kind of probe molecular motion: i) rotational diffusion of a chromophore; ii) lateral diffusion of two independent probes;   iii) conformational transition in a molecule with two photochemically-independent chromophores, and iv) internal rotational torsion in a probe molecule. A parameter characteristic of the motion is determined experimentally.   Then, the effective viscosity (η_{eff}) of the probe microenvironment is estimated by comparing the measured parameter with values obtained in a series of reference solvents.   A wide selection of dyes and methods have been employed to estimate η_{eff} in the case of surfactant micelles and phospholipid bilayers [8, 9], but there are only a few reports in the case of polymer/surfactant systems. They are based on either fluorescence depolarization or intramolecular excimer formation.

a.   Fluorescence depolarization
For a chromophore to absorb light, a component of its transition moment must be parallel to the electric vector of the incident light. As a consequence, irradiating a sample of randomly oriented molecules with plane-polarized light results in an optical selection of which molecules are excited. If the motion of these molecules is slow compared to the timescale of fluorescence, the emitted light will also be polarized, and its degree of polarization will be related to the extent of motion of the

molecule. In this way, fluorescence depolarization can be used to determine the apparent microviscosity of a medium containing a probe. In practice, a steady-state experiment is carried out by measuring the fluorescence intensities detected through polarizers oriented parallel ($I_\parallel$) and perpendicular ($I_\perp$) to the plane of polarization of the excitation beam. From these, one calculates either the steady-state polarization parameter P or, more meaningfully, the emission anisotropy r, defined in Eq. (1):

$$P = \frac{I_\parallel - I_\perp}{I_\parallel + I_\perp} \qquad r = \frac{I_\parallel - I_\perp}{I_\parallel + 2I_\perp} \tag{1}$$

To interpret the results, one normally treats the probe as a spherical rotor, so that its motion can be described in terms of a single correlation time. It is generally assumed that the rotational diffusion of the fluorophore is the only significant process which causes fluorescence depolarization. Data are then analyzed in terms of a modified version of the Perrin equation [22], (Eq. 2),

$$\frac{r}{r_0} = \frac{\frac{1}{P} - \frac{1}{3}}{\frac{1}{P_0} - \frac{1}{3}} = 1 + \frac{kT\tau}{\eta V_0} \tag{2}$$

where P_0 and r_0 are the limiting values of the polarization and emission anisotropy obtained in the absence of rotational motion, τ is the average lifetime of the fluorophore excited state, k the Boltzmann constant, T the absolute temperature, and V_0 the effective molecular volume of the probe. It is possible to estimate the effective viscosity η_{eff} of the probe microenvironment from calibration curves of r_0/r vs. $T\tau/\eta$ in a series of reference solvents. The assumptions underlying this data treatment have been discussed in detail elsewhere [8,9]. Well characterized chromophores useful in depolarization studies of micellar aggregates include diphenylhexatriene, perylene, 2-methylanthracene, and octadecyl rhodamine B (Figure 1) [23]. Only a few applications of fluorescence depolarization in the area of polymer/surfactant interactions have been reported to date. They include studies of the interactions of SDS with polysoaps [24] and poly-(N-isopropylacrylamide) [25].

b.   Intramolecular excimer formation
Bis(1-pyrenylmethyl) ether (dipyme, Figure 1) is a hydrophobic probe that exhibits intramolecular pyrene excimer emission (intensity I_E) in competition with fluorescence from isolated excited Py monomer emission, (intensity I_M) [26, 27]. Excimer formation requires a change in the conformation of the molecule (Figure 4). The extent of excimer emission depends upon the rate of conformational change,

the motion being resisted by the local friction imposed by the environment. As a consequence, the excimer-to-monomer intensity ratio, I_E/I_M, provides a measure of the effective viscosity sensed by the probe (the larger the ratio, the less viscous is the environment where dipyme is located). Representative emission spectra are shown in Figure 5 for dipyme solubilized in n-octyl β-D-thioglucopyranoside (OTG) micelles and in hydrophobic microdomains formed in water by a copolymer of N-isopropylacrylamide and N-n-octadecylacrylamide. The emission spectrum of dipyme within OTG micelles presents a large contribution from excimer (I_E/I_M = 0.88), while for dipyme in polymeric micelles the ratio takes a value of 0.11. This striking difference is a strong indication that the polymeric micelles (low I_E/I_M) are significantly more rigid than the OTG micelles. In mixed solutions of the polymer and OTG the contribution of the excimer is enhanced (I_E/I_M = 0.60), reflecting the increased fluidity of the mixed micelles, compared to the polymeric micelles. The vibrational fine structure in the dipyme monomer emission, quantified as $[I_1/I_3]^{DP}$ is sensitive to the polarity of the probe environment. The $[I_1/I_3]^{DP}$ ratios for the probe in the two situations in the example cited are typical for dipyme in hydrophobic environments. Other dichromophoric probes used to estimate the microviscosity of micellar systems include 1,3-di(1-pyrenylpropane) [28], 1,6-di(1-pyrenylhexane) [29], and 1,3-di(1-naphthylpropane) [30]. These do not exhibit polarity-induced spectral changes, hence the information they provide is limited to microviscosity. Their application in the area of polymer/surfactant interactions has not been reported as yet.

B.   Determination of Mean Aggregation Numbers

Fluorescence probe methods are among the most reliable for the measurements of micellar aggregation numbers, offering many advantages over classical techniques, such as light and X-ray scattering or gel permeation chromatography. They are based on the quenching of emission from a fluorescent probe (F) by a quencher (Q) added in increasing amounts to a solution containing a known amount of surfactant and F. Both F and Q are chosen to have a high affinity for the micelles. In the data analysis, one assumes a Poisson distribution of F and Q among the micelles. Thus, some micelles contain F but not Q, some contain F and Q and may have various levels of multiple occupancy. Quenching occurs only in micelles containing both F and Q. Steady-state measurements and time-resolved experiments can be used to determine aggregation numbers. The reader is referred to Chapter 10 of this volume, where both techniques are described in detail.

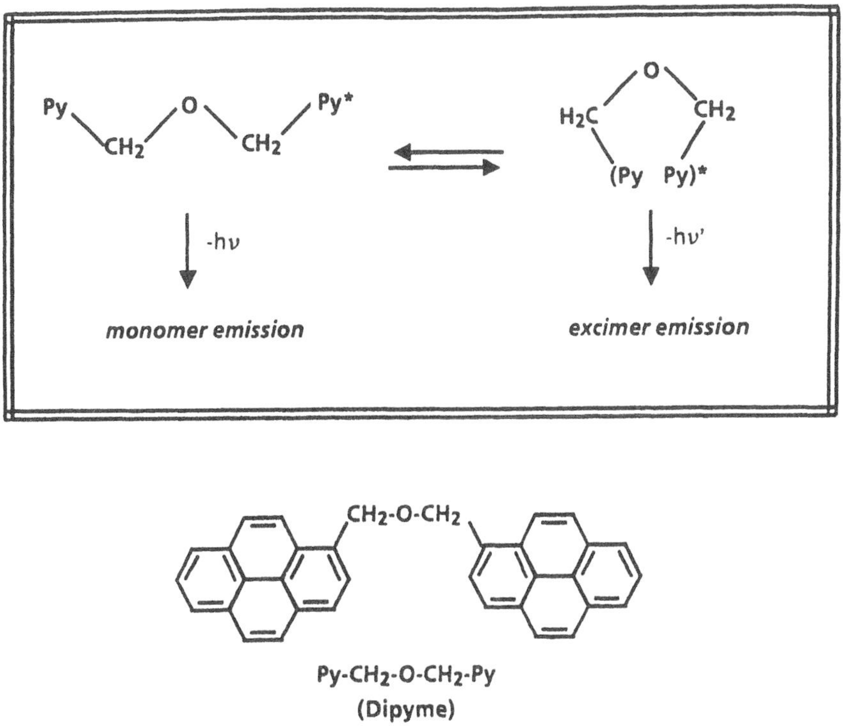

Py-CH$_2$-O-CH$_2$-Py
(Dipyme)

FIG. 4   Representation of intermolecular excimer formation in dipyme [bis(1-pyrenylmethyl)ether].

C.   Other Fluorescence Techniques

The application of non-radiative energy transfer (NRET) to probe interpolymeric association and interactions was pioneered by Nagata and Morawetz [31]. The process of NRET between two chromophores originates in dipole-dipole interactions between an energy donor in its excited state and an acceptor in its ground state [32]. The probability of energy transfer between two chromophores depends sensitively on their separation distance and, to a lesser extent, on their orientation. The distance dependence of the energy transfer efficiency, E, is a well defined function of the distance R between the donor and the acceptor as shown in Eq. (3):

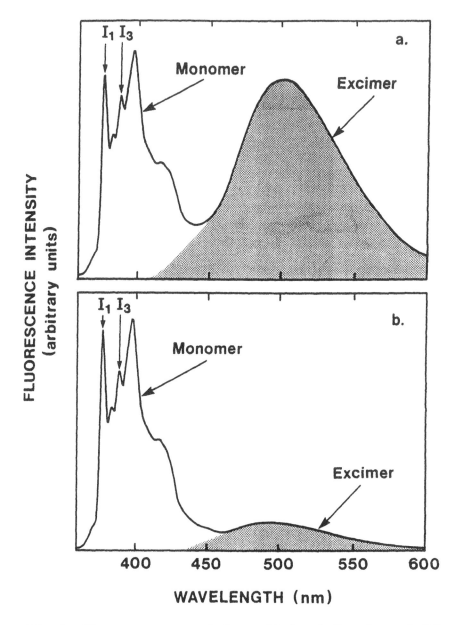

FIG. 5 Fluorescence spectra of dipyme (a) in micelles of n-octyl β-D-thioglucopyranoside ([OTG] = 1.67 x 10^{-2} M) and (b) in an aqueous solution of PNIPAM-C_{18}/100 (1.87 g L^{-1}); 20°C,λ_{exc} = 348 nm; [dipyme]= 2 x 10^{-7} M. From ref. 81 with permission.

$$E = \frac{R_0^6}{R_0^6 + R^6} \qquad (3)$$

where R_0 for rotationally averaged pairs is the donor/acceptor distance for which 50% of energy transfer takes place [33]. In the case of small molecules in fluid solution, NRET takes place when the two interacting chromophores diffuse towards each other during the lifetime of the donor. When the donors and acceptors are attached to a polymer chain, the distribution of donor/acceptor separations is related to polymer chain dimension and conformation. In mixed solutions of polymers carrying either donor or acceptor chromophores, the extent of NRET between the two labels can be related to the extent of interpolymeric association. The application of the NRET process to study polymer/surfactant interactions has been quite limited to date. The few examples presented in section IV of this chapter may generate ideas for future applications.

III. THE STUDY OF POLYMER-SURFACTANT INTERACTIONS WITH FLUORESCENTLY-LABELLED POLYMERS

A. Preparation of Fluorescently-Labelled Polymers

For studies of most industrially relevant polymers, the labelling technique of choice is to introduce a fluorophore by chemical modification of commercial or natural polymers. Alternatively, labelled polymers can be obtained by adding a dye-substituted co-monomer during the polymerization. A third approach is to prepare a polymer carrying at random a small number of "reactive" functional groups, which are reacted with an appropriate dye in a subsequent step.

1. Polymer postmodification

Protocols for the chemical modification of water-soluble polymers fall into two broad categories: 1) "statistical chemistry," which involves the use of non-selective chemical reagents for random derivatization; and 2) specific modifications, accomplished by reacting a functionalized dye with polymer sites having intrinsic selective chemical reactivities (e.g. COOH, NH_2). This strategy is often useful for the modification of polysaccharides. The reagents commonly used in random derivatization include strong alkali and acylating or alkylating agents. These can trigger partial degradation of the polymer. Nonetheless, a large number of polymers have been labelled without significant perturbation of their molecular weight and molecular weight distribution. From the gamut of reactions used to modify water-soluble polymers, the following methods have been applied to prepare fluorescently labelled polymers:

Ether formation: This reaction is applicable if the polymer to be labelled has free hydroxyl groups and if it is stable in strong base. The polymer is treated with a base (NaH, Na_2CO_3) to generate alkoxides, and subsequently reacted with a dye bearing a primary alkyl halide or a tosylate group. The technique has been employed to label polyethylene oxide [34] and polysaccharides, such as hydroxypropyl cellulose [35], tylose [36] and methyl cellulose [36] (Figure 6).

Esterification: The reaction has been employed to label polysoaps [23] and polyethylene oxide [37]. The resulting ester linkage between the polymer and the label is hydrolyzed easily in alkaline medium, a circumstance that may prove limiting in certain applications.

Amide formation: Amides may be formed by nucleophilic substitution of esters or anhydrides with suitable amines, as in the case of copolymers of maleic anhydride and alkyl vinyl ethers [38]. Amide linkages can also be prepared by carbodiimide-mediated coupling of amines to carboxylic acid groups in organic or mixed aqueous/organic media, as demonstrated in the labelling of alginic acid with nitroxide probes for EPR analysis [39].

2.   Postmodification of reactive polymers

This technique is useful in the preparation of polymers having no functionality readily amenable to reacting with available dye derivatives. The idea is to prepare a polymer carrying at random a small number of "reactive" functional groups. These are sites where fluorescent groups can be introduced later. The method offers several advantages over direct copolymerization of a dye-substituted co-monomer: 1) the free-radical polymerization does not have to be performed in the presence of chromophores which, like many aromatic compounds, often act as chain-transfer agents; 2) the polymerization and subsequent polymer modification can be carried out in homogeneous solutions and in the absence of surfactants, contrary to the case of the micellar polymerization of water-soluble polymers; and 3) from a single reactive polymer it is possible to prepare polymers, identical with respect to their molecular characteristics, but labelled with different chromophores or various amounts of the same chromophore. Examples of labelling by post modification of a reactive polymer include the preparation of pyrene- and naphthalene-labelled poly-(N-isopropylacrylamides) [40], obtained by reaction of 4-(1-pyrenyl)butylamine and/or 1-aminoethyl-1-naphthalene with a copolymer of N-isopropylacrylamide and N-(acryloxy)succinimide. The techniques are modifications of reactions reported earlier for the modification of PNIPAM with proteins and other biomolecules [41].

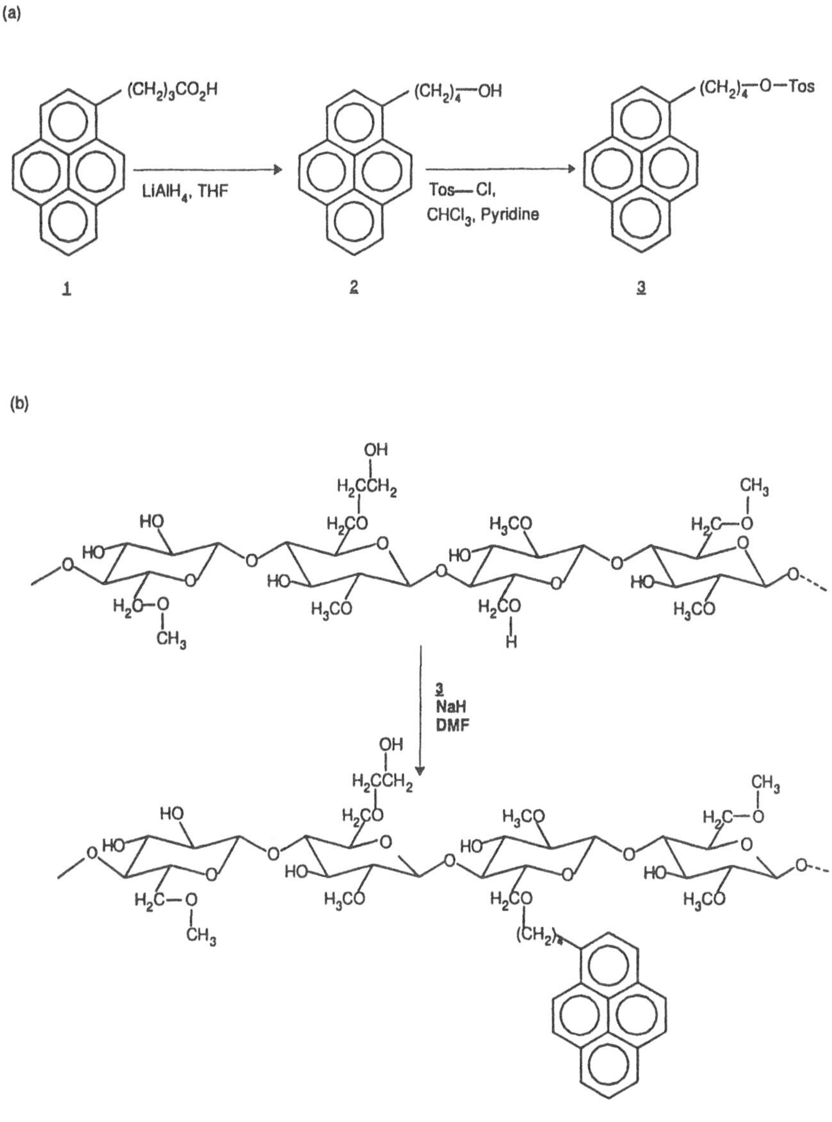

(a)

(b)

FIG. 6 Synthetic scheme for the post-modification by ether formation: preparation of pyrene-labelled tylose. From ref. 36 with permission.

3. Copolymerization of a dye-substituted monomer

Most labelled polyelectrolytes employed in polymer/surfactant studies were prepared
by this method. They include (Figure 7) a) pyrene-labelled polyacrylic acids,
obtained by free-radical copolymerization of acrylic acid and a comonomer, such as
2-[[4-(1-pyrenyl)butanoyl]-amino]propenoic acid [42], (1-pyrenyl)acrylic acid

PAA-Py
Mn = 40,000

$$-(-CH_2-CH-)_{98.5}-(-CH_2-\underset{NH}{\overset{CO_2H}{\underset{|}{\overset{|}{C}}}}-)_{1.5}-$$

with CO_2 substituent, and amide linkage:

$$O=\overset{(CH_2)_3}{\underset{Py}{}}$$

PAA-vPy
Mn = 45,000, Mw/Mn = 2

$$-(-CH_2-\underset{CO_2H}{\overset{|}{CH}}-)_{99.5}-(-CH_2-\underset{Py}{\overset{|}{CH}}-)_{0.5}-$$

PAMSA-Py

$$-(-CH_2-CH-)_n-(-CH_2-CH-)_m-$$
$$O=\overset{|}{C} \qquad O=\overset{|}{C}$$
$$\overset{|}{NH} \qquad \overset{|}{NH}$$
$$(CH_2)_{10} \qquad (CH_2)_2$$
$$CO_2Na \qquad NHSO_2Py$$

P(A/Ad/Py)AMPS
Mn 11,000

Ad: adamantyl

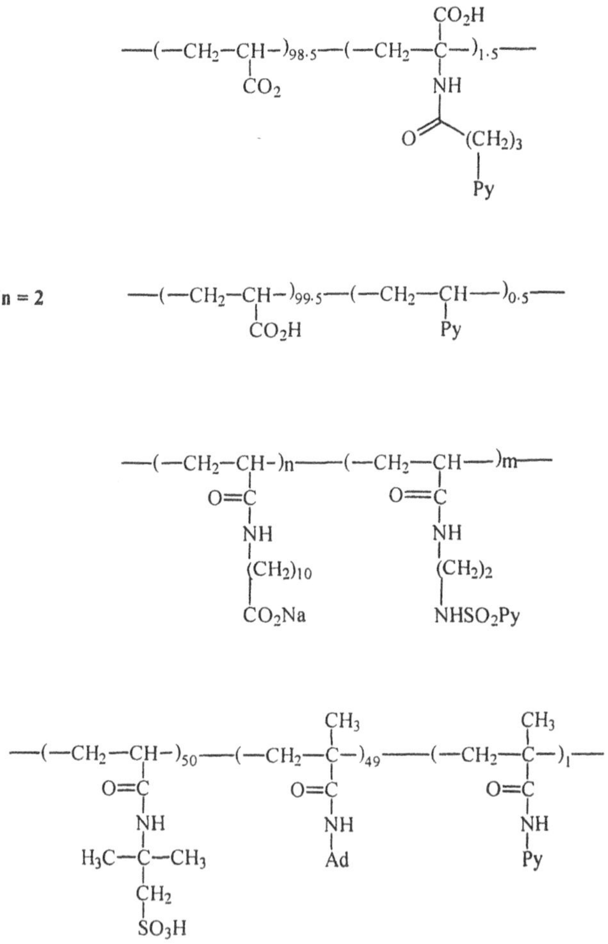

$$-(-CH_2-CH-)_{50}-(-CH_2-\overset{CH_3}{\underset{|}{\overset{|}{C}}}-)_{49}-(-CH_2-\overset{CH_3}{\underset{|}{\overset{|}{C}}}-)_1-$$
$$O=\overset{|}{C} \qquad O=\overset{|}{C} \qquad O=\overset{|}{C}$$
$$\overset{|}{NH} \qquad \overset{|}{NH} \qquad \overset{|}{NH}$$
$$H_3C-\overset{|}{C}-CH_3 \qquad Ad \qquad Py$$
$$\overset{|}{CH_2}$$
$$\overset{|}{SO_3H}$$

FIG. 7 Chemical structures of the labelled polyelectrolytes discussed in the text.

(PAA-Py) [43], or 1-vinylpyrene (PAA-vPy) [44]; b) various amphiphilic polyelectrolytes, such as terpolymers of sodium 2-(acrylamido) 2-methylpropane sulfonate, N-(1-adamantyl)methacrylamide, N-(1-naphthylmethyl)-methacrylamide, and/or N-(1-pyrenylmethyl)methacrylamide [P(A/Ad/Py)AMPS] [45]; and c) copolymers of sodium 11-(acrylamido)undecanoate acrylamide, and [2-(1-pyrenylsulfonamido)ethyl]acrylamide (PAMSA-Py) [46]. The methacrylamide comonomers used by Morishima *et al.* [45] were polymerized in N,N-dimethylformamide, a solvent for all comonomers. Under these conditions the methacrylamides employed provided "ideal" copolymerization systems: the copolymer composition was found to be proportional to the monomer composition in the feed, allowing the preparation of completely random terpolymers of well-defined composition. In contrast, the preparation of labelled amphiphilic acrylamides is more difficult, as pointed out by Kramer *et al.* [46] In order to overcome solubility problems they had to perform the polymerization in the presence of a surfactant (SDS) under conditions known as "micellar polymerization" [47]. This technique often leads to a non-random incorporation of the co-monomers, as a result of the inherent heterogeneity of the medium. Thus, the labelled copolymers may possess a "blocky" microstructure. The presence along the polymer backbone of sequences of hydrophobic substituents and/or fluorescent labels presumably affects the interactions of the copolymer with surfactants, but the importance of the effect is hard to appreciate.

The synthesis of labelled neutral water-soluble copolymers can often be performed in an organic solvent, such as dioxane or tetrahydrofuran, in which all comonomers are soluble. Problems associated with incomplete dissolution of the comonomers are therefore prevented. The label distribution is random and the copolymer composition is dictated by the initial monomer feed. Examples of labelled neutral aqueous polymers include pyrene- and/or naphthalene-labelled poly-(N-isopropylacrylamides) (PNIPAM-Py, PNIPAM-C$_{18}$Py, Figure 8) [48].

B. Spectroscopic Parameters for the Study of Labelled Polymer-Surfactant Systems

Variables monitored in studies of the interactions of surfactants with labelled polymers include: quantum yields of fluorescence, fluorescence depolarization, pyrene excimer emission and the fine structure of pyrene monomer emission. The techniques are similar to those used in probe experiments; there are, however, some key differences between the photophysics of dyes free to diffuse in solution and those of chromophores attached to macromolecules. These will be reviewed briefly here.

PNIPAM-Py
Mv = 1.1 x 10⁶

$$-(-CH_2-CH-)_{200}-(-CH_2-CH-)_{\overline{1}}-$$

with $C=O$, NH, CH (H_3C, CH_3) on the first repeat unit and $C=O$, NH, $(CH_2)_4$, Py on the second.

PNIPAM-C₁₈-Py
Mv = 390,000

$$-(-CH_2-CH-)_{200}-(-CH_2-CH-)_{\overline{1}}-$$

with $C=O$, NH, CH (H_3C, CH_3) on the first repeat unit and $C=O$, N, $(CH_2)_4$ / $(CH_2)_{17}$, Py / CH_3 on the second.

PyO₂C-PEO-CO₂Py
Mn = 4,800 Mw/Mn < 1.10

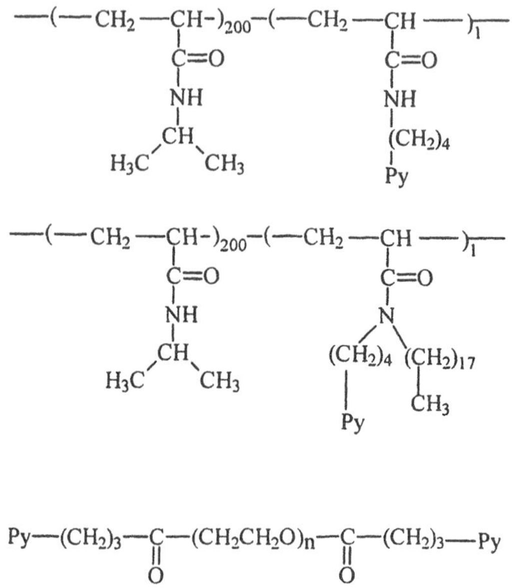

$$Py-(CH_2)_3-\underset{O}{\overset{\|}{C}}-(CH_2CH_2O)_n-\underset{O}{\overset{\|}{C}}-(CH_2)_3-Py$$

FIG. 8 Chemical structures of neutral polymers discussed in the text.

1. Fluorescence quantum yield

The dansyl (1-dimethylaminonaphthalene-5-sulfonyl) group is sensitive to the polarity of its environment, as discussed in detail in the case of the analogous probe ANS. The response of the dansyl label to its microenvironment has been exploited by U. P. Strauss in his extensive studies of polysoaps [49].

2. Pyrene excimer emission

The ratio of pyrene excimer to monomer emission intensities has been used to follow the interactions of surfactants with labelled polyelectrolytes, neutral water-soluble polymers, and hydrophobically-modified polymers.   In many of these systems excimers originate from pre-formed pyrene dimers or higher aggregates, rather than from the usual dynamic mechanism involving encounter of an excited pyrene with a pyrene in its ground state.   The occurrence of pyrene aggregates in solution can be established from observations such as: 1) hypochromism and wavelength shifts in the UV absorption spectra; 2) differences in the excitation spectra monitored for the excimer and for the monomer; and 3) the absence of a rising component in the time dependent excimer emission measured in the nanosecond timescale [50].

3. Pyrene monomer emission

The emission spectrum of substituted pyrenes usually lacks the micropolarity sensitivity characteristic of the pyrene emission.   There are a few specific examples of 1-substituted pyrene groups which exhibit changes in their emission as a function of solvent polarity, albeit not as pronounced as those presented by pyrene itself.   The N-(1-pyrenyl)methylamido group employed by Morishima *et al.* [45] is one such chromophore (see P(A/Ad/Py)AMPS, Figure 7).

4. Fluorescence depolarization

Labels covalently attached to macromolecules usually do not satisfy the criteria of the Stokes-Einstein equation, which would permit calculation of the microviscosity of the probe environment from a knowledge of the parameters r_0, r, V_0, and t in Eqs (1) and (2).   It is still meaningful to follow trends in the fluorescence anisotropy as a function of labelled polymer composition and of the relative amounts of polymer and surfactant.   This was demonstrated in the elegant work of McGlade and Olufs who investigated the interactions of SDS with various copolymers of α-olefins and maleic acid, labelled with 9-hydroxymethylanthracene [24].   The anisotropy ratio can be used to calculate the rotational diffusion coefficient of a label, D_r, using the Perrin-Webber equation:

$$\frac{r}{r_0} = 1+\frac{kT\tau}{V_0\eta} = 1+6D_r\tau \tag{4}$$

The changes in D_r as a function of added surfactant can give indications on the mobility of the label as it experiences changes in its microenvironment.   The technique has been applied extensively to study polymers in solution as well as gels [51], but not to investigations of polymer-surfactant systems.

C.   Polymer-Labelled Surfactant Systems

Polymer/surfactant interactions may be monitored by fluorescence not only with extrinsic probes or labelled polymers, but also by using fluorescent surfactants. This strategy has not been employed frequently, since few such surfactants are available readily.   Schild and Tirrell have reported the use of sodium 2-(N-dodecylamino)naphthalene-6-sulfonate (SDNpS) to study the interactions between ionic surfactants and poly(N-isopropylacrylamide) [52]. The sodium salt of this probe exhibits a blue shift in the wavelength of maximum emission with decreasing polarity (from 430 nm in water to 408 nm in 1-butanol).   When incorporated within SDS micelles, SDNpS exhibits a fluorescence with maximum intensity at 422 nm. The similarity in structure of SDNpS with sodium n-alkyl sulfates coupled with the environmental sensitivity of the naphthyl sulfonate emission enables one to monitor closely the mechanism of adsorption of a surfactant to a polymer chain.   The underlying assumption that alkyl sulfonates can be modelled by probes such as SDNpS remains to be studied in detail.

D.   Labelled Polymers and Labelled Surfactants: a Word of Caution

As pointed out in the previous section, in any study of labelled materials, concern always arises about whether or not the label disturbs the system under investigation. Probes are believed to be relatively unobtrusive at the low concentrations (usually < 10^{-6} M) required for fluorescence measurements. This is not necessarily the case for labels. Aromatic chromophores are by nature hydrophobic, and the attachment of such groups to a water-soluble polymer converts the substrate into a "hydrophobically-modified" (amphiphilic) polymer. From the body of evidence on this important class of materials gathered in recent years, it is clear that even minute amounts of hydrophobic substituents on an aqueous polymer may have large consequences on its solution properties. In a recent note, Morawetz reviews the case of poly(methacrylic acid) in water [53].   He concludes his historical perspective by pointing out that, while models based on fluorescence labelling studies are convincing for the labelled polymer, they "cannot be used to infer the state of the unlabelled PMA."

IV. RECENT APPLICATIONS OF FLUORESCENCE MEASUREMENTS IN STUDIES OF POLYMER-SURFACTANT INTERACTIONS

A. Neutral Polymers-Ionic Surfactants

The association between nonionic polymers, such as poly(ethylene oxide), PEO, or poly(vinylpyrrolidone), PVP, and SDS has been studied by a variety of techniques. The exact nature of the interaction between PEO and SDS still remains open to debate. At issue are the location of polymer segments with regard to the micelle and the nature of the attractive forces. The aggregation numbers (N_{agg}) of micelles bound to the   polymers were determined by steady-state and time-resolved fluorescence measurements. A few key results include: 1) N_{agg} is independent of the molecular weight of PEO (from 6,000 to 100,000); 2) N_{agg} increases with salt concentration; 3) N_{agg} increases with increasing SDS concentration and with decreasing polymer concentration; and 4) the cac is independent of polymer concentration.

Kamenka et al. [54] have reported a comparative study of the interactions of two anionic surfactants, SDS and copper didodecylsulfate [$Cu(DS)_2$] with PEO and PVP, in order to test the influence of counterion charge on polymer/surfactant interactions. This is an important issue to be solved in order to explain the differences in the association of polymers with cationic and anionic surfactants. The ratio I_1/I_3 of pyrene solubilized in $Cu(DS)_2$ micelles and SDS micelles was found to be 1.12 and 1.21, respectively, implying that the microenvironment of pyrene is less polar in $Cu(DS)_2$ micelles than in SDS micelles. The intensity of pyrene fluorescence in $Cu(DS)_2$ micelles is significantly lower than in SDS micelles, as a result of the known quenching effect of Cu^{2+} [55]. When PVP or PEO was added to $Cu(DS)_2$ micelles, the ratio I_1/I_3 increased as the polymer concentration increased (up to 10 g L^{-1}), indicating that the pyrene microenvironment is more polar in the mixed micelles than in free micelles. The effect of PVP was larger than that of PEO, suggesting a stronger interaction. The overall emission intensity also increased with polymer concentration, especially in the case of PVP, suggesting that the PVP-bound aggregates are more ionized than free micelles, thus alleviating the quenching effect of Cu^{2+}.

Maltesh and Somasundaran have conducted extensive studies of the interactions between PEO and SDS. Most of their early studies were carried out at concentrations such that SDS micelles and polymer-bound SDS aggregates co-existed in solution, thus only average properties of the two species could be obtained [56]. Recently, they probed the effects of different cations on the size and number of SDS/PEO aggregates, under conditions of dilute polymer and surfactant concentrations, where there were no unbound SDS micelles [57]. They used the dynamics of Py excimer formation to determine the aggregation numbers in the presence of cations of decreasing affinity for PEO, viz. $Cs^+ > Na^+ > Li^+ > Mg^{2+}$ [58].

The SDS aggregates were large in the presence of Li^+ (N_{agg} = 48) and Mg^{2+} (N_{agg} = 51), and much smaller (N_{agg} = 22) in the presence of Cs^+ and Na^+. The number of aggregates per polymer chain was found to decrease with increasing ionic strength, but it was not dependent on the type of salt used. The authors concluded that the size of SDS aggregates confined in strict geometrical limits by a polymer chain is controlled by the nature of the interactions, rather than by the number of sites available.

A cationic derivative of poly(ethylene glycol), was prepared recently by Koyama and co-workers [59]. The polymer, which carries approximately 15 mol % of pendant amino groups linked at random along the main chain, was shown to form complexes with the anionic surfactant lauric acid. A fluorescence study using Py as a microenvironment probe indicated that the cac of this system occurs at a surfactant concentration of 7×10^{-4} M, a concentration much lower than the cmc of sodium laurate (20×10^{-3} M). Similar experiments were carried out with a poly(ethylene glycol) derivative having terminal amino groups at both ends. This polymer was shown to interact with lauric acid as well, but with a cac of 5×10^{-3} M in the case of polymer/surfactant solutions containing the same amine group concentration as in the measurements performed with the PEG derivative having amino-pendant groups. From the value of the ratio I_1 /I_3 of Py emission in the mixed aggregates it can be concluded that the lauric acid/polymer aggregates are less polar in the case of the PEG derivative with pendant amino-groups, compared to the amino-terminated PEG.

Changes in polymer conformation upon association with SDS were monitored by following the response of the fluorescence of a pyrene end-labelled sample (PyO_2C-PEO-CO_2Py, Figure 8). Studies of both labelled and unlabelled polymers (pyrene probe) pointed to the fact that, in this specific case (PEO, MW 5,000), the perturbations due to pyrene labelling were negligible. Initial binding of SDS ([SDS] = 5×10^{-5} M) causes an increase in Py excimer emission as a result of a contraction of the PyO_2C-PEO-CO_2Py chain, whereas saturation of the polymer by SDS causes the PyO_2C-PEO-CO_2Py chain to stretch out, with concomitant increase in Py monomer emission intensity relative to excimer emission. Using different samples of PyO_2C-PEO-CO_2Py, both Quina et al. [60] and Hu et al. [61] detected a lower cac, compared to unlabelled polymer. This observation was taken as a sign that a PEO chain labelled with two pyrene groups is in fact a model of a hydrophobically-modified polymer, rather than a true representation of PEO.

A recent study by means of static and dynamic fluorescence measurements of the interactions between SDS and fully hydrolyzed poly(vinyl alcohol) (PVA) or PEO indicates that the interaction starts at a critical aggregation concentration below the cmc both at 20 °C and at 40 °C[62]. The aggregation numbers are affected in the same way by the two polymers at the lowest SDS concentrations, but they increase more steeply with surfactant concentration in the case of PVA. This is attributed to the reduced flexibility of PVA compared to PEO.

The interactions of perfluorinated surfactants with several neutral polymers are the focus of several current studies, prompted by the well-known greater surface activity of fluorocarbons compared to the hydrocarbon analogues. The application of fluorescence techniques to study these systems is limited by the poor compatibility of the typical probes with fluorocarbons. Both Thomas [63] and Asakawa et al. [64] have reported that the I_1/I_3 of Py in the presence of fluorocarbon micelles is the same as that measured in water, a result which led them to conclude that pyrene in micellar solutions of perfluorinated surfactants resides in the water phase and not within the micelles. Contradictory results were reported by Muto et al. [65] and by Kalyanasundaram [66], who observed a significant decrease in the I_1/I_3 ratio at perfluorinated surfactant concentrations above the cmc.

The same ambiguity exists in the limited number of reported investigations of polymer/perfluorinated surfactant systems. In a study of the interactions between perfluorooctanoic acid and PNIPAM, Schild and Tirrell [67] were unable to detect any change in the I_1/I_3 ratio of the pyrene emission with surfactant concentration. In contrast, a response of the pyrene emission as a function of solution composition was detected in two studies on the interactions between lithium perfluorooctane sulfonate (LiFOS) and PVP [68] or a copolymer of vinylpyrrolidone and vinylacetate [69]. A comparative study of the interactions of LiFOS and lithium dodecylsulfate (LiDS) with PVP was carried out in the presence of two fluorescence probes, Py and PCA. The I_1/I_3 ratio of pyrene in aqueous solutions of PVP decreased upon addition of both surfactants, indicating a decrease in the polarity sensed by the probe in the following order: H_2O > LiFOS micelles > PVP/LiFOS complexes > PVP/LiDS complexes > LiDS micelles. From the work of Muto et al. [65] it is known that in the case of simple LiFOS micelles, Py is solubilized near the LiFOS/water interface, close to the sulfonate head groups. Hence, the observations reported in the case of the PVP/LiFOS complexes would indicate that the main binding site for LiFOS to PVP involves the head group of the surfactant. The wavelength of maximum emission of PCA dissolved in PVP/LiFOS and PVP/LiDS is also sensitive to surfactant concentration: it undergoes a shift to shorter wavelengths, followed by a rapid shift to longer wavelength as the surfactant concentration exceeds the cmc. Effects of the copolymer composition on the micropolarity reported by Py were measured as well, in the absence of surfactant and with increasing concentration of both LiFOS and LiDS. The overall trends described for the PVP/surfactant systems were still apparent; however, further data are needed to clarify several aspects of the mechanism of complex formation.

The aggregation of surfactants and cellulose ethers continues to be of special interest in view of the industrial importance of this class of polymers. Zana et al. [70] had shown earlier that ethyl(hydroxyethyl)cellulose (EHEC) interacts with cationic surfactants, such as CTAX, (X= Cl, Br). They demonstrated that the polymer-bound micelles have a larger degree of ionization and a lower aggregation number than free micelles. The strength of the EHEC/surfactant interaction was shown to increase with temperature, as reflected by a decrease with increasing temperature of both cac and N_{agg}. Zana and coworkers proceeded to study the interactions of EHEC with an anionic surfactant (SDS) [71]. Fluorescence quenching measurements carried out by time resolved spectroscopy with the Py/DoPC pair demonstrated that with SDS, as was the case with cationic surfactants, the polymer-bound micelles are much smaller ($N_{agg} = 20$, at 20 °C) than free SDS micelles ($N_{agg} = 60$, at 20 °C). Compared to other neutral polymers, EHEC has a particularly pronounced effect on the size of the SDS aggregates, a reflection of the strength of the EHEC/SDS interaction. Thuresson and co-workers did a comparative study of the interactions between SDS and either EHEC or HM-EHEC, a hydrophobically modified derivative containing 1.7 mol % nonylphenol[72]. The aggregation numbers of the HM-EHEC/SDS complexes increase with increasing SDS, reaching a maximum value of 590 at 40 mM SDS. Quenching experiments of the Py emission with the hydrophobic quencher DoPC were performed for HM-EHEC/SDS and EHEC/SDS. At high surfactant concentrations, the quenching rate constant k_q is about the same in the two systems ($\sim 1.5 \times 10^7$ s^{-1}). This can be expected since the aggregation numbers are similar. At lower concentrations, however, there are substantial differences. In the micelles made up predominantly of HM-EHEC hydrophobic tails, the quenching rate is much smaller than in the micelles with high SDS content, and rather independent of the SDS concentration. As the fraction of SDS increases, k_q increases. At high SDS concentration, when the mixed aggregates consist mostly of SDS the quenching rate becomes less dependent on SDS concentration. For EHEC/SDS, k_q decreases monotonically with SDS concentration. The quencher escape frequency is nearly zero in most cases, indicating that the quencher remains associated with micelles during the lifetime of the probe.

The hydroxypropyl methyl cellulose (HPMC)/SDS system has been investigated over an extended composition interval from infinite dilution to a polymer concentration above c*, the critical overlap concentration [73]. Fluorescence measurements with the Py probe were used to monitor the onset of interactions and to determine N_{agg} of the bound clusters. SDS adsorbs in a cooperative mechanism to form clusters of N_{agg} ranging from 50 for solutions of 2.0 g L^{-1} HPMC to 20-25 for infinitely dilute solutions.

Low-molecular weight poly(propylene oxide) (PPO, MW < 1000) is soluble in water, with a cloud point of *ca.* 32 °C. It forms complexes with cationic surfactants, as demonstrated by Brackman and Engberts [74], Sierra and Rodenas [75], and DeSchryver and co-workers [76,77] in the case of CTAX (X= Cl$^-$, Br$^-$, and NO$_3^-$). Association is usually accompanied by an increase in the cloud point. Using the Py probe technique, in tandem with conductivity and viscosity measurements, it was established in all cases that i) the association onset, the cac, is lower than the cmc; ii) the bound micelles have aggregation numbers smaller than free micelles, and iii) the bound aggregates possess a higher micellar ionization degree. A detailed study by static and dynamic fluorescence suggests that the polymer chains are wrapped around the surfactant aggregates, with their hydrophobic segments penetrating into the aggregates Stern layer by displacing water molecules. In the case of the system PPO/SDS, quenching experiments with the Py/CPC pair revealed that the aggregation numbers of the mixed micelles increase sharply with increasing SDS concentration [78]. At a critical concentration, higher than the SDS cmc, the aggregates become large enough to incorporate the polymer chains into the surfactant micelles, forming SDS/PPG mixed micelles in aqueous solution, similar to the CTAB/PPG and surfactant medium chain alcohols mixed micelles.

Neutral water-soluble polymers containing a small number of long alkyl chains, often called hydrophobically-modified (HM) polymers, interact particularly strongly with surfactants. Since the contact between the hydrophobic groups and water is unfavorable, these polymers have a strong tendency to self-associate and/or to associate with surfactants. The rheology of these fluids has been investigated in great detail. In most cases progressive addition of surfactant gives rise to an increase in the viscosity of the solution, followed by a decrease at higher surfactant concentration. The enhancement in viscosity is ascribed to the formation of mixed micelles between the polymer alkyl substituents and the surfactant molecules, thus reinforcing interpolymeric cross-links. Upon further surfactant addition, the number of polymer alkyl chains in the mixed micelles decreases, reducing the number of interpolymeric bridges. Molecular aspects of the interactions have been examined by fluorescence spectroscopy in a few systems. Dualeh and Steiner have reported a study by fluorescence (depolarization) on the solution properties of a C_{12}-grafted hydroxyethylcellulose (HM-HEC) in the presence of SDS [79]. This polymer, undergoes strong association with SDS, in contrast with the "parent" polymer HEC [80]. Winnik et al. [81] reported a study of the interactions between surfactants and a series of hydrophobically-modified poly(N-isopropylacrylamides) (HM-PNIPAM), namely copolymers of NIPAM and N-n-alkylacrylamides (n-alkyl = decyl, tetradecyl, and octadecyl groups). Fluorescence probe experiments with Py and dipyme revealed a strong association between the copolymers and ionic surfactants (SDS, TTAC, CTAC). The association of surfactants and polymers results in the formation of mixed clusters consisting of the n-alkyl substituents of the polymers surrounded by surfactant molecules (ca. 30 per n-alkyl group in the case of the C_{12} and C_{14} surfactants and ca. 15 per n-alkyl chain for CTAC). The formation of mixed clusters, as detected by changes in the ratio I_1 / I_3 of the Py emission, occurred by a non-cooperative mechanism over a large range of surfactant concentrations. Experiments with analogous Py-labelled PNIPAM and Py-labelled HM-PNIPAM samples confirmed this conclusion [82]. Association of the surfactants to the polymers was detected by a decrease in the Py excimer emission intensity with concomitant increase in the Py monomer emission intensity (Figure 9). The association occurred at a critical concentration of surfactants below the cmc in the case of PNIPAM-Py. The hydrophobically-modified labelled polymers also

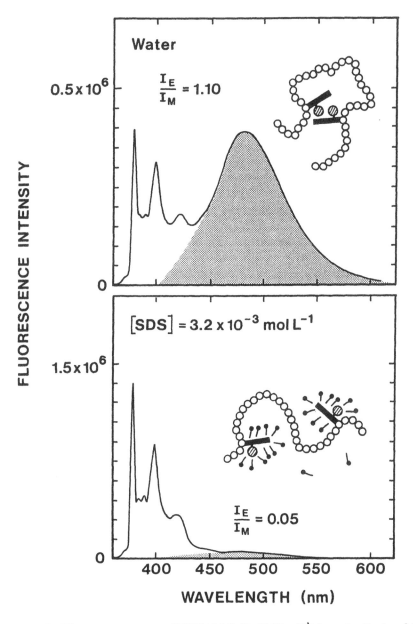

FIG. 9 Fluorescence spectra of PNIPAM-C$_{18}$Py (0.05 g L^{-1}) in water (top) and in the presence of SDS (3.2 x 10^{-3} M) (bottom); temperature: 20°C, λ_{em} = 330 nm. From ref. 82, with permission.

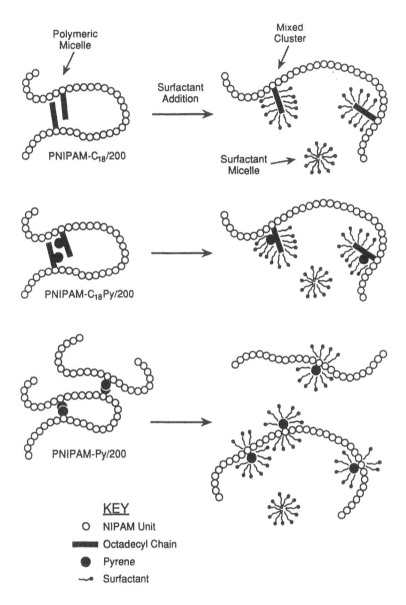

FIG. 10 Schematic representation of the interactions between surfactants and HM-PNIPAM, PNIPAM-C$_{18}$/Py, and PNIPAM-Py. The structures represented include (intramolecular) polymeric micelles and the polymer/surfactant mixed clusters formed in each case. From ref. 82 with permission.

underwent association with anionic and cationic surfactants, but via a non-cooperative mechanism.   A stylized illustration of the interactions between surfactants and HM-PNIPAM, PNIPAM-Py and Py-labelled HM-PNIPAM is presented in Figure 10.

Effing *et al.* [83] have examined the association of SDS and a series of copolymers of acrylamide and N-n-decylacrylamide using a battery of analytical techniques, including fluorescence measurements with the Py probe.   The copolymers in the absence of surfactants form micellar aggregates, which solubilize Py, as evidenced by values of I_1/I_3 *ca* 1.3 to 1.47 in 2% solutions of polymers with high and low decyl contents, respectively. A continuous small decrease from 1.47 to 1.27 in the ratio I_1/I_3 was observed upon addition of increasing amounts of SDS. Interestingly, this system exhibits associative phase separation in concentrated polymer/surfactant solutions although the polymers themselves do not exhibit a cloud point in aqueous solution.   Copolymers of acrylamide and the hydrophobic monomer N-(4-ethylphenyl)acrylamide also form mixed micelles in the presence of SDS [84] via a non-cooperative association process.

B.   Neutral Polymers-Nonionic Surfactants

Interactions between nonionic polymers and nonionic surfactants are usually very weak, being driven primarily by the tendency towards a reduction in free energy of the total system.   They can occur only between surfactants and sufficiently hydrophobic polymers. Extremely sensitive tools are needed to detect such weak interactions. Brackman *et al.* [85] have presented evidence of complex formation between poly(propylene oxide) (PPO) and n-octyl β D- thioglucopyranoside (OTG) on the basis of microcalorimetry and EPR probe measurements.   Fluorescence measurements with Py-labelled polymers also gave indications of association between hydroxypropyl cellulose (HPC) and nonionic surfactants, such as OTG, but only at surfactant concentration nearing the cmc [86]. The effect of an increase in hydrophobicity of the polymer on the strength of its interactions with nonionic surfactants was demonstrated clearly in a fluorescence study of the interactions between HM-PNIPAM and OTG.   In this system the association took place by a cooperative mechanism at surfactant concentrations lower than the cmc.   This mechanism is different from the gradual association characteristic of anionic and cationic surfactants.   That two distinct mechanisms occur can be appreciated by comparing the changes as a function of added surfactant, SDS or OTG, in the ratios $[I_1/I_3]^{DP}$ and I_E/I_M for dipyme solubilized in PNIPAM-C$_{18}$. The addition of SDS is accompanied by gradual increases in the ratios I_E/I_M and $[I_1/I_3]^{DP}$ until the cmc is reached, indicating the formation of mixed micelles of lower microviscosity than the polymeric micellar core. In contrast, upon addition of OTG, the values of both I_E/I_M and $[I_1/I_3]^{DP}$ undergo significant transitions, as shown in Figure 11. Aggregation

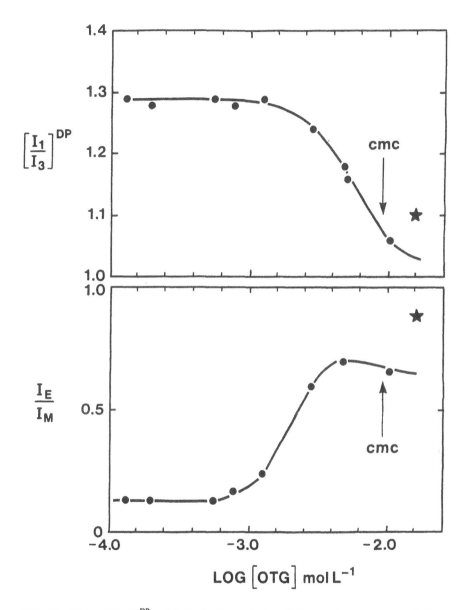

FIG. 11 Plots of $[I_1 / I_3]^{DP}$ and I_E/I_M for the emission of dipyme in an aqueous solution of PNIPAM-C_{18} (1.87 g L^{-1}) as a function of [OTG]. Values for dipyme in SDS micelles are indicated by a start in each plot; temperature 20°C, λ_{exc} = 348 nm. From ref. 81 with permission.

onset takes place well below the cmc. Note that in this case too, the mixed micelles appear to offer a less rigid microenvironment to the probe (I_E/I_M increases).

The interaction between the nonionic surfactant $C_{12}E_5$ and a high molecular weight PEO (5.94×10^5) was examined recently by fluorescence probing methods, with the Py/DMBP probe/quencher pair [87]. The quenching experiments gave evidence of a moderate micellar growth from 155 to 203 $C_{12}E_5$ molecules in the absence of PEO over a wide concentration range at 8 °C. In the presence of PEO, the micellar aggregation number shows a strong dependence on surfactant concentration. At low $C_{12}E_5$ concentration (< 0.25 %) the aggregation number remains low (ca 160). It increases with surfactant concentration to reach a plateau value of 270 at 1.2 % (30 nM) $C_{12}E_5$. At high surfactant concentrations the aggregation number is higher in the presence of polymer than in its absence. The large size of the aggregate may account for the marked lowering of the cloud point of PEO for these compositions.

Unlike most neutral polymers, hydrophobe-modified ethoxylated urethanes, HEUR have been found to interact strongly with nonionic surfactants, leading to an increase in solution viscosity. A study by fluorescence probing techniques with Py and the Py/DMBP probe/quencher pair has elucidated some of the phenomena involved in the interactions of a polymer of the HEUR family ($C_{12}EO_{460}C_{12}$) with the neutral surfactant $C_{12}E_8$ [88]. The ratio I_1/I_3 of the Py emission in the mixed system $C_{12}EO_{460}C_{12}/C_{12}E_8$ decreases at concentrations much lower than the cmc of the pure surfactant.   The decrease in the ratio is gradual, showing a less cooperative aggregation process. Fluorescence quenching measurements indicate that in the absence of surfactant the polymer aggregation number is about 15 to 30 end groups per micelle [89]. In the presence of $C_{12}E_8$ the surfactant tails and the polymer end-groups form mixed micelles. Using fluorescence quenching methods, the authors established that the average number of polymer ends per aggregate increases almost in proportion to the polymer concentration. The total aggregation numbers of the hydrophobic domains at different surfactant concentrations increase slightly with surfactant concentration in solutions of constant polymer concentration. Temperature was found not to influence the total aggregation number in contrary to pure nonionic surfactant micelles. Thus it was possible to estimate the aggregation numbers for the micelles formed at the compositions which correspond to the minima in cloud point curves. The surfactant/polymer aggregation numbers were found to be nearly independent on polymer concentrations (from 57/5 at the lowest polymer concentration (5 g L^{-1}) to 67/13 for a polymer concentration of 40 g L^{-1}).

C.   Polyelectrolytes and Surfactants

Charged polysaccharides interact strongly with oppositely-charged surfactants. A high cooperativity of binding is usually accompanied by the formation of large surfactant clusters. Examples of systems studied by fluorescence include: i) anionic branched polysaccharides, such as sodium dextran sulfate (NaDxS) and DTAB [90]; ii) linear anionic polysaccharides, such as sodium hyaluronate (NaHA) and DeTAB or DTAB [91] iii) sodium carboxymethylamylose (NaCMA) and SDS or CTAC[93] and iv) sodium carboxymethylcellulose and DTAB or CTAB [92]. The addition of salts to a NaCMA solution triggers supramolecular organization of the macromolecules which can form inclusion complexes with a variety of organic molecules. Zhen and Tung [93] carried out an extensive study by fluorescence of the interactions of NaCMA with surfactants in water and in the presence of added salts. By monitoring changes in the ratio I_1 /I_3 of Py as a function of added surfactant in aqueous solution of NaCMA they were able to detect interactions of the polyelectrolyte with both a cationic surfactant, CTAC, and an anionic surfactant, SDS, with a cac of 1×10^{-4} M for CTAC and 3×10^{-3} M for SDS. Dynamic fluorescence quenching measurements indicate that in both cases the mixed micelles are smaller than free micelles ($N_{agg} \approx 61$, NaCMA = 0.4 %, [CTAC] = 1×10^{-3} M; $N_{agg} \approx 47$; NaCMA = 1.6 %, [SDS] = 5×10^{-3} M). Contrary to the trend exhibited by salt solutions of surfactants in the absence of polymer, *i.e.* a cmc decreasing with added salts, the cac of NaCMA/CTAC is larger than the cmc of the free micelles and aggregation numbers for the surfactant aggregates in salt polymer solutions are identical to those of free micelles. Similar trends are observed in the NaCl/NaCMA/SDS system. These results point to the formation of NaCMA inclusion complexes with both CTAC and SDS. The cavities within the NaCMA supramolecular structures are highly hydrophobic, thus able to accommodate surfactant molecules. From the experimentally measured association constant of CTAC it was estimated that each NaCMA molecule includes about 3.5 CTAC molecules. The balance of CTAC molecules form micelles which do not interact.

Starburst dendrimers are macromolecules constructed from various low molecular weight multifunctional cores by attachment of radially branched layers, termed generations [94]. In the case of poly(amidoamine) dendrimers, the first ones to be synthesized and characterized in detail, the exterior surface consists either of carboxylate groups ("half-generation") or of amine groups ("full generation"). The interactions of surfactants with a series of dendrimers of this family have been studied by Caminati, Turro, and Tomalia using the Py probe as an indicator of the extent of surfactant aggregation [95]. The association process of carboxyl terminated dendrimers with cationic surfactants such as DTAB leads to the formation of dendrimer/surfactant aggregates of two different types, depending on the size of the dendrimers. In the case of the earlier generations (0.5 to 3.5) the association occurs by a non-cooperative mechanism, resulting in random condensation of surfactant molecules on the negatively charged dendrimer surfaces. Addition of DTAB to the later generations (4.5 to 9.5) dendrimers leads first to a random, non-cooperative binding of DTAB on the dendrimer surfaces, followed by a cooperative condensation of the cationic surfactant on the surfaces. These results can be correlated to the change in the morphology of the dendrimers from an open, branched structure (generations 0.5 to 3.5) to a closed, increasingly compact surface of the higher generations. In contrast, cationic starburst dendrimers do not perturb noticeably the micellization of the anionic surfactant SDS.

D. Hydrophobically Modified Polyelectrolytes and Oppositely Charged Surfactants

In these systems very strong association is expected, since both electrostatic and hydrophobic forces promote the formation of mixed micelles. A key factor affecting the interactions is the actual molecular architecture of the polymer, in particular the distribution of hydrophobic groups along the polymer chain. The fluorescence probe technique has been an important tool in the study of these systems, as described by Zana in chapter 10 of this volume for hydrophobically-modified polyacids/cationic surfactant systems. There have been only a few studies of labelled polycation/surfactant systems. This approach may prove very fruitful in the study of hydrophobically-modified polycations, as revealed by a preliminary report on the interactions of SDS with a hydrophobically-modified polycationic cellulose ether.

The hydrophobically modified cationic polyelectrolyte, Quatrisoft® LM200, is known to form cationic hydrophobic microdomains via intra- and interpolymeric association of the side chains. Upon addition of anionic surfactants, mixed micelles form readily due to both electrostatic and hydrophobic forces of interaction. Associative phase separation occurs near charge neutralization, and resolubilization is known to proceed in the presence of excess surfactant. Goddard and Leung [96] have observed LM200/SDS systems to form gels in the resolubilization zone, as well as the preprecipitation zone at high polymer concentrations. In the dilute regime, Guillemet and Piculell [97] have shown by fluorescence probe and viscosity determinations that although gels do not form, interaction does indeed occur at surfactant concentrations as low as 10^{-5} M. Winnik, Regismond, and Goddard [98] have performed comparative experiments on a pyrene labelled Quatrisoft® LM200 (LM200-Py), and reported its response to the addition of SDS and to changes in ionic strength.   Excitation spectra, UV absorbance and quantum yield determinations of solutions of the labelled polymer in water indicate that the broad fluorescence excimer emission originates from preformed pyrene dimers or higher aggregates which are stabilized by hydrophobic bonding. Fluorescence labelling was performed via ether linkages. In the case of LM200, the pyrenyl substituents can be introduced on two chemically distinct positions along the polymer: either the primary hydroxyl groups along the cellulosic backbone or the hydroxyl groups in close proximity to the hydrophobic substituents (Figure 12a). The pyrene linked to those two sites are expected to experience different environments and exhibit

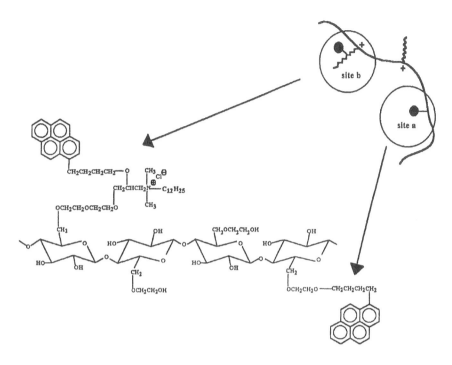

FIG. 12a Chemical structure of LM200-Py indicating the two sites of pyrene incorporation.

different photophysical properties. Addition of SDS to LM200-Py solutions has an affect on the emission spectrum in the following way: above a SDS concentration of 1.0×10^{-5} M, there is a gradual increase in intensity of the monomer emission until a plateau value of 2×10^{-4}M is reached. This concentration coincides with the neutralization value, where all the charges on the polymer are compensated by surfactant sulfate groups. An unusual spectroscopic feature occurred in a narrow concentration range around [SDS] ~ 0.008M (see Figure 12b). The monomer emission intensity exhibited a minimum and the excimer emission intensity displayed a weak maximum, which may signal the existence of a narrow surfactant concentration range where LM200-Py/SDS mixed micelles entrap at least two pyrene groups capable of forming excimers. Upon further increase of SDS concentration, the pyrene groups are solubilized as separated entities within larger or more numerous mixed micelles.

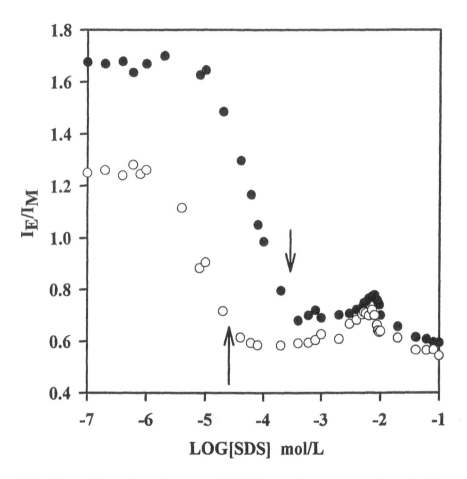

FIG. 12b I_E/I_M ratio of aqueous LM200-Py solutions as a function of SDS concentration. Open circles: [LM200-Py] 0.1 g L^{-1}; full circles 1 g L^{-1}. The arrows indicate the SDS concentration corresponding to charge neutralization. From ref. 98 with permission.

E. Polycations-Cationic Surfactants

Cationic surfactants have been shown to interact with hydrophobically-modified polycations, but there has been no report of their interactions with "simple" polycations. Bakeev *et al.* have studied the interactions between the surfactants DTAB or CTAB and copolymers of N-ethyl-4-vinylpyridinium bromide and small amounts of either the N-dodecyl- or N-cetyl-4-vinylpyridinium bromide [99]. The formation of copolymer/surfactant complexes in dilute solution was established on the basis of studies by potentiometry, sedimentation velocity, viscometry, and fluorescence spectroscopy with the PCA probe. The probe was added to solutions of the various polymers in water, and the wavelength of maximum emission (λ_{em}) of PCA was recorded as a function of added surfactant. In solutions of copolymers bearing 3 to 6 mol % of dodecyl or cetyl groups, the following trends were observed: a) at low surfactant concentrations λ_{em} (470 nm) is identical to that observed for PCA in an aqueous 0.01 M NaCl solution; b) at surfactant concentrations 2 to 3 times lower than the cmc, an abrupt blue shift is observed, followed by c) saturation at concentrations near the cmc ($\lambda_{em} = 450$ nm). The transitions were not as sharp as those occurring at the cmc of surfactants (in the absence of polymer), hinting at an association mechanism of lower cooperativity. The specific fluorescence technique selected in this study did not provide information on the interactions of surfactants with the polycations containing more than 3-6 mol % of alkyl side groups, since these copolymers form hydrophobic domains in water even in absence of surfactants. The hydrophobic PCA is solubilized within these domains and therefore λ_{em} is already blue-shifted. This limitation may be overcome using a different fluorescence technique.

The interactions between a hydrophobically-modified cationic hydroxyethylcellulose, Quatrisoft® LM200 (see structure, Figure 13a) and various cationic surfactants, (DTAB, DTAC, TTAB, CTAC), have been studied by Winnik, Regismond, and Goddard, using pyrene probe studies [100]. The occurrence of polycation/cationic surfactant interactions was suspected to occur, on the basis of precipitation and redissolution effects noted by Goddard and Leung for systems near the cmc of the surfactant [101,96]. Fluorescence probe experiments with Py confirmed that there is a definite interaction in these systems. The decrease in the ratio of I_1/I_3 was observed well before the cmc of the surfactants (Figure 13b). The relatively large concentration span from the onset of the decrease in I_1/I_3 to the cmc, indicates that the association between surfactant and Quatrisoft® LM200 takes place via a mechanism of relatively low cooperativity. Here, at low surfactant concentrations, intermolecular hydrophobic interactions exist between polymeric side chains, creating hydrophobic domains were the surfactant and pyrene molecule will preferentially aggregate (Figure 14). Above the cac, the hydrophobic substituents of the polymers are separated from each other and incorporated within

R = C$_{12}$H$_{25}$: Quatrisoft LM200

R = CH$_3$: JR400

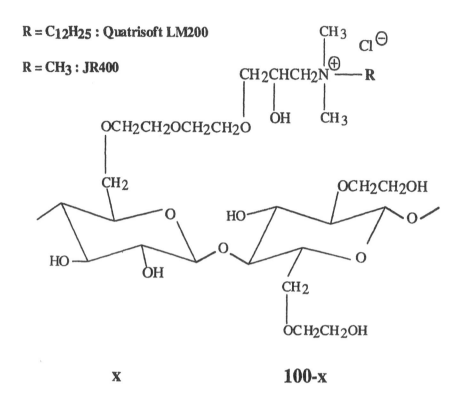

X   100-x

FIG. 13a   Chemical structures of Quatrisoft® LM200 and Polymer JR400.

surfactant micelles.   This behavior is very different from that exhibited by the unmodified cationic cellulose ether Polymer JR400 in the presence of surfactant: in all cases, the sharp transition in I_1/I_3 as a function of surfactant concentration was centered at a concentration corresponding exactly to the cmc, indicating that the surfactants have no interaction with the polymer, but rather that they proceed to form micellar structures freely diffusing in solution.

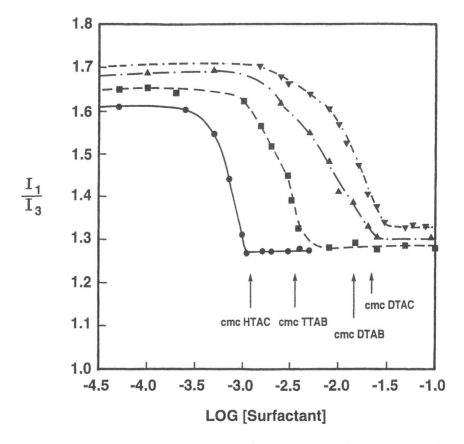

FIG. 13b I_1/I_3 ratios of pyrene emission in aqueous Quatrisoft®LM200 (1.0 g L^{-1}) solution as a function of surfactant concentration. ▲: DTAB; ▼: DTAC; ■: TTAC; ●: CTAC. Arrows indicate surfactant cmc values. From ref. 100 with permission.

F.   Proteins and Surfactants

It is well-known that SDS is a very potent denaturant of water-soluble proteins [102]. Since hydrophobic effects are the major driving force for the folding of water-soluble proteins into their native forms, it is generally agreed in the literature that the unfolding of proteins by SDS is caused by binding of SDS molecules to the hydrophobic portions of the polypeptide chain. However there is still no unanimity about the precise structure of the resultant protein/surfactant complexes. Studies of

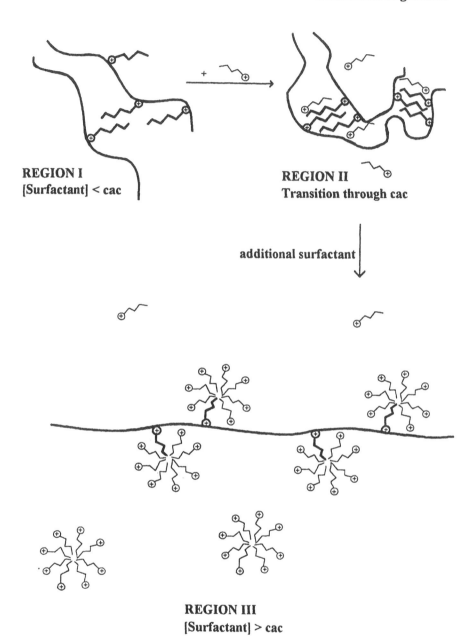

REGION I
[Surfactant] < cac

REGION II
Transition through cac

additional surfactant

REGION III
[Surfactant] > cac

FIG. 14   Schematic representation of the interactions between cationic polymer
Quatrisoft®LM200 and cationic surfactants. From ref. 112 with permission.

protein/SDS systems by fluorescence spectroscopy and SANS (see below) brought convincing evidence in favor of a model proposed in 1974 by Shirahama *et al.*[103], described in detail in chapter 4 of this volume. From free electrophoresis mobility determinations, they put forward a "necklace" model, which assumed that the polypeptide chains are flexible in solution and that micelle-like clusters of SDS are distributed along the unfolded polypeptide chains. This model presents many analogies with Cabane's subsequent description of the PEO/SDS complexes in solution [104]. It is worth noting that not all proteins are denatured by anionic surfactants. These proteins do not bind significant amounts of SDS.

Turro *et al.* have analyzed in detail the SDS/bovine serum albumin (BSA) system using a gamut of spectroscopic techniques, including fluorescence spectroscopy [105]. The fluorescence probes employed were Py and 6-*p*-toluidino-2-naphthalenesulfonate (TNS). Pyrene is partially solubilized within hydrophobic sites of BSA, as indicated by the value of I_1/I_3 (1.19) for Py in BSA solutions (1 wt %). Addition of SDS results in a slight decrease of the ratio (I_1/I_3 *ca* 1.0) for [SDS] > 5 mM. Aggregation numbers of the BSA-bound SDS aggregates were determined via pyrene excimer decay measurements. They ranged from 33 to 82, depending on the SDS concentration.

Unlike Py, TNS is hydrophilic and possesses an anionic sulfonate group which can compete with the SDS sulfonate headgroups for binding to cationic sites on the protein. The photophysics of TNS is very similar to that of the closely related probe ANS. It exhibits a very weak fluorescence in water, but when it is solubilized in a hydrophobic environment its emission increases by several orders of magnitude [106]. In a micellar SDS solution, the fluorescence emission of TNS remains weak, implying that the probe is not solubilized within the micelles but rather remains dissolved in water. The TNS emission is enhanced markedly when the probe is added to aqueous BSA (1 wt%). The changes of the TNS fluorescence intensity as a function of [SDS] shown in Figure 15 exhibit a complex pattern. The same trends are followed also by the fluorescence of Py, although not as pronounced. The initial increase in the hydrophobicity (increase in TNS emission intensity) is attributed to the cobinding of SDS and probe molecules near the hydrophobic regions of the protein. The reduction in hydrophobicity experienced by the probe upon incremental addition of SDS indicates the release of a fraction of the TNS molecules into a more hydrophilic environment, possibly as a result of competitive binding. A reversal of this second trend is observed at total [SDS] > 2.7 x 10^{-3}M, a concentration corresponding to the onset of a massive binding of the surfactant on the protein, leading to an uncoiling of the protein. As the protein unfolds, the number of available hydrophobic binding sites increases again, as suggested by the increase in TNS fluorescence intensity. At [SDS] > 10^{-3} M, protein-bound TNS is displaced by SDS and expelled into the aqueous environment where its fluorescence is markedly quenched. Taken together with the data gathered with the Py probe, the spectroscopy of TNS supports the formation of micelle-like aggregates along the

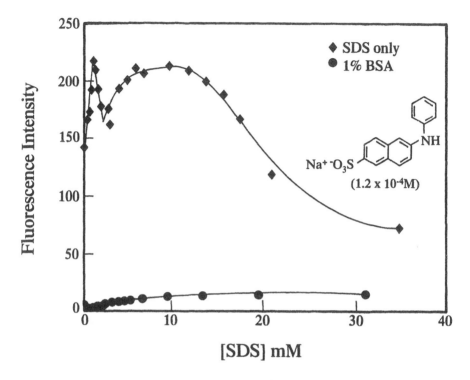

FIG. 15 Fluorescence intensity of TNS (1.25×10^{-6} M) in the presence and absence of 1% BSA as a function of total SDS concentration. From ref. 106 with permission.

protein chain, in the "necklace" fashion suggested by Shirahama. Corroborative evidence from ESR and ^{2}H NMR spectroscopy, also provided in this study, led Turro *et al.* to propose a model for the structure of the protein/surfactant complex in which the protein actually wraps around the SDS micelles.

Gelatin, a denatured protein, is the coating medium for most photographic products. Complex formation between anionic surfactants and gelatin has been known for many years [107]. The general consensus emerging is that the gelatin/anionic surfactant complex is yet another example of the "necklace" structure. These conclusions have been confirmed by several recent fluorescence studies using the PCA, ANS, and Py probes. The onset of gelatin/SDS interaction has been monitored by changes in the emission wavelength of PCA for gelatin solutions (0.5 wt %, 1 wt %, 5 wt %) [108]. The measurements indicate that the cmc of SDS is reduced in the presence of gelatin and appears to be insensitive to gelatin concentrations higher than 1 wt % (cac *ca.* 4×10^{-4} M). Pyrene probe studies carried out with gelatin from a different source confirmed the generality of the interaction mechanism [109]. The Py probe is incorporated into hydrophobic microdomains of high rigidity which may be attributed to residual triple helix segments. Steady state emission spectra (I_1/I_3) and fluorescence quenching by oxygen indicate that these microdomains become less polar and more fluid with surfactant association.

The interactions of photographic gelatin and SDS were studied also by ingenious NRET experiments which confirmed that small micellar aggregates are formed above the cac and that large aggregates capable of solubilizing probes are not formed at lower concentration [110]. Coumarin 1 was the energy donor and Rhodamine 6G the energy acceptor. The R_0 of the Coumarin 1/Rhodamine 6G pair is only a few nm [111]. When the donor and acceptor are confined to a single micelle, their separation is constrained by the dimensions of the micelles. If those are sufficiently small the energy transfer efficiency will markedly increase, compared to the situation of an isotropic solution of the two dyes of similar concentration ($< 10^{-5}$ M). The effect was illustrated in a plot depicting the changes of the ratio of emission from the donor to that of the acceptor (I_{Rh6G}/I_{C1}) with [SDS]. The sharp increase in the ratio at [SDS] $\sim 10^{-3}$ M corresponds to the onset of aggregation (cac). The ratio decreases smoothly with surfactant concentration above the cac, since, as the number of micelles grows, there is a smaller chance that both probes reside in the same micelle.

Further aspects of the interactions in SDS/gelatin mixtures were probed with ANS [111]. At low SDS concentration the fluorescence intensity of ANS is nearly indistinguishable from the background emission. When [SDS] reaches the value of 10^{-3} M, there is a sharp increase in the emission intensity. The intensity rises to reach a plateau value, then it increases again as the SDS concentration reaches the cmc value. The point at which the intensity deviates substantially from the background fluorescence is taken as the onset of formation of SDS micelles complexed to gelatin. The complex pattern of the ANS spectroscopy was interpreted in terms of a simple thermodynamic model based on electrostatic effects.

V. CONCLUSIONS

This review has presented many illustrations of the key features of fluorescence spectroscopy that make this technique particularly well-suited to clarify, on a molecular scale, many aspects of polymer/surfactant interactions. The breadth of systems studied by fluorescence spectroscopy has increased steadily over the last ten years. Probe measurements can be performed easily with readily available instrumentation and materials. The most important types of information the technique provides are the mean aggregation numbers of the complexes formed and measures of the polarity and internal fluidity of these structures. Probe and labelling experiments provide different and often complementary information about these systems. Labelling experiments offer the special advantage in that they allow one to follow changes in polymer/polymer association and in the conformation of the polymer chain upon interacting with surfactants. Moreover fluorescently labelled polymers are often excellent models of the corresponding hydrophobically-modified polymers, a class of materials of increasing practical importance.

ACKNOWLEDGEMENTS

A grant from the Natural Sciences and Engineering Research Council of Canada has supported this project.

GLOSSARY

1. Abbreviations and full names for the surfactants discussed in the text

Anionic surfactants:

SDS	sodium dodecylsulfate
$Cu(DS)_2$	copper didodecylsulfate
LiFOS	lithium perfluorooctanesulfonate
LiDS	lithium dodecylsulfate
SOS	sodium octanesulfonate
SDNpS	sodium 2-(N-dodecylamino)naphthalene-6-sulfonate

Cationic surfactants:

CTAC	cetyltrimethylammonium chloride
CTAB	cetyltrimethylammonium bromide
TTAC	tetradecyltrimethylammonium chloride
DeTAB	decyltrimethylammonium bromide
DTAB	dodecyltrimethylammonium bromide
DDAB	didodecyldimethylammonium bromide
DEFUMAC	diethanol heptadecafluoro-2-undecanol methylammonium chloride

Nonionic surfactants:

OTG	n-octyl β-D-thioglucopyranoside
Triton X-100	Octylphenolethoxylate

2. Abbreviations and full names for the polymers discussed in the text

Nonionic Polymers.

EHEC	ethyl(hydroxyethyl)cellulose
HEC	(hydroxyethyl)cellulose
HM-EHEC	hydrophobically-modified ethyl(hydroxyethyl)cellulose
HM-PNIPAM	hydrophobically-modified poly(N-isopropylacrylamide)
HPC	hydroxypropyl cellulose
MC	methyl cellulose
PEG	poly(ethylene glycol)
PEO	poly(ethylene oxide)
PPO	poly(propylene oxide)
PNIPAM	poly(N-isopropylacrylamide)
PVA	poly(vinyl alcohol)
PVP	poly(vinylpyrrolidone)
Ty	tylose

Polyelectrolytes.

BSA	bovine serum albumin
gelatin	
HM-PAA	hydrophobically-modified poly(acrylic acid)
NaCMA	sodium carboxymethylamylose
NaDxS	sodium dextransulfate
NaHA	sodium hyaluronate
NaPA	sodium polyacrylate
NaPMA	sodium polymethacrylate
NaPVS	sodium polyvinylsulfate
PAA	poly(acrylic acid)
PAAC	poly(allylammonium) chloride
PDMAAC	poly(dimethyldiallylammonium chloride)
PSS	polystyrenesulfonate
PSX	alt-co-(maleic acid-alkylvinyl ether) (X: number of carbon atoms in the alkyl group)
	poly(amidoamine) dendrimer
Polymer JR400	Cationic hydroxyethylcellulose (see Structure Figure 13a)
Quatrisoft ® LM200	hydrophobically-modified cationic hydroxyethylcellulose (see structure, Figure 13a)

3. Symbols and abbreviations

cac	critical aggregation concentration
cmc	critical micelle concentration
ε	dielectric permittivity
ε_{eff}	effective dielectric constant
N_{agg}	average aggregation number
η_{app}	apparent microviscosity
Y_o	electrostatic surface potential
[M]	total micelle concentration
M_n	a micelle with n quenchers
K	$K=k_+/k_-$ association equilibrium constant of the quencher to the micellar phase
k_+	entrance rate constant for a quencher into a micelle
k_-	exit rate constant for a quencher from a micelle
\<n\>	average number of quencher per micelle
$[Q_T]$	total fluorescence quencher concentration
$[Q_m]$	quencher concentration in the micellar phase
$[Q_w]$	quencher concentration in the aqueous phase
[S]	total surfactant concentration

4. Photophysical parameters

I	fluorescence intensity observed with continuous excitation
I_0	fluorescence intensity observed with continuous excitation in the absence of added quencher
I_1/I_3	ratio of the intensity of the I and III bands in the emission of pyrene; also known as the *py* scale
I_M	fluorescence intensity of pyrene monomer emission
I_E	Fluorescence intensity of pyrene excimer emission
F_f	quantum yield of fluorescence
k_f	radiative fluorescence decay constant
k_q	first-order intramicellar quenching rate constant
k_{IC}	internal conversion rate constant
k_{ISC}	intersystem-crossing rate constant
k_P	rate constant for photochemical reaction
k_{ET}	rate constant for non-radiative energy transfer
k_o	rate constant in the absence of quencher
λ_{em}	wavelength of maximum emission
λ_{exc}	excitation wavelength
P	steady-state polarization parameter
Q	fluorescence quencher

R	Förster radius for non radiative energy transfer
R_o	critical Förster radius for non-radiative energy transfer
r	emission anisotropy
τ	average fluorescence lifetime
V_o	effective molecular volume of a fluorescence probe

5. Probes

ANS	1-anilinonaphthalene-8-sulfonate
Dipyme	bis(1-pyrenylmethyl)ether
DPH	diphenylhexatriene
	9-methylanthracene
2-NpS	2-naphthalenesulfonate
	octadecylrhodamine
PCA	1-pyrenecarboxaldehyde
Py	pyrene
1-PyS	1-pyrenesulfonate
TNS	2-*p*-toluidinylnaphthalenesulfonate

6. Quenchers

DPC	Decylpyridinium chloride
DoPC	Dodecylpyridinium chloride
CPC	Cetylpyridinium chloride

REFERENCES

1. J.K. Thomas, *Chem. Rev. 80*: 283 (1980).
2. N.J. Turro, B.H. Baretz, P. L. Kuo, *Macromolecules 17*: 1321 (1984).
3. E.B. Abuin and J.C. Scaiano, *J. Am. Chem. Soc. 106*: 6274 (1984).
4. K.P. Ananthapadmanabhan, P.S. Leung, and E.D. Goddard, *Coll. Surf. 13*: 63 (1985).
5. K. Kalyanasundaran, in *Photochemistry in Microheterogeneous Systems*, Academic press, New York, 1987.
6. J.K. Thomas, *J. Phys. Chem. 92*: 5580 (1988).
7. J.H. Fendler, *Chem. Rev. 87*: 877 (1987).
8. F. Grieser and C.J. Drummond, *J. Phys. Chem. 92*: 5580 (1988).
9. R. Zana. in *Surfactants in Solution: New methods of Investigation* (R. Zana, ed.), Marcel dekker, New York, 1987, Ch. 5.
10. M. van de Auweraer, E. Roelants, A. Verbeeck, and F.C. De Schryver, in *Surfactants in Solution* (K.L. Mittal, ed.), Plenum Press, New York, 1989, Volume 7, p. 141.

11. F.M. Winnik, in *Interactions of Surfactants with Polymers and Proteins* (E.D. Goddard and K.P. Ananthapadmanabhan, eds.), CRC Press, Boca Raton, Fla., 1993, Ch. 9.
12. K. Kalyanasundaram and J.K. Thomas. *J. Am. Chem. Soc. 99*: 2039 (1977).
13. D.C. Dong and M.A. Winnik, *Can. J. Chem. 62*: 2560 (1984); D.C. Dong and M.A. Winnik. *Photochem. Photobiol. 34*: 17 (1982).
14. D.S. Karpovich and G.J. Blanchard, *J. Phys. Chem. 99*: 3951 (1995).
15. F.M. Winnik, M.A. Winnik, and S. Tazuke, *J. Chem . Phys. 91*: 594 (1987).
16. K. Nakashima and I. Tanaka, *Langmuir 9*: 90 (1993).
17. J.K. Thomas, *Chem. Rev. 80*: 283 (1980).
18. J.C. Dederen, L. Coosemens, F.C. De Schryver, and A. van Dormael, *Photochem. Photobiol. 30*: 443 (1979).
19. H. Trauble and P. Overpath, *Biochim. Biophys. Acta. 307*: 491 (1973); L Stryer, *J. Mol. Biol. 13*:482 (1965).
20. N.J. Turro, *Modern Molecular Photochemistry*, Benjamin/Cummings, Menlo Park, 1978.
21. J. Slavik. *Biochim. Biophys. Acta. 694*: 1 (1982); G.R. Fleming and G. Porter. *Chem. Phys. Letters 52*: 28 (1977).
22. F. Perrin, *Ann. Phys. (Paris), 12*: 169 (1929).
23. K. Nakashima, T. Anzai, and Y. Fujimoto, *Langmuir 10*: 658 (1994).
24. M.J. McGlade and J.L. Olufs, *Macromolecules 21*: 2346 (1988).
25. J. Ricka, M. Meewes, Ch. Quellet, and Th. Binkert, *Prog. Coll. Polym. Sci. 91*: 156 (1993).
26. F.M. Winnik, M.A. Winnik, H. Ringsdorf, and J. Venzmer, *J. Phys. Chem. 95*: 2583 (1991).
27. D. Georgescault, J.P. Desmasèz, R. Lapouyade, A. Babeau, H. Richard, and M.A. Winnik, *Photochem. Photobiol. 31*: 539 (1980). K.A. Zachariasse, W.L.C. Vaz, C. Sotomayor, and W. Kühnle, *Biochim. Biophys. Acta 688*: 323 (1986).
28. K.A. Zachariasse, W. Kühnle, and A. Weller *Chem. Phys. Lett. 73*: 6 (1980).
29. M.-L. Viriot, M. Bouchy, M. Donner, and J.C. André, *Photochem. Photobiol. 5*: 293 (1983).
30. N.J. Turro, M. Aikawa, and A. Yekta, *J. Am. Chem. Soc. 101*: 772 (1979).
31. I. Nagata and H. Morawetz, *Macromolecules 14*: 87 (1981).
32. T. Förster, *Discuss. Faraday Soc. 7*: 27 (1959).
33. J.R. Lakowicz, *Principles of Fluorescence Spectroscopy*, Plenum Press, New York, 1983, Ch. 10.
34. S.T. Cheung, A.E.C. Redpath, and M.A. Winnik, *Makromol. Chem. 183*: 1815 (1982).
35. F.M. Winnik, M.A. Winnik, S. Tazuke, and C.K. Ober, *Macromolecules 20*: 38 (1987).
36. F.M. Winnik, in *Hydrophilic Polymers: Performance with Environmental Acceptance*, (E. Glass, ed.) Adv. Chem. Ser. 248, American Chemical Society, Washingtom, 1996, pp 409-424.

37. H.T. Oyama, D.J. Hemker, and C.W. Frank, *Macromolecules 22*:1255 (1989) and references therein.
38. U.P. Strauss and G. Vesnaver, *J. Phys. Chem. 79*: 1558 (1975).
39. M. Yalpani and L.D. Hall, *Can. J. Chem. 59*: 3105 (1981).
40. F.M. Winnik, *Macromolecules 23*:233 (1990); F.M. Winnik, *Polymer 31*: 2125 (1990).
41. P. Ferruti, A. Betteli, and A. Feré, *Polymer 13*: 426 (1972); C. A. Cole, S. M. Schreiner, J. H. Priest, N. Monji, and A.S. Hoffman, *ACS Symposium Ser. 350*:245 (1987).
42. N.J. Turro and K.S. Arora, *Polymer 27*: 783 (1986).
43. D.-Y. Chu and J.K. Thomas, *Macromolecules 117*: 2142 (1984).
44. R.D. Stramel, C. Nguyen, S.E. Webber, and M.A. Rodgers, *J. Phys. Chem. 92*: 2934 (1988).
45. Y. Morishima, Y. Tominaga, M. Kamachi, T. Okada, Y. Hirata, and N. Mataga, *J. Phys. Chem. 95*: 6027 (1991).
46. M.C. Kramer, C.G. Welch, J.R. Steger, and C.L. McCormick, *Macromolecules 28*: 5248 (1995) and references therein.
47. K.D. Branham, D.L. Davis, J.C. Middleton, and C.L. McCormick, *Polymer 35*: 4426 (1994).
48. H. Ringsdorf, J. Venzmer, and F.M. Winnik, *Macromolecules 24*: 1678 (1991).
49. U.P. Strauss, in *Interactions of Surfactants with Polymers and Proteins* (E.D. Goddard and K.P. Ananthapadmanabhan, eds.), CRC Press, Boca Raton, Fla., 1993, Ch. 6.
50. F.M. Winnik, *Chem. Rev. 93*:587 (1993).
51. M. Asano, F.M. Winnik, T. Yamashita, and K. Horie, *Macromolecules 28*:5861 (1995) and references therein.
52. H.G. Schild and D.A. Tirrell *Langmuir 6*: 1676 (1990).
53. H. Morawetz, *Macromolecules 29*: 2689 (1996).
54. N. Kamenka, I. Burgaud, C. Treiner, and R. Zana, *Langmuir 10*: 3455 (1994).
55. J.C. Dederen, M. Van der Auweraer, and F.C. DeSchryver, *J. Phys. Chem. 85*: 1198 (1981).
56. C. Maltesh and P. Somasundaran, *Langmuir 8*:1926 (1992)and references therein.
57. C. Maltesh and P. Somasundaran, *J. Colloid Interface Sci. 157*: 14 (1993).
58. R. Sartori, L. Sepulveda, F. Quina, E.A. Lissi, and E. Abuin, *Macromolecules, 23*: 3878 (1990).
59. Y. Koyama, M. Umehara, A. Mizuno, M. Itaba, T. Yasukouchi, K. Natsume, A. Suginaka, and K. Watanabe, *Bioconjugate Chem. 7*: 298 (1996).
60. F. Quina, E. Abuin, and E.A. Lissi, *Macromolecules 23*: 5173 (1990).
61. Y. Hu, C. Zhao, M.A. Winnik, and P.R. Sundararajan, *Langmuir 6*: 880 (1990).
62. J. van Stam, N. Wittouck, M. Almgren, F.C. DeSchryver, and M. Da Graça Miguel, *Can. J. Chem. 73*: 1765 (1995).
63. J.K. Thomas, in *The Chemistry of Excitation at Interfaces*, American Chemical Society, Washington, 1984.

64. T. Asakawa, M. Mouri, S. Miyagishi, and M. Nishida, *Langmuir 5*: 343 (1989).
65. Y. Muto, K. Esumi, K. Meguro, and R. Zana, *J. Colloid Interface Sci. 120*: 162 (1987).
66. K. Kalyanasundaram, *Langmuir 4*: 942 (1988).
67. H.G. Schild and D.A. Tirrell, *Makromol. Chem., Makromol. Symp. 64*: 159 (1992).
68. T. Nojima, K. Esumi, and K. Meguro, *J. Am. Oil Chem. Soc. 69*: 64 (1992).
69. K. Esumi and T. Maekawa, *J. Am. Oil Chem. Soc. 72*: 145 (1995).
70. R. Zana, W. Binana-Limbele, N. Kamenka, and B. Lindman, *J. Phys.Chem. 96*: 461 (1992).
71. N. Kamenka, I. Burgaud, R. Zana, and B. Lindman, *J. Phys. Chem. 98*: 6785 (1994).
72. K. Thuresson, O. Söderman, P. Hansson, and G. Wang, *J. Phys. Chem. 100*: 4909 (1996).
73. S. Nilsson, *Macromolecules 28*: 7837 (1995).
74. J.C. Brackman and J.B.F.N. Engberts, *Langmuir 7*:2097 (1991).
75. M.L. Sierra and E. Rodenas, *J. Phys. Chem. 97*: 12387 (1993).
76. M.H. Gehlen and F.C. De Schryver, *Chem. Rev. 93*: 199 (1993)
77. S. Reekmans, M. Gehlen, F.C. De Schryver, N. Boens and M. van der Auweraer, *Macromolecules 26*: 687 (1993).
78. E. Rodenas and M.L. Sierra, *Langmuir 12*: 1600 (1996).
79. A.J. Dualeh and C.A. Steiner, *Macromolecules 23*: 251 (1990).
80. E.D. Goddard and R.B.Hannan, in *Micellization, Solubilization, and Micro-emulsions* (K.L. Mittal, ed.), Plenum Press, New York, 1977, Vol. 2, p. 835.
81. F.M. Winnik, H. Ringsdorf, and J. Venzmer, *Langmuir 7*: 905 (1991).
82. F.M. Winnik, H. Ringsdorf, and J. Venzmer, *Langmuir 7*: 912 (1991).
83. J.J. Effing, I.J. McLennan, and J.C.T. Kwak, *J. Phys. Chem. 98*: 2499 (1994).
84. S. Biggs, J. Selb, and F. Candau, *Langmuir 8*: 838 (1992).
85. J.C. Brackman and J.B.F.N. Engberts, *Chem. Soc. Rev. 85* (1993); J.C. Brackman and J.B.F.N. Engberts, in *Structure and Flow in Surfactant Solutions* (C.A. Herb and R.K. Prud'homme, eds.), ACS Symp. Ser. 578, American Chemical Society, Washington, 1994, Ch. 24.
86. F.M. Winnik, *Langmuir 6*: 522 (1990).
87. E. Feitosa, W. Brown, M. Vasilescu, and M. Swanson-Vethamuthu, *Macromolecules 29*: 6837 (1996).
88. E. Alami, M. Almgren, and W. Brown, *Macromolecules 29*: 5026 (1996).
89. E. Alami, M. Almgren, W. Brown, and J. François, *Macromolecules 29*: 2229 (1996).
90. P. Hansson and M. Almgren, *J. Phys. Chem. 99*: 16694 (1995).
91. K. Thalberg and B. Lindman, *J. Phys. Chem. 93*:1478 (1989).
92. P. Hansson and M. Almgren, *J. Phys. Chem. 100*: 9038 (1996).
93. Z. Zhen and C.-H. Tung, *Polymer 33*: 812 (1992).
94. D.A. Tomalia, A.M. Naylor, and W.A. Goddard, *Angew. Chem. 29*: 139 (1990).

ort

ort2

95. G. Caminati, N.J. Turro, and D.A. Tomalia, *J. Am. Chem. Soc. 112*: 8515 (1990).
96. E.D. Goddard and P.S. Leung, *Coll. Surf. 65*: 211 (1992).
97. F. Guillemet and L. Piculell, *J. Phys. Chem. 99*: 9201 (1995).
98. F.M. Winnik, S.T.A. Regismond and E.D. Goddard, *Langmuir 13*: 111 (1997).
99. K.N. Bakeev, E.A. Ponomarenko, T.V. Shishkanova, D.A. Tirrell, A.B. Zezin, and V.A. Kabanov,. *Macromolecules 28*: 2886 (1995).
100. F.M. Winnik, S.T.A. Regismond, and E.D. Goddard, *Colloids Surf. A: 106*: 243 (1996).
101. E.D. Goddard and P.S. Leung, *Langmuir 8*: 1499 (1992)
102. C. Tanford, *Adv. Protein Chem. 23*: 121 (1968).
103. K. Shirahama, K. Tsujii, and T. Takagi, *J. Biochem (Tokyo), 75*: 309 (1974)
104. B. Cabane, *J. Phys. Chem. 81*: 1639 (1977).
105. N.J. Turro, X.-G. Lei, K.P. Ananthapadmanabhan, and M. Aronson, *Langmuir 11*: 2525 (1995).
106. S. Niu, K.R. Gopidas, and N.J. Turro, *Langmuir 3*:1271 (1992).
107. K.G.A. Pankhurst and R.C.M. Smith, *Trans. Faraday Soc. 40*: 565 (1944)
108. J. Greener, B.A. Contestable, and M.D. Bale, *Macromolecules 20*:2490 (1987).
109. M. Henriquez, E. Abuin, and E.A. Lissi, *Colloid Polym. Sci. 271*: 960 (1993).
110. G.H. Whitesides and D.D. Miller, *Langmuir 10*: 2899 (1994).
111. G.A. Kenney-Wallace, J.H. Flint, and S.C. Wallace, *Chem. Lett. 32*: 71 (1975).
112. F.M. Winnik and S.T.A. Regismond, *Coll. Surf A. 118*: 1 (1996).

8

Isothermal Titration and Temperature Scanning Calorimetric Studies of Polymer - Surfactant Systems

GERD OLOFSSON and GENG WANG Division of Thermochemistry, Center for Chemistry and Chemical Engineering, Lund University, Box 124, S-221 00 Lund, Sweden

SYNOPSIS

Two modern microcalorimetric techniques applicable to the study of polymer–surfactant interaction are presented, *i.e.* isothermal titration calorimetry and high sensitivity differential scanning calorimetry, HSDSC. Principal design features and properties of the instruments are reviewed and the interpretation of the primary results are discussed. The calorimetric titration curves obtained from experiments with addition of concentrated surfactant solution to dilute polymer solution are compared with titration curves from surfactant dilution experiments without polymer. Differences between the two sets of curves are ascribed to polymer–surfactant interaction. The evaluation of the critical aggregation concentration cac and the saturation concentration C_2 from the titration curves is presented. Examples of studies of various polymer–surfactant systems using titration calorimetry are reviewed. Thermally induced events in dilute aqueous polymer and polymer–surfactant solutions can be studied using HSDSC. Micelle formation of block copolymers and the suppression of this aggregation by surfactants is one example. Another is the self-association preceding phase separation and its relation to the formation of thermoreversible gels in EHEC systems containing low concentrations of ionic surfactants.

I. INTRODUCTION

As evidenced by recent review articles [1-4], until now calorimetry has been little used in the study of polymer–surfactant interactions and not contributed significantly to the development in the field. One reason for this lack of use of various forms of calorimetry may be that commercial instruments suitable for the study of polymer–surfactant systems have not been obtainable until fairly recently.   The situation has changed now that modern high sensitivity microcalorimeters are readily available. For instance, we have found that by using titration microcalorimeters, interactions between nonionic polymers and ionic surfactants can be mapped quite in detail as a function of concentration in dilute and semidilute aqueous solution [5]. These instruments operate at a constant, preset temperature and measurements are made under isothermal conditions.   This technique is sometimes referred to as ITC (Isothermal Titration Calorimetry). High sensitivity differential (temperature) scanning calorimetry, HSDSC, is a technique where the thermal response, *i.e.* the apparent heat capacity, of samples of constant composition is studied as a function of temperature. This method is useful for studies, for instance, of how the clouding behaviour of polymers that give liquid-liquid phase separation is affected by the addition of surfactants [6,7]. Both types of instruments have been developed by biologically oriented calorimetrists. Titration calorimetry is used, for instance, to study bio-thermodynamic model systems, ligand binding to proteins and hydrophobic hydration–hydrophobic interaction. A practical application is its use for ligand binding studies in structure-based drug design programs. HSDSC is used for the study of *e.g.* protein unfolding and lipid bilayer transitions. The application of the methods to macromolecular systems outside the biological/biochemical interest area has been much more limited. We believe that much useful information can be gained by including calorimetric methods among the experimental techniques used to study polymer–surfactant systems.

In this chapter we will concentrate our discussion on isothermal titration microcalorimetry and HSDSC and only briefly mention other related techniques that have been used. We do not present any general treatment of basic thermodynamic relations for the systems and measurements which we discuss. Instead, we refer interested readers to the chapter in a previous volume in this series by Desnoyers *et al.*[8] which deals with the related topic of calorimetric measurements on surfactant systems. Methods applicable to polymer–surfactant systems not discussed by us may be found in this reference and in the monograph "Solution Calorimetry"[9] which deals with various calorimetric techniques applied to both reacting and non-reacting systems.

II. ISOTHERMAL TITRATION CALORIMETRY (ITC)

A. Principal Design Features and Properties

Calorimeters used when making isothermal titration experiments are usually called microcalorimeters, which is not a well-defined term. We will use the prefix micro to indicate calorimeters with a power sensitivity of 1 μW or better. Such calorimeters usually have reaction vessels of 1 to 3 mL volume but larger vessels may be used. Solution microcalorimeters are normally arranged as twins and function as differential instruments. One of the calorimetric vessels contains the reaction system and the other, called the reference vessel, is an inactive dummy preferably having approximately the same heat capacity and heat conductance. The most commonly used, commercially available isothermal microcalorimeters are heat conduction calorimeters, also called heat-leak, heat-flow or heat-flux calorimeters. In a heat conduction calorimeter, heat released (or taken up) in the reaction vessel flows to (or from) a surrounding heat sink, usually an aluminium block. Normally, the heat flow is registered by a thermopile positioned between the sample container and the heat sink. The temperature difference between the vessel and the heat sink, which is the driving force for the heat flow, will generate an electrical potential, U, over the thermopile. Provided that the temperature in the vessel and the heat sink is uniform, the thermal power (the rate of heat production) released in the reaction vessel is given by the Tian equation:

$$P = \varepsilon \, (U + \tau \, dU/dt) \qquad (1)$$

where $P = dq/dt$ is the thermal power, ε is the calibration constant, τ the time constant of the instrument and dU/dt the time derivative of the thermal potential. For a steady-state process equation (1) reduces to

$$P = \varepsilon \, U \qquad (2)$$

The heat quantity q released in the reaction vessel by the process is given by the potential–time integral

$$q = \varepsilon \int U \, dt \qquad (3)$$

Semiconducting thermopiles, often called thermocouple plates, are used as sensors. They have a relatively large thermal conductance and the temperature difference between the calorimetric vessel and the heat sink is usually small and of the order of 1 mK. Such instruments can therefore normally be considered as isothermal calorimeters. However, if measurements are made on concentrated solutions with large reaction enthalpies, the temperature difference may be larger, of the order of tenths of degrees, which under certain circumstances needs to be taken into account. The time constant for heat conduction calorimeters is usually fairly large, typically in the range 100 s to 1000 s. This means that for fast processes the measurement period needed to reach thermal equilibrium is much longer than the time for the reaction to go to completion. Although the reaction may be fast, a titration calorimetric experiment with 10 consecutive injection steps may require several hours. However, the time needed can be reduced by an order of magnitude by using a dynamic correction method. Sample injections are made at 5 to 7 min intervals before the calorimeter signal has returned to the baseline [10]. After the experiment, the curve is deconvoluted using the Tian equation (1) with a value for the time constant determined from electrical calibration experiments. Since we are only interested in determining the heat quantity (and not the kinetics of the process) one time constant is sufficient to achieve the required resolution.

The heat conduction microcalorimetric system marketed by Thermometric (Järfälla, Sweden) is described in refs. 11 and 12 while the system marketed by CSC (earlier Hart Scientific) (Provo, Utah, USA) does not seem to have been described in any detail. Microcal's (Northampton, MA, USA) titration microcalorimeter has a different design and is in principle an adiabatic calorimeter. Two coin-shaped cells are permanently mounted on either side of a semiconducting thermocouple plate. The thermopile functions as a differential thermometer and the instrument has no heat sink. Instead, the temperature is allowed to slowly increase during an experiment by heating the reference cell by a small constant power. The temperature difference between the sample and reference cells is monitored and a proportional power fed to the sample cell is adjusted to keep the temperatue of the two cells the same (power compensation calorimeter). The heat generated or taken up in a process is proportional to the integral of the differential power. This instrument has a small time constant. The totally filled cells have a volume of 1.3 mL and are permanently mounted. This microcalorimeter is fully described in ref. 13.

The titrant solution is usually added stepwise in small portions in order to have a good control of the concentration of the reactants and to achieve the best accuracy in the measurements. High-quality syringes with normally 100 to 500 µL volume are combined with accurate, computer-controlled syringe drives for the delivery of the titrant solution.

Titration microcalorimeters have been developed and used primarily for biochemical titrations, particularly, protein–ligand binding processes and for the study of hydration of small hydrophobic molecules. This means that the stirring in the calorimetric vessel has been optimized for dilute aqueous solutions. When designing the stirrers and adjusting the stirring speed, the aim has been to minimize the mechanical wear of dissolved proteins and to reduce the heat introduced by keeping the stirring speed as low as possible. This means that even small changes in the viscosity of the calorimetric solution can have a detrimental effect on the stirring efficency. An efficient turbine stirrer is needed to achieve satisfactory mixing in the cylindrical 3 mL vessel used with the TAM microcalorimeter [12]. It is important that the stirrer induces a vertical movement in the liquid. Horizontal stirring resulting in liquid layers is easy to achieve but will not give a homogenous solution. In the disc-shaped vessel of the Microcal microcalorimeter, the tip of the injection needle is modified to act as a stirrer and the syringe rotated to achieve stirring. The stirring blades cannot extend further into the sample than the width of the cell and a large fraction of the liquid is out of direct reach. We have found it very valuable to perform bench experiments using transparent plastic vessels to check the stirring effiency. The flow pattern in the solution to be studied can then be examined by injecting colored solutions. One should keep in mind that during a series of injections where surfactant is added to a polymer solution, the viscosity may change significantly during the titration experiment. For instance, the reduced viscosity at 20 °C of a 0.29 wt % solution of EHEC A (a sample of ethyl(hydroxyethyl)cellulose that has been studied by various methods by different groups, see Appendix) was found to start to increase at 2 mmol/L SDS from approximately 1000 mL/g in pure water to reach a pronounced maximum of 3850 mL/g at 4 mmol/L SDS and then to decrease to become lower than in water at SDS concentrations above 7 mmol/L [14]. Such a large viscosity change puts a great demand on the stirring efficiency. We recommend bench experiments.

Kresheck and co-workers were the first to use titration calorimetry in the study of polymer–surfactant interactions [15]. The main focus of their studies was the interaction between surfactants and biological macromolecules such as proteins and polypeptides [16,17]. They used a semi-adiabatic calorimeter consisting of a 25 mL sample cup which was inserted in a Dewar flask of approximately 50 mL volume and the whole calorimeter immersed in a thermostatted bath [18]. The temperature in the calorimeter was monitored by a 2000 ohm thermistor. The titrant was delivered continuously through the experiment by means of a buret with a constant speed actuator. This calorimeter did not reach the sensitivity and resolution of the microcalorimeters described above.

Usually microcalorimeters are calibrated electrically. This is a convenient method which also is highly accurate if the electrical heater is properly placed. However, in order to check the overall performance of the calorimetric system including the auxilliary equipment we recommend the use of a suitable test reaction. The dissolution of propan-1-ol in water is a convenient reaction for this purpose [12,19].

B. Experimental Considerations

When studying polymer–surfactant interaction using titration calorimetry, the experiments can be made in two different ways: a concentrated surfactant solution is added in small portions to a dilute polymer solution or a concentrated polymer solution is added to a (dilute) surfactant solution. We have found the first approach to give more informative titration curves as it allows the direct comparison of the (de)aggregation process of the surfactant in solutions with and without polymer, see section II.C. The choice of surfactant concentration will be limited by the requirement that it should be possible to handle the solution in the syringe used for the injections. As the injection needle should be narrow to limit diffusion, the solution cannot be too viscous and it must not contain crystals. This means that when working with ionic surfactants, the Krafft temperature should be well below the temperature at which the injection system is kept. Although some surfactant solutions are easily supersaturated, in our experience the results from measurements using supersaturated or close to saturated solutions tend to show bad reproducibility. Besides, there is the risk of clogging the injection needle. It is desirable to keep the surfactant concentration reasonably high in order to minimize dilution effects and to allow the surfactant monomer concentration to be neglected when interpreting the results. However, the final choice will also depend on the polymer concentration that is chosen and how detailed the titration curve should be at low surfactant concentration and how high the total surfactant concentration should be at the end of the titration experiment. The sensitivity of the calorimeters will allow injections down to a few tenths of μmol of surfactant per step. The titration experiment is repeated with the same titrant solution but with pure water (or solvent) in the calorimeter vessel to determine the dilution behavior of the surfactant solution in the absence of polymer. The viscosity is one important factor when chosing the polymer concentration. In order to obtain meaningful results the stirring must be efficient throughout the titration experiment. The enthalpic effect from the dilution of the polymer solution in the calorimeter vessel is usually insignificant but is easily checked by separate dilution experiments injecting pure water or the solvent used into the calorimetric liquid.

In vessels with a gas phase which will allow the volume of the calorimetric liquid to increase, the amount of polymer is constant throughout the experiment. In totally filled vessels like the ones used in the Microcal calorimeter, there will be an overflow following each injection and the amount of polymer will decrease during the titration series. This must be taken into account when evaluating the calorimetric results. Bloor *et al.*[20,21] circumvented the problem by using the same concentration of polymer in the injectant solution as in the calorimetric vessel. In this way the polymer concentration was kept constant during the titration experiment. The presence of polymer in the titrant solution makes the description of what is happening during the titration more complicated and the overflow correction is still needed in order to keep track of the total surfactant concentration in the reaction vessel.

The enthalpy of micelle formation varies strongly with temperature and so will the enthalpy of surfactant aggregation in the presence of polymer. This means that far-reaching conclusions based on the shape of calorimetric titration curves and the magnitude of enthalpic effects should not be drawn from measurements at only one temperature.

C. Interpretation of Calorimetric Titration Curves

When interpreting the calorimetric titration curves it may be helpful to consider first the course of events in an experiment consisting of the dilution of a concentrated amphiphile solution in water. The amphiphile will be in micellar form in the titrant solution and in the injections giving final concentrations below the cmc, the micelles will break up to give monomers. The measured enthalpy change, ΔH_{obs} (usually calculated per mole of injected surfactant) consists of the enthalpy of dilution of the micellar solution, ΔH_{dil}, and the enthalpy change for demicellization, ΔH_{demic}, equal to ($-\Delta H_{mic}$), the enthalpy of micelle formation. In the cmc region, part of the added micelles will break up and at higher final concentrations the added micelles are only diluted. The enthalpy of micelle formation can be calculated from the difference between the observed enthalpy changes below and above the cmc. The plot shown in Figure 1a of the fraction f_{mic} of incremental amphiphile that aggregates against the total amphiphile concentration C_{tot}, normalized to the cmc, illustrates the development of the titration curve in the cmc region [22]. The variable f_{mic} = $d(N_s C_{mic})/dC_{tot}$ is calculated from the mass action law model applied to a monodisperse nonionic amphiphile; N_s denotes the aggregation number and C_{mic} the concentration of micelles. The curve for $N_s = \infty$ represents the behavior at true phase separation. The convention used in Figure 1a is that 1% of the surfactant is in micellar form at the cmc. The concept of a critical micellization concentration, cmc, is not well defined but has its most precise interpretation within the (pseudo)phase separation model of micelle formation. The definition of cmc within the framework of other models is less clear [23]. For a further discussion of cmc see the IUPAC Appendix II, Ref. 24. Reported values can have different meanings depending on experimental methods and models used. A calorimetric titration curve recorded from consecutive additions of small portions of concentrated surfactant solution starting with pure water and where the observed enthalpy change per injection, ΔH_{obs}, is plotted against total surfactant concentration will reflect the shape of Figure 1a. This is exemplified by the calorimetric titration curve for 10 wt % C_8E_4 in water shown in Figure 1b [25]. If the change in surfactant concentration in each titration step is reasonably small, the following is valid to a good approximation:

$$\Delta H_{obs} = (1 - f_{mic}) \, \Delta H_{demic} + \Delta H_{dil} \qquad\qquad (4)$$

In the beginning of the titration all added micelles will break up to monomers and f_{mic} is zero. In the cmc region a fraction $(1 - f_{mic})$ will change into monomers while a fraction f_{mic} will remain in micellar form. This means that the steepness of the titration curve (normalized to the cmc) is an indication of the cooperativity of the aggregation process, i.e. of the aggregation number N_s. For large values of N_s the cmc can easily be derived from the calorimetric curve as the crossing point between extrapolated pretransition and linear ascent lines as shown in Figure 1b. Values of cmc derived in this way agree well with values derived by other methods. This procedure is easy to apply for

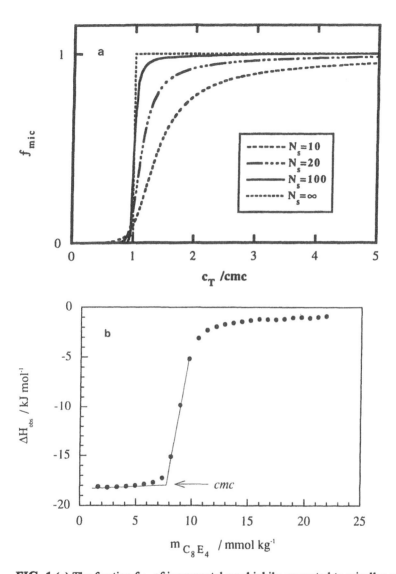

FIG. 1 (a) The fraction f_{mic} of incremental amphiphile converted to micelles versus the total surfactant concentration normalized to the cmc, c_T /cmc. The variable f_{mic} = $d(N_S C_{mic})/dc_T$ is calculated from the mass action law model. The curve for $N_S = \infty$ represents the behavior at true phase separation. Cmc is defined as the concentration where 1% of the surfactant is in micellar form. From ref. 22 with permission;   (b) differential enthalpies of addition of 10 wt % C_8E_4 as a function of molality. The change in molality was 8.3×10^{-4} mol kg^{-1} in each step. At 25 °C the cmc for C8E4 is 7.3 mmol/kg, ΔH_{mic} = 16.75 ± 0.10 kJ/mol and N_S = 23 ± 3 [25].

pronounced cooperative aggregation with N_s above say 20 for which the (pseudo)phase separation model gives a reasonable description of changes in physical properties. However, the cmc can be defined and derived in other ways from calorimetric titration curves. For instance, Desnoyers and co-workers [26] chose to use the inflection point in the titration curve as a measure of the cmc. For highly cooperative micellization with N_s above say 50, the difference between cmc defined as the start of the aggregation or as the inflection point is small and without practical significance. In cases of less cooperative aggregation where the association process extends over a considerable concentration range, cf. the curve for $N_s = 10$ in Figure 1a, the concentration where aggregates start to form is often also denoted cmc. The relevance of such values is, however, unclear.

If the temperature of the measurements happens to be very close to or coincides with the temperature where ΔH_{mic} is zero, the titration curve will show little or no change from ΔH_{dil} when the concentration passes the cmc. This is, for instance, the case for SDS in water at 25°C as seen from Figure 2a which shows the differential enthalpies of dilution of 10 wt % SDS solution. The value for ΔH_{mic} (at the cmc) is -0.20 ± 0.05 kJ/mol at 25.00 °C [27] so the dilution curve in water shows only a change of slope. However, micelle formation in water is characterized by large negative heat capacity changes [28] and a change of temperature will change ΔH_{mic} and the titration curve will show the expected jump at the cmc. A typical titration curve from the addition of concentrated SDS solution to a dilute polymer solution is included in Figure 2a which shows the results of the calorimetric titration of 0.10 wt % (0.023 mol repeat unit /kg) solution of PEO 1.5m with 10 wt% SDS solution at 25°C [29,30]. Differences between the titration curve in polymer solution and the dilution curve in water are ascribed to SDS–polymer interaction. The curve in the PEO solution starts to deviate from the dilution curve above 3.5 mmol/kg SDS to give a pronounced endothermic peak followed by a broad shallow exothermic one. Then the curve joins the dilution curve at about 20 mmol/kg. The onset of aggregation of surfactant in the presence of polymer is characterized by a critical aggregation concentration cac. At a second critical concentration, denoted C_2, the influence of the polymer on the aggregation of surfactant ceases. These characteristic concentrations are seen somewhat clearer in a plot of the incremental enthalpy changes against surfactant concentration, $[\Delta H_{obs}(k) - \Delta H_{obs}(k-1)]/ \Delta m$, where $\Delta H_{obs}(k)$ is the observed enthalpy change in the kth injection and Δm is the change in molality, see the insert in Figure 2. We find it easiest and most reproducible to identify cac with the concentration for the maximum of the first peak in the difference plot which corresponds to the concentration for the inflection point in the leading edge of the endothermic peak. We define C_2 as the concentration where the

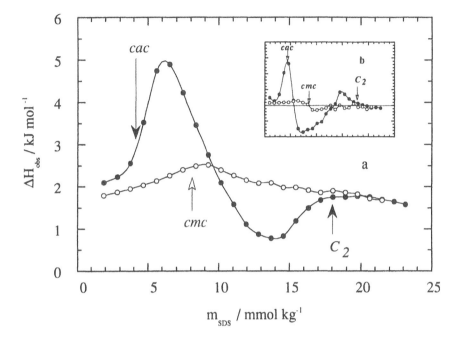

FIG. 2 (a) Calorimetric titration curve from the addition of a 10.00 wt % SDS to (●) 0.10 wt % PEO1.5m solution and (o) dilution curve in water at 25 °C; (b) difference curve of 10.00 wt % SDS (●) in 0.10 wt % PEO 1.5m solution and (o) in water.

slope of the difference curve becomes zero, within uncertainty limits, that is the concentration where the titration curve in the polymer solution joins the dilution curve in water. Thus, in addition to the differential enthalpy of surfactant–polymer interaction, the critical aggregation concentration cac and the saturation concentration C_2 and concomitantly the extent of binding, can be derived from the calorimetric titration curves.

A qualitative description of the calorimetric curve showing the differential enthalpy of dilution of SDS in 0.10 % PEO solution in Figure 2 could be the following. The micelles added in the first couple of injections break up to monomers. The slightly more positive dilution enthalpy in the polymer solution, compared to water, indicates a weak endothermic interaction between PEO and SDS monomers. When the cac region is reached, aggregates start to form and EO segments are incorporated to give mixed micelles containing about 30 SDS monomers [31]. At this first stage of SDS aggregation, with polymer in excess, there is an intimate contact between polymer segments and surfactant in the mixed micelles leading to dehydration of EO segments. The enthalpy change for the dehydration of EO groups is positive, $\Delta H_{dehyd} = 7$ kJ/[mol (-C_2H_4O-)] at 25°C [32,33], so ΔH_{obs} becomes more positive as the concentration passes the cac. Then ΔH_{obs} goes through a maximum and drops to become exothermic relative to the dilution curve in water. The number of SDS monomers per micelle will increase as the total concentration increases and the aggregates will more and more resemble free micelles. The exothermic contribution to ΔH_{obs} could arise from the rehydration of EO segments that are expelled in this reorganization of the mixed micelles. It is also possible that the enthalpy of formation of the primary SDS aggregates differs from the full-grown micelles which could contribute to the endothermic peak observed close to the cac [34]. As the saturation concentration C_2 is approached, the EO groups will be found in the outer headgroup region where to a large extent they will stay hydrated. While the cac and the maximum of the endothermic peak are almost independent of PEO concentration, the peak height, the downward slope, the location and size of the exothermic peak (relative to the dilution curve) and C_2 are directly related to the polymer content [30]. The titration curves from the addition of SDS to solutions of other nonionic polymers such as ethyl(hydroxyethyl)cellulose, EHEC, and polyoxypropylene PPO1000 show the same features, see below. The more hydrophobic the polymer, the lower the cac, the higher the endothermic peak and the deeper the following exothermic hump. For the short chain PPO1000 the cac varies with polymer concentration, but the general shape of the curve is the same as for the long-chain polymers.

Titration curves from the addition of, for instance, SDS to PEO solutions of varying concentration and to EHEC solutions [5,30] have features in common with curves observed from the addition of SDS to 1-pentanol solutions shown in Figure 3 [35]. The similarities indicate that it may be fruitful to regard the interaction in polymer solutions as the solubilization of polymer segments and/or substituents in surfactant aggregates to form mixed micelles. Surfactant micellization is well understood and the solubilization of (small) uncharged molecules in ionic micelles can be modelled theoretically [35-38]. Micelle formation of SDS in the presence of pentanol could quantitatively be

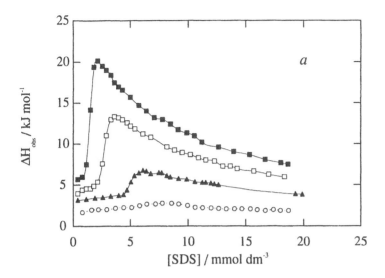

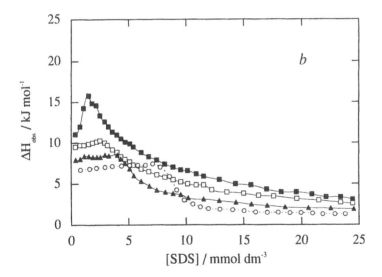

FIG. 3 Differential enthalpies of dilution of 10 wt % SDS solution as a function of total SDS concentration. Initial calorimeter solution: (○) pure water, (▲) 0.059, (□) 0.118, and (■) 0.176 mol dm⁻³ pentanol solution; (a) at 25 °C and (b) at 35 °C. From refs. 5 and 35, with permission.

described and enthalpic titration curves were calculated that satisfactorily agreed with the experimental curves at various pentanol concentrations at 25 and 35°C [35]. The fitting of the experimental curves needed only one adjustable parameter. At 25°C, ΔH_{mic} for SDS in water is close to zero so the observed endothermic enthalpy changes in the pentanol solutions above the cmc, Figure 3a, predominantly stems from the dehydration of pentanol that becomes solubilized in the mixed micelles. At 35 °C, ΔH_{mic} of SDS is -5.1 kJ/mol and therefore gives a significant contribution to ΔH_{obs}. As the total concentration of SDS increases, the fraction of SDS monomers in the mixed micelles increases and the shape of the curves reflects the changing composition of the micelles [35]. In solutions of small molecules such as pentanol, the cmc will vary with the concentration of solubilizate. The major contributing factors to the depression of the cmc are i) a decrease of the electrostatic headgroup repulsion on the surface of the micelle due to the reduction of the surface charge density when solubilized molecules are incorporated into the micelle, ii) an increased micellar entropy due to the mixing of SDS and solubilized molecules in the aggregates. In polymer–surfactant aggregates the reduction of the interfacial area between hydrophobic polymer segments and exposed hydrophobic parts of the aggregating surfactant is considered an additional factor favoring micellization.

D. Examples of Results

Describing a surface tension study of the PEO/SDS system, Jones formalized the concept of two critical concentrations that is used today to discuss polymer–surfactant interaction [39]. The first concentration, here denoted cac, (Jones used the notation T_1) represents the concentration at which interaction between surfactant and the polymer first occurs. In general, cac is less than the cmc of the surfactant and varies only little if at all with polymer concentration. Only for short-chain polymers such as PPO 1000, does the cac decrease significantly with increasing polymer content. Values of the cac for the same system, determined by different methods, including calorimetry, usually show good agreement. Thus, the cac is experimentally well defined and its meaning generally well understood.

The second critical concentration, denoted C_2 by us but T_2 by Jones and others, is much more elusive. It is considered to indicate the surfactant concentration where the polymer has become saturated with surfactant aggregates or the concentration where free surfactant micelles start to form or where the aggregation of surfactant is no longer influenced by the polymer. Figure 4 illustrates the difficulties in reaching a generally valid experimental definition of C_2. Bloor et. al. [20,40] have studied the system SDS and PVP using potentiometric measurements with a surfactant ion selective electrode and titration calorimetry. The results from the emf measurements are plotted as a function of SDS concentration (on a logaritmic scale) in Figure 4a and the concentration of monomeric SDS deduced from these results in Figure 4b. The two critical concentrations are indicated as the start and end of the deviation between the emf curves with and without polymer. The SDS monomer concentration increases all the way up to C_2 (T_2) and after that decreases. The results of the titration calorimetric measurements in the form of differential enthalpies of addition of concentrated SDS solution (containing 1 w/v% PVP) in 1 % w/v solution of PVP and in water are shown in Figure 4c. It is clear that C_2 determined from the emf measurements occurs well below the concentration where the enthalpy curves merge. The existence of free SDS micelles above C_2 was confirmed by gel filtration experiments [40]. Figure 4 depicts the situation in 1 % w/v solution (and 25 °C) but it changes if the polymer content is halved to 0.5 % w/v. The emf measurements give a C_2 value of 30 mmol/L (from Figure 3 in [40]) while the calorimetric titration curve in 0.5 wt % solution joins the dilution curve in water at 31 mmol/kg SDS [29,30] which is a fully satisfactory agreement beween the two ways to define C_2. The values of C_2 derived from the emf measurements in 0.5, 1 and 2 w/v % PVP are 30, 41 and 48 mmol/L SDS [40]. They do not increase in proportion to the polymer content and, thus, the amount of bound SDS per gram PVP decreases with increasing concentration. Bloor *et al.*[21] have also studied the SDS–PPO and SDS–EHEC systems by potentiometric and calorimetric measurements. For the PPO system containing 0.5 % polymer, the emf and calorimetric titration curves in the polymer solution join the curves in water at about the same SDS concentration, 100 mmol/L, that is, the two ways of defining C_2 agree. This gives the extent of binding to 19 mmol of SDS per gram PPO which agrees very well with the amount of SDS bound to PPO in 0.1 wt % solution of 18.6 ± 1 mmol/kg that we derive from titration calorimetric results [29]. However, the picture of the SDS–PPO system is made very complex by the reported appearance in the gel filtration results for the 0.5 % solution of a peak ascribed to free SDS micelles already in the concentration range (20–30) mmol/L [21]. The surfactant ion selective electrode senses the activity of monomeric SDS, while the calorimetric measurements probe the environment of the added SDS micelles. As long as there is a significant difference between the calorimetric titration curves in polymer solution and in the solvent, the freshly added micelles are affected by the presence of the polymer (or the polymer–surfactant complex). It is not possible to identify the changes in the micelles or in the micellar surroundings leading to noticeable differences in the dilution enthalpies but they probably stem either from changes in electrostatic interaction between the micelles or from the reorganization of surfactant aggregates involving

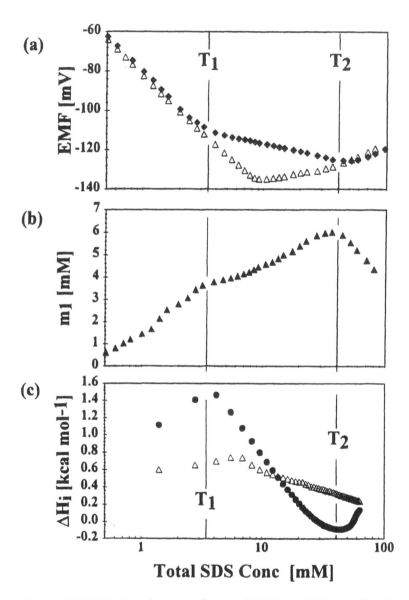

FIG. 4 SDS-PVP (1% w/v) system in 0.1 mM NaBr at 25 °C as a function of total SDS concentration: (a) surfactant electrode EMF results where the PVP content is (△) 0 % w/v or (◆) 1 % w/v; (b) SDS monomer concentration ; (c) enthalpies of addition of (△) pure SDS solution into water and (●) SDS + 1% PVP solution into 1 % PVP solution. From ref. 20 with permission.

changes in hydrophobic hydration of surfactant molecules and/or polymer segments. As the extent of counterion binding is almost unaffected by the presence of polymer near C_2 [40], the electrostatic contribution to the dilution enthalpy will be unchanged. Therefore, we conclude that as long as the enthalpic titration curves differ, the added micelles experience a reorganization when introduced into the polymer-containing solution. This is our motivation for denoting the merging of the dilution curves as C_2 and to consider this point the concentration where the aggregation of the surfactant is no longer influenced by the polymer. The comparison made above between C_2 determined from emf and calorimetric measurements show that for some systems and some compositions the two methods agree while at other compositions (higher polymer content?) C_2 determined from emf measurements is lower than the calorimetric value. This indicates that while the added micelles in the calorimetric measurements still feel a difference between solutions with and without polymer, the SDS monomer activity is the same. Clearly, our understanding of polymer–sufactant systems approaching C_2 is still poor and needs to be developed before a satisfactory description of this and other experimental results can be made.

The calorimetric titration curves, as exemplified by SDS–PEO in Figure 2, show that the aggregation of surfactant in the presence of polymer goes through different phases depending on the polymer/surfactant ratio. When surfactant is added in increasing amounts to a polymer solution, there is first an endothermic contribution starting at the cac. This is seen in the titration curves as an initial endothermic peak. In most of the cases we have studied, an exothermic contribution (relative to the dilution curve in water) follows [5, 29, 30]. The height of the initial peak increases with increasing polymer content and the crossing towards the exothermic side of the dilution curve in water takes place at increasingly higher surfactant concentration. The two dilution curves in water and polymer solution will merge at a surfactant concentration C_2 that depends on the polymer content. The existence of a transitional concentration between cac and C_2 has been inferred from results of conductance measurements. Results on the SDS–PEO system in the form of plots of the slope of conductance curves as a function of SDS concentration clearly show three different critical concentrations [41]. We observe that values for the intermediate concentration, denoted c' by François et al. [41], coincide with the concentration where the differential enthalpy curve in PEO solution crosses the dilution curve in water [30]. Indications of such intermediate concentration regions have also been seen in conductance curves from measurements on SDS–PVP [42] and SDS–PVA (polyvinyl alcohol) solutions [43]. Thus, the character of the surfactant–polymer interaction and the properties of the resulting aggregates will vary with the polymer/surfactant ratio. The interaction of polymer and the involvment of polymer segments in the initial "imperfect" mixed micelles, with an aggregation number about half the value for free micelles, will differ considerably from the interaction with the fully grown aggregates. When presenting models for polymer–surfactant interaction and descriptions of polymer–surfactant complexes it is necessary to clearly state the compositon of the systems.

Noteworthy is that the shape of the titration curves is the same for long-chain PEO 100,000 with more than 15 aggregates per chain at C_2 [31] and PEO 1.5m, Figure 2, as for short-chain PEO 4000, which can accommodate less than one SDS aggregate on the average per unimer. The growth of the aggregates shows the same enthalpic pattern with increasing total SDS concentration independent of possible electrostatic repulsion between aggregates on the same chain. However, for PEO 8000 the hump on the exothermic side of the dilution curve is separated into two, fairly well resolved peaks [30]. The reason for this change of shape is unknown but may be related to a possible aggregate–aggregate repulsion. As we are using polydisperse samples, a significant fraction of the unimers will carry on the average two SDS aggregates at C_2 [31].

Brackman et al. were the first to observe that there can be a significant response in titration-calorimetric measurements from polymer–surfactant interaction although the decrease in cmc is small or may even be nonexistent [44]. These authors studied the interaction between the nonionic surfactant n-octyl b-D-thioglucopyranoside, OTG, and PPO 1000. Their calorimetric results are shown in Figure 5. The differential enthalpies of dilution of 10 wt % OTG solution in 0.5 wt % PPO and in water are plotted against the final OTG concentration. In the presence of 0.5 wt % PPO, the dilution enthalpy in the premicellar region is slightly less exothermic than in water and the transition region is located in the same concentration range. However, above the cmc there is an endothermic effect of +4.3 kJ/mol while the dilution enthalpy in water at the same OTG concentration is close to zero. Evidently, the aggregation of the nonionic OTG is affected by PPO. The authors conclude that the Gibbs energy of micellization, i.e the cmc, of OTG is unchanged by the presence of PPO [44,45]. However, it is questionable if this conclusion is supported by the results shown in Figure 5. The bends in the enthalpy curves do appear in the same concentration range, but the curve in the PPO solution is shifted about 1 mmol/L towards lower concentration. This is a small but not insignificant change. Measurements on the OTG–PPO system made by us confirm the results by Brackman et al., however, we did not observe any splitting of the calorimetric response peaks into an endothermic peak followed by an exothermic peak. A possible reason could be a more efficient stirring in our calorimetric vessel. In all systems for which we observe a well-defined endothermic peak in the enthalpy curves as shown in Figure 2 we also

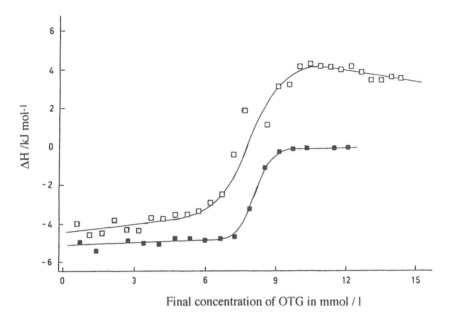

FIG. 5 Titration curves from the addition of concentrated OTG solution in (■) water and (Δ) in 0.5 wt % PPO 1000 at 25 °C. The cmc for OTG is 8.05×10^{-3} mol/L. Adapted from ref. 44.

see a decrease in the cmc. Admittedly, in some cases this shift is small as seen in Figure 6a, which shows the results from addition of DoAC solution to 0.10 and 0.20 wt % PEO at 25°C [5]. The cac is about 1 mmol/kg below the cmc of 14.0 mmol/kg for DoAC in water. In some systems the endothermic peaks seems to have flattened out but the perturbed enthalpy curve indicates significant interaction as seen in Figure 6b which shows the enthalpic titration curves for TTAB in 0.25 wt% EHEC D and in water at 25°C [5]. A cursory glance may give the impression of an increase in the cac compared to the cmc but we believe, based on the change of shape of the calorimetric curves with increasing temperature discussed below, that the initial endothermic peak has become so flat that the cac is not seen. Another system for which the calorimetric titration curve at 25°C does not give a clear indication of the cac is SDS–PVP [30]. Addition of SDS to 1.00 wt % PVP gives a steady increase in

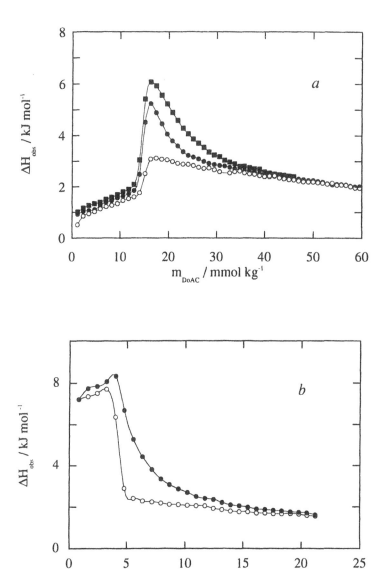

FIG. 6 (a) Calorimetric titration curves from the addition of 10 wt % DoAC solution to (●) 0.10 and (■) 0.20 wt % solution of PEO 8000 and (b) addition of 10 wt % TTAB solution to (●) 0.25 wt % solution of EHEC D at 25 °C. Open circles denote dilution of surfactant solutions in water. From ref. 5 with permission.

ΔH_{obs} to a well-defined maximum at 3.5 mmol/kg which is higher than the cac of 2.0 to 2.6 mmol/kg determined from surface tension [46, 47], conductance [42] and gel filtration experiments [40]. At 35°C, the endothermic peak has disappeared and there is a sudden drop in ΔH_{obs} at 2 mmol/kg which is the expected cac [29]. In general, an increase in temperature has a profound influence on the shape of the enthalpy curves. Figure 7 shows the titration curves from addition of SDS to 0.25 wt % EHEC D solution at 35 and 45°C [5]. The titration curve at 25°C closely resembles the one for PEO shown in Figure 2. As the temperature increases, the first endothermic peak decreases and at the same time the exothermic hump (relative to the water curve) fades away. At 45°C there is no crossing of the polymer and water curves. Although the shape of the enthalpy curves changes drastically, cac and C_2 were not significantly affected by the temperature increase. An important contribution to

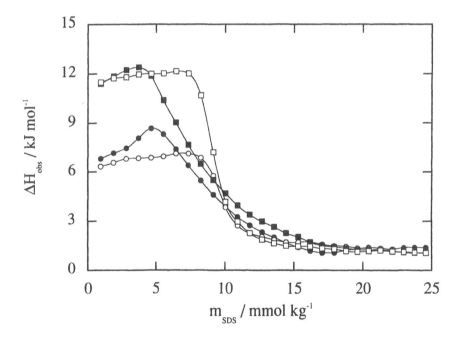

FIG. 7 Enthalpic titration curves from the addition of 10 wt % SDS to 0.25 wt % EHEC D at (●) 35°C and (■) 45°C and in water at (○)35 °C and (□) 45 °C. From ref. 5 with permission.

the change of shape of the enthalpy curves is the increasingly more exothermic micellization enthalpy shown by the increasing jump in the titration curve in water. The heat capacity change for micelle formation of SDS is -500 J/K mol [27]. In addition, the change with temperature of the hydrophobic hydration of polymer segments or substituents involved in the surfactant aggregates may contribute to the change of shape similar to what was observed for the solubilization of pentanol in SDS micelles, see Figure 3. There may also be contributions from changes in the polymer–surfactant interaction with temperature [5].

Calorimetric titration curves at 25 °C from the addition of micellar CTAC and DoAC solution to 0.10 wt % PEO solution give prominent endothermic peaks but only slightly reduced cac, *cf.* Figure 6a [29]. In solutions containing extra NaCl, the interaction peaks vanished at different salt concentrations in different systems depending on the strength of interaction as indicated by the decrease in the cac. The extra salt lowers the cmc of the surfactant and in these weakly interacting systems the polymer loses its stabilizing effect. Also in systems with stronger interaction such as SDS–PEO, the addition of extra salt decreases the strength of polymer–surfactant interaction as measured by the change in Gibbs energy but still has a strong influence on the behavior of the system. One effect of extra salt is an increase in C_2 and this increase is more pronounced for more hydrophobic polymers [29]. In solutions with extra salt, the electrostatic repulsion beween neighboring aggregates in the surfactant–polymer assemblies is screened and the polymer chains can accommodate more aggregates [48]. The titration curve from addition of $Mg(DS)_2$ solution to 0.10 wt % PEO in water looks quite similar to the curve from addition of SDS to 0.10 % PEO in 0.10 mol/kg NaCl at the same temperature, 35°C [29]. This is consistent with the more effective screening by the double-charged magnesium ion compared to the univalent counterions.

The interaction beween SDS and a hydrophobically modified EHEC, HM-EHEC, has been studied recently using various physico-chemical methods [49, 50]. HM-EHEC contains 1.7 mol % branched nonylphenol groups per mole anhydroglucose unit equivalent to approximately 6.5 groups per EHEC unimer. Hydrophobic microdomains of the micellar type form in solutions of HM-ECEC by the association of the nonylphenol groups. The calorimetric titration curves from the addition of SDS to solutions of HM–EHEC and the unmodified EHEC C (see Appendix) at 25°C are shown in Figure 8. The difference between the two curves in the beginning of the titration shows that already at very low concentrations SDS solubilizes in the hydrophobic domains to form mixed micelles. This solubilization is non-cooperative and the concentration of microdomains is low so they soon become saturated. On further increase of concentration, SDS starts to interact cooperatively with the polymer in the same way as with the unmodified EHEC. Analogous behavior was seen in the

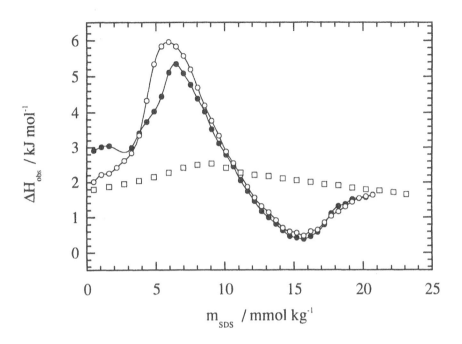

FIG. 8 Calorimetric titration curves at 25°C from the addition of 10 wt % SDS to 0.25 wt % solution of (●) HM-EHEC and (○) EHEC C and to (□) water. The solid lines are only guides for the eyes. From ref. 49 with permission.

enthalpy curves from the titration with SDS of solutions of PEO end-capped with dodecyl chains and the unmodified PEO [51]. SDS added in the beginning became solubilized in the aggregates formed by the dodecyl chains but as the aggregate concentration was low they soon became saturated. At higher concentration SDS interacted with main PEO chain in the same way as with unmodified PEO. Titration of end-capped PEO with DTAB showed deviation of the enthalpy curve at low surfactant concentration indicating solubilization of DTAB monomers in the hydrophobic domains. At higher DTAB concentration the titration curve became similar to the dilution curve in water showing that DTAB did not interact significantly, as expected, with the PEO backbone.

Recently, Engberts and co-workers have studied the interaction between single and double chain alkyl phosphates and alkylmethylammonium bromide surfactants with hydrophobically modified polyacrylates and polyacrylamides using titration calorimetry and cryo electron microscopy [52]. They observed, for instance, that the single chain surfactants are solubilized in hydrophobic microdomains, that the anionic surfactants interact with anionic, hydrophobically modified polymers despite the unfavorable electrostatic forces, and that mode of interaction between hydrophobically modified polymers and the single chain and the vesicle-forming double chain surfactants is different.

Skerjanc et al. [53,54] measured enthalpies of binding of dodecyl- and hexadecylpyridinium cations to poly(styrenesulfonate) at 25°C using a batch microcalorimeter. The binding isotherms were determined separately and the enthalpy of binding of the surfactant ions was derived. The binding enthalpy increases sharply at a low, well defined concentration, which agrees with the cac determined from potentiometric measurements using surfactant ion selective electrodes. Beyond this concentration the enthalpy stays constant. The observed binding enthalpy is considered to largely stem from aggregate formation by hydrophobic interaction.

Isothermal titration calorimetric measurements reflect the enthalpic effects resulting from the interaction between polymer and surfactant as a function of composition. It is also possible to use heat capacity measurements to probe polymer–surfactant interactions. Partial molar heat capacities and partial molar volumes have been shown to be sensitive indicators of structural changes in aqueous systems [55], and for instance micellar and microemulsion systems have been extensively studied using flow heat capacity microcalorimeters [8]. The study of heat capacities and volumes in aqueous PEO and PVP solutions as functions of concentration of ionic surfactants at 25°C showed that this type of thermodynamic mesurements can successfully be extended to polymer-surfactant solutions [56]. The method is described in ref. 8.

III. HIGH SENSITIVITY DIFFERENTIAL SCANNING CALORIMETRY, HSDSC

A. Principal Design Features

The measurement principle of high sensitivity differential scanning calorimeters, HSDSC, is basically the same as for ordinary power compensated DSC instruments. They are twin instruments with two identical cells, the sample cell containing the sample solution and a reference cell containing an inert reference solution. A sensitive temperature sensor is mounted between the sample and reference cells and the cells are mounted in an adiabatic shield. Other arrangements may also be found. In an experiment the two cells are heated at the same predetermined rate, usually between 30 to 90 K/h. Initially, the temperature of the two cells increases linearly with time and the temperature difference is maintained at zero. If a temperature difference starts to develop due to a thermally induced event in the sample, the differential control system supplies more (or less) power to the sample cell to eliminate the temperature difference. The recorded parameter is the differential power, proportional to the heat capacity difference between sample and reference cells, as a function of temperture. A DSC experiment consists of two temperature scans, one with the sample cell containing the sample solution and a second with an inert reference solution, usually the solvent, in the sample cell. The cell volume varies between 0.5 and 1.2 mL depending on the instrument. Commercial HSDSC instruments are available from CSC (Provo, Utah, USA), Institute for Biological Instrumentation, Russian Academy of Sciences, (Pushchino, Moscow Region, Russia), Microcal (Northampton, MA, USA) and Setaram, (Caluire, France). Cells in the instruments from the Institute for Biological Instrumentation and Microcal are permanently mounted, while the cells in the other instruments are removable.

B. Measuring Principle

In a temperature scanning calorimeter, one measures the specific heat of a system as a function of temperature. The apparent specific heat capacity of the solute, $c_{p,\phi}$, in a binary solution is given by

$$c_{p,\phi} = c_p{}^* + (c_p - c_p{}^*)/w_2 \tag{5}$$

where c_p is the specific heat of the solution, $c_p{}^*$ is that of the solvent, and w_2 is the weight fraction of the solute. Since the quantity $c_p - c_p{}^*$ is small for dilute solutions, for example approximately -0.7% for a 1% aqueous solution of an organic solute, calorimeters with a high sensitivity are needed. The apparent molar heat capacity of the solute at constant pressure, $C_{p,\phi}$, is given by

$$C_{p,\phi} = M_2 c_p + 1000(c_p - c_p{}^*)/m_2 \tag{6}$$

where M_2 is the molar mass of the solute and m_2 the molality. These expressions are valid for temperature scanning experiments with constant mass of solution and solvent in the cells. In order to avoid evaporation effects,

HSDSC instruments usually work with totally filled cells. Such calorimeters measure the difference in heat capacity between fixed volumes of solution and solvent as a function of temperature. The term excess apparent specific heat, $C_{p,ex}$, is used when describing the results. It is calculated as a function of temperature from the difference between the sample and reference scans. For accurate determinations of partial molar specific heats of solutes etc., density measurements of comparable accuracy are needed over the temperature range of interest.

Most HSDSC measurements are made to study thermally induced events such as phase changes or conformational transitions. Absolute values for the heat capacity are of secondary importance for this type of study. The enthalpy change for the thermally induced process can be calculated from the area under the transition peak. For ternary systems consisting of polymer, surfactant and water, thermodynamic equations for multicomponent systems should be used. However, in many cases solvent 1 plus component 2 can form a mixed solvent with fixed composition which can be considered as a solvent for solute 3 [8]. Normally, HSDSC studies are used to determine excess apparent heat capacities. In such studies the reference cell can contain either pure water or the mixed solvent (water + solute 2) when working with dilute solutions.

C. Interpretation of HSDSC Curves

The interpretation of DSC data on systems with structural changes such as protein denaturation or helix-coil transitions of polynucleotides is discussed, for instance, in the review article by Sturtevant [57]. This review contains literature references to the original articles presenting the evaluation methods. Structural transitions of small molecules cannot be studied by DSC unless they form aggregates of high cooperativity. Molecules or aggregates having masses of thousands of g/mol are required to give transitions sufficiently sharp for useful DSC observation [57].

The temperature-induced event most commonly encountered in polymer–surfactant systems is phase separation. Aqueous solutions of polymers such as PEO, PPO and various cellulose ethers show liquid-liquid phase separation upon heating. This event can be picked up by DSC if it takes place within the operational temperature (and pressure) range of the instrument. Usually, polymer solutions are not binary systems in a strict meaning as in most cases the polymers are not pure homologues but have a more or less wide distribution of molar masses. This should be taken into account in the thermodynamic treatment of, for instance, phase changes. However, the heat capacity behavior of truly binary systems will give an indication about what to expect from a polymer solution. Thermodynamic relations for the heat capacity of heterogeneous systems have been derived by Filippov [58] and the application of these relations to the interpretation of DSC results has been presented [59, 60]. When the temperature of a homogenous solution is raised and no phase transition occurs, the heat absorbed by the sample is spent on the heating of this single phase. The signal from the DSC instrument will be proportional to the heat capacity of the sample solution. On transition of the system from a one-phase to a two-phase state there is a jump in the heat capacity to a higher value. When the temperature of a heterogenous system is raised, the heat absorbed by the sample is spent on heating of the coexisting phases and on the transformation of one phase into the other. The apparent heat capacity of the system is the sum of the heat capacity of the two phases plus a heat capacity contribution from changes in composition of the phases with temperature.

HSDSC traces showing this simple heat capacity behavior upon liquid - liquid phase separation in aqueous polymer solution have been recorded for PEO (in salt solution?) [61] and for a rather hydrophilic EHEC, sample EHEC D, with a clouding temperature, CP, at 65°C in 1 wt% solution, see Figure 9a [7]. There is a modest increase of the specific excess heat over a limited temperature range covering the clouding temperature. In other cases, a large endothermic peak, located in the temperature region where clouding occurs, dominates the DSC traces [61-63].

Figure 9b shows the traces of solutions of two rather hydrophobic EHEC samples A and B, with clouding temperatures of 35 and 34°C. Characteristic data for the EHEC samples are given in the Appendix. The trace for the more hydrophilic sample EHEC D is included in the figure for comparison. The endothermic peaks observed in DSC traces of dilute solutions of EHEC and other clouding polymers such as poly(N-isopropylacrylamide) (PNIPAM) and poly(vinylmethylether) [62], (hydroxypropyl)cellulose [63] and PPO [61] appear at temperatures close to the cloud point. The temperature for the peak maxima, T_{max}, observed for various PNIPAM samples were very close to CP determined from optical density measurements [62]. It can be noted that these samples gave very narrow DSC peaks with half-height widths less than 1°C. For other polymers giving broader peaks, CP is lower than T_{max} but well above the onset temperature of the peak. The peaks observed for polymers such as PPO and EHEC have the same shape as the endothermic transition peaks ascribed to micelle formation observed in solutions of ethylene oxide - propylene oxide, EO-PO, block copolymers [61,64-67]. In analogy with this observation, the endotherms are attributed to aggregation of polymer chains or segments preceding phase separation [7,61].

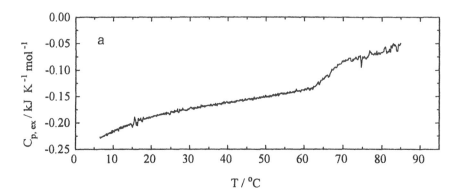

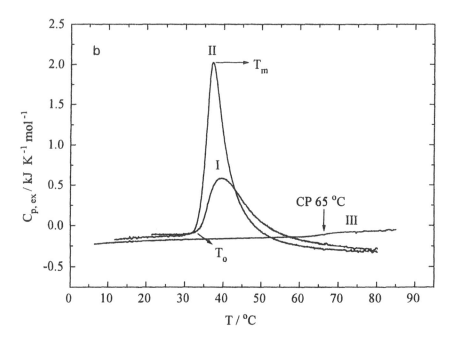

FIG. 9   (a) DSC trace of 2.00 wt % EHEC D (CP 65 °C) in water; (b) traces of 1.00 wt% (I) EHEC A (CP 35°C), (II) EHEC B (CP 34°C) and (III) 2.00 wt % EHEC D (CP 65°C) in water. $C_{p, ex}$ is expressed per mole of anhydroglucose unit.

For block copolymers forming micelles the excess heat capacity is lower after the transition peak than before. This means that the heat capacity change is negative which is expected as large negative heat capacity changes characterize, for instance, micelle formation of low molar mass amphiphiles in aqueous solution [28]. The hydrophobic groups in the interior of the micellar aggregates of normal surfactants or block copolymers will to a large extent lose their contact with water and this dehydration gives rise to the large negative heat capacity change. In contrast, the heat capacity change accompanying for instance the thermal denaturation of globular proteins is positive [57]. When the protein chain unfolds, hydrophobic groups from the water-free interior are exposed to water and the increased hydrophobic hydration gives an increase in heat capacity. Most polymer solutions for which endothermic peaks appear in DSC traces at the clouding temperature also show a pronounced decrease in heat capacity after the peak. The negative heat capacity change supports the assumption that the peak arises from aggregation and accompanying dehydration of hydrophobic substituents or segments. Various parameters such as the temperature for the beginning and the maximum of the peak, the peak width and the size, expressed as an enthalpy change, can be derived from the recorded peaks [7,62]. The parameter values can be used to characterize polymer samples. However, further studies are needed to elucidate the molecular mechanism giving rise to the endothermic peaks at about the clouding temperature in some polymer solutions.

D. Examples of Studies of Polymer-Surfactant Systems

Block copolymers of the type $EO_xPO_yEO_x$, also called poloxamers, have a wide industrial application. The compounds are surface active and form micelles and lyotropic liquid crystalline phases. Above the critical micellization temperature, cmt, or above the critical micelle concentration, cmc, they form micelles with a core containing the hydrophobic PO blocks. The aggregation number depends on the size of the PO block and the x/y ratio and is in the range 30-70 for many commercial compounds. In many applications the block copolymers are used in combination with "normal" surfactants. Hecht *et al.* [64,65] have made a detailed study of block copolymer/surfactant interaction using various techniques including HSDSC. They were particularly interested in finding out how addition of surfactants affected the micellar size and what the polymer/surfactant complexes looked like. The polymer $EO_{97}PO_{69}EO_{97}$, in the following denoted as F127, with a molar mass of 12 500 g/mol was used. Their study showed that the presence of even small amounts of ionic surfactants interfere with the micelle formation of the polymer and eventually at high enough surfactant concentration, it is totally

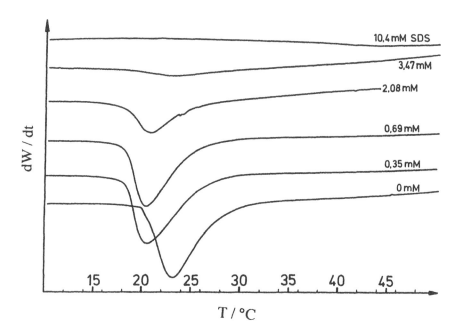

FIG. 10 DSC traces of 1 wt% F127 containing varying concentrations of SDS. The endothermic peaks indicate micelle formation. Scanning rate = 0.2 K/min (no added NaCl). From ref. 64 with permission.

suppressed. This process can be followed by DSC. Some of the curves showing results from 1 wt% F127 solutions containing up to 10.4 mmol/L of SDS are shown in Figure 10. Note that in this figure the endothermic peaks point downward, opposite to the convention used in the other figures. At very low concentrations of surfactant the micellization peak of F127 is not influenced, but above 0.6 mmol/L SDS, the peak decreases with increasing SDS concentration to completely disappear at 4.0 mmol/L of SDS. This last concentration corresponds to 4.3 mol of SDS per mole of F127. The micellization enthalpies derived from the peak areas are plotted against SDS concentration in Figure 11 [64]. An increase of the F127 concentration to 3 wt% shifts the changes to higher surfactant concentration but the DSC peak disappears at about the same surfactant to F127 molar ratio. As seen from Figure 11 the addition of the cationic surfactant CTAB has closely the same suppressive effect on the micellization of F127 as the anionic SDS. Mean aggregation numbers of F127 in 3 wt% solution were determined as a function

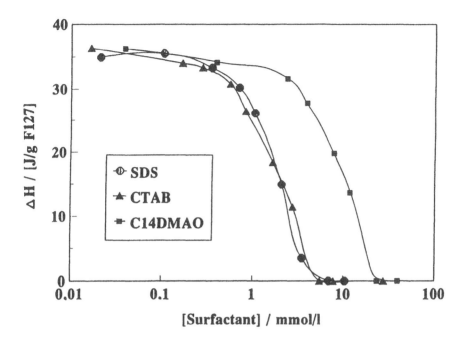

FIG. 11   Enthalpies of micelle formation of F127 in 1 wt% solution determined from DSC scans as function of SDS, CTAB and C$_{14}$DMAO concentrations.   From ref. 65 with permission.

of SDS concentration at 27°C from SANS experiments.   Without SDS, the aggregation number was determined to 69 and was constant up to an SDS concentration of 1 mmol/L but above this concentration it decreased continously to become 1 at 100 mmol/L. The authors conclude that the suppression of the F127 micelles by ionic surfactants is a continous process without marked onset or saturation.   DSC results for C$_{14}$DMAO included in Figure 11 show that this zwitterionic surfactant also supresses the micelle formation of block copolymers. However, the underlying mechanism may be different as the concentration needed is about 10 times higher than for the ionic surfactants although C$_{14}$DMAO has the lowest cmc (= 0.16 mmol/L).

Some hydrophobic forms of ethyl(hydroxyethyl)cellulose, EHEC, with cloud points, CP, of 30 to 35°C in 1 wt% solution, form thermoreversible gels in the presence of ionic surfactants [68,69]. This means that a liquid solution will convert to a transparent gel upon warming to slightly below its CP. The behavior is fully reversible as the gel liquifies upon cooling.   In a recent study,

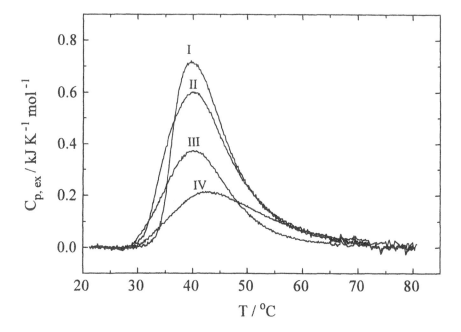

FIG. 12 DSC traces of 1.00 wt% EHEC A showing the effect of added SDS: (I) 0, (II) 0.5 mmol/kg, (III) 2 mmol/kg and (IV) 4 mmol/kg SDS. From ref. 7 with permission.

DSC experiments and phase studies were combined to investigate thermal events possibly related to thermoreversible gelation in EHEC systems containing SDS or CTAC [7,70]. Maps depicting the phase properties of 1 wt% solution of two different EHEC samples as functions of SDS or CTAC concentration at temperatures bewteen 20 and 80 °C show that the two surfactants have a similar influence on the phase behaviour. The turbidity boundary (cloud-point curve) increased steadily with surfactant concentration. Above a surfactant concentration of 1.5 to 2 mmol/kg a stiff gel formed at temperatures above 37–39 °C. The gel region extended to 6 to 7 mmol/kg of surfactant for one of the EHEC samples while with SDS it was significantly larger for the other sample. The part of the gel region situated below the turbidity boundary gave a clear, stable gel, while the gel was cloudy above the

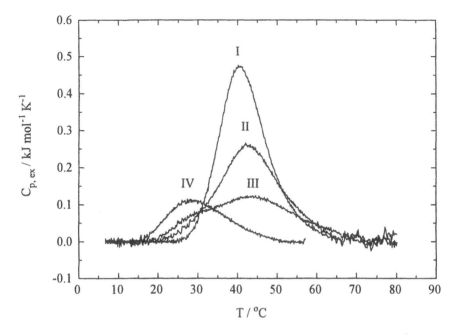

FIG. 13 The effect of addition of CTAC on DSC traces of 1.00 wt% EHEC A : (I) 2 mmol/kg, (II) 4 mmol/kg, (III) 6 mmol/kg, and (IV) 12 mmol/kg CTAC. From ref. 70 with permission.

boundary but showed no tendency to phase separate up to approximately 15°C above the cloud-point line. At higher temperatures syneresis took place. EHEC samples giving gel formation in the presence of surfactants gave large endothermic peaks in DSC traces of solutions in pure water as depicted in Figure 9b. The effect of addition of SDS is seen in Figure 12 which shows DSC curves for 1 wt % EHEC in pure water, in 0.5mmol/kg SDS which is well below the gel region, in 2 mmol/kg SDS just below- and 4 mmol/kg SDS well inside the region. As the SDS concentration further increased within the gel region, the peak moved to higher temperature and became smaller and broader. At SDS concentrations above the gel region, the peak started to increase in size and became less broad. The temperature for peak maxima moved with SDS concentration roughly in the same way as the cloud point. However, this was not the case in CTAC solutions as seen in Figure 13. Solutions I-III form gels above 37, 39 and 45°C, respectively. The peaks in the DSC curves broaden with increasing CTAC content but the maximum temperatures change only little. Solution IV containing a higher CTAC concentration is above the gelling region. It has a clouding temperature above 70°C but the peak in the DSC curve is centered around 30°C. This lack of correlation between the clouding temperature and the peak in DSC curves becomes more pronounced when a small amount of extra electrolyte is added. The cloud point of EHEC/CTAC solutions containing 4 mmol/kg NaCl drops below 20 °C for CTAC concentrations between 2.5 and 5.5 mmol/kg while the region for gel formation is not much changed. Without extra salt the cloud point increases from 40 to 56°C over the same CTAC concentration range. However, the DSC curves of an EHEC solution containing, for instance, 6 mmol/kg CTAC looked closely the same with and without 4 mmol/kg NaCl. Likewise, the DSC curve for EHEC solution containing 4 mmol/kg SDS remained almost unchanged upon addition of 4 mmol/kg NaCl (Figure 14), although the cloud point dropped from 45 to 27°C. The addition of extra salt separates the gelling interval from the one-phase region and gel formation occurs in macroscopically phase-separated samples [70]. The composition of the coexisting phases are not known but the polymer-rich phase still has a high water content. The phenomeon behind the peaks in the DSC curves appears to be determined by temperature. The increase in polymer content does not seem to have any significant influence. The drastic drop in CP caused by the addition of small amounts of additional electrolytes to aqueous solutions of nonionic polymers, like EHEC, containing low concentrations of ionic surfactants was first reported by Carlson et al. [71]. The effect is described as synergistic since the electrolytes alone at such low concentrations have no influence on the cloud point of the polymer solution and the surfactant itself would give (a more or less) monotonic increase of CP. The DSC results indicate different origins of the clouding/phase separation process at low surfactant concentration in solutions with or without extra electrolyte. Without salt, the cloud points and the peaks in the DSC curves were seen at temperatures in the vicinity or above the CP of the pure polymer solution. In the presence of salt, phase separation occurred at significantly lower temperatures without any detectible thermal effect. However, the gel points still were located in the vicinity or above the CP of the pure polymer solution where also the DSC peaks appeared. Thus, the temperature-induced aggregation of polymer, which is thought to give the thermal effect registered by DSC, seems to be needed for gelation but not for phase separation. The two phenomena could be separated by the addition of electrolyte. These findings are compatible with the explanation for the synergistic salt effect proposed by Piculell and Lindman [72].

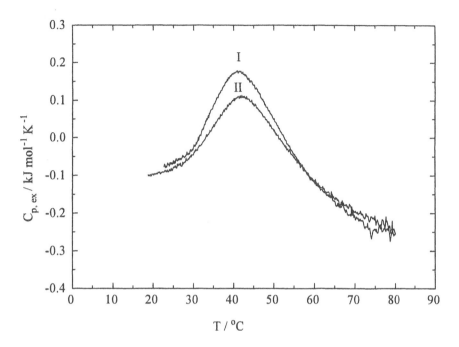

FIG. 14 DSC traces of 1.00 wt% EHEC A in 4 mmol/kg SDS (I) without extra salt and (II) in 4 mmol/kg NaCl. CP without salt is 45°C and in 4 mmol/kg NaCl 27°C. From ref. 70 with permission.

IV. CONCLUSIONS

Much interesting information about polymer–surfactant interaction can be gained from studies using isothermal titration and temperature scanning calorimetry. Both are microcalorimetric methods with a high sensitivity allowing measurements on dilute and semidilute solutions. Enthalpic titration curves from the addition of surfactants to aqueous polymer solutions give information about the influence of polymers on the aggregation of surfactants. For instance, the concentration for the start of aggregation can be derived as well as the concentration where the influence of the polymer ceases, and the method can allow conclusions on whether the binding is cooperative or not. From measurements at varying temperatures, the effect of temperature on the aggregation behavior can be inferred. High sensitivity differential scanning calorimetry, HSDSC, can be used to study thermally induced events in polymer solutions and how they are affected by the addition of surfactants. Examples of such events are micelle formation of block copolymers and the self-association preceding liquid-liquid phase separation in aqueous solutions of polymers containing hydrophobic groups or segments.

GLOSSARY

Symbols and notation

P	thermal power
U	electrical potential
q	heat quantity
ε	calibration constant
τ	time constant
cmc	critical micelle concentration
cac	critical aggregation concentration
T_1	critical aggregation concentration
C_2	"saturation" concentration
T_2	"saturation" concentration
C_{mic}	concentration of micelles
C_{tot}	total amphiphile concentration
f_{mic}	fraction of added surfactant in micellar form
N_S	aggregation number
ΔH_{obs}	observed enthalpy change
ΔH_{dil}	enthalpy of dilution (of micelles)
ΔH_{mic}	enthalpy of micelle formation
$c_{p,\phi}$	apparent specific heat of solute
c_p	specific heat of solution
$c_p{}^*$	specific heat of pure solvent
$C_{p,\phi}$	apparent molar heat capacity of solute
$C_{p,ex}$	apparent excess specific heat
CP	cloud point
M_2	molar mass of solute
m	molality
w	weight fraction

T_{max}	temperature for peak maximum
DSC	differential scanning calorimetry
HSDSC	high sensitivity differential scanning calorimetry

Polymers and surfactants

PEO	polyoxyethylene
PPO	polyoxypropylene
PVP	polyvinylpyrrolidone
EHEC	ethyl(hydroxyethyl)cellulose
HM-EHEC	hydrophobically modified EHEC
PNIPAM	poly(N-isopropylacrylamide)
F127	EO-PO-EO block copolymer
SDS	sodium dodecylsulfate
DTAB	dodecyltrimethylammonium bromide
TTAB	tetradecyltrimethylammonium bromide
CTAC	hexadecyltrimethylammonium chloride
DoAC	dodecylammonium chloride
OTG	n-octyl b-D-thioglucopyranoside
$C_{14}DMAO$	dimethyl-1-tetradecanamine N-oxide

APPENDIX

1. Samples of polyoxyethylene

$HO(CH_2CH_2O)_nH$

n=90	Mw = 4000	PEO 4k
n=180	Mw = 8000	PEO 8k
n=450	Mw = 20 000	PEO 20k
n=2250	Mw = 100 000	PEO 100k
n=33800	Mw = 1 500 000	PEO 1.5m

2. Characteristic parameters of investigated samples of EHEC (according to the manufacturer)

Batch name	Notation	DS(ethyl)	MS_{EO}	CP	M_n
CST 103	EHEC A	1.5	0.7	35 °C	100 000
DVT 89017	EHEC B	1.9	1.3	34 °C	100 000
EHM 0	EHEC C	0.6-0.7	1.8	65 °C	100 000
E230G	EHEC D	0.8	0.8	65 °C	100 000

DS(ethyl) is the average number of ethyl groups and MS_{EO} the number of ethylene oxide groups per anhydroglucose unit. CP values refer to 1.0 % aqueous solution.

REFERENCES

1. E.D. Goddard, *Colloids Surf. 19*: 255 (1986).
2. S. Saito, in *Nonionic Surfactants* (M.J. Schick, ed.), Marcel Dekker, New York 1987, pp. 881-926.
3. K. Hayakawa and J.C.T. Kwak, in *Cationic Surfactants* (D.N. Rubingh and P. M. Holland, eds.), Marcel Dekker, New York, 1991, pp.189-248.
4. B. Lindman and K. Thalberg, in *Interactions of Surfactants with Polymers and Proteins* (E.D. Goddard and K.P. Ananthapadmanabhan, eds.), CRC, Boca Raton, Fla, 1993, pp. 203-276.
5. G. Wang and G. Olofsson, *J. Phys. Chem. 99*: 5588 (1995).
6. H. G. Schild and D. A. Tirrell, *Langmuir 7*: 665 (1991).
7. G. Wang, K. Lindell, and G. Olofsson, *Macromolecules, 30*: 105 (1997).
8. J.E. Desnoyers, G. Perron, and A.H. Roux, in *Surfactant Solutions. New Methods of Investigation* (R. Zana ed.), Marcel Dekker, New York, 1987, pp. 1-55.
9. *Solution Calorimetry. Experimental Thermodynamics.* Vol. IV (K.N. Marsh and P.A.G. O'Hare, eds.) Blackwell, London, 1994.
10. M. Bastos, S. Hägg, P. Lönnbro, and I. Wadsö, *J. Biochem. Biophys. Methods, 23*: 255 (1991).
11. J. Suurkuusk and I. Wadsö, *Chem. Scr. 20*:155 (1982).
12. I. Wadsö, in *Solution Calorimetry. Experimental Thermodynamics*, Vol. IV. (K.N. Marsh and P. A. G. O'Hare, eds.), Blackwell, Oxford, 1994, pp. 267-301.
13. T. Wiseman, S. Williston, J.F. Brandts, and L.-N. Lin, *Anal. Biochem. 179*: 131 (1989).
14. C. Holmberg, S. Nilsson, S.K. Singh, and L.-O. Sundelöf, *J. Phys. Chem. 96*: 871 (1992).
15. G.C. Kresheck and W.A. Hargraves, *J. Colloid Interface Sci. 83*: 1 (1981).
16. G.C. Kresheck, W.A. Hargraves, and D.C. Mann, *J. Phys. Chem. 81*: 532 (1977).

17. K.M. Kale, L. Vitello, G.C. Kresheck, G. Vanderkooi, and R.J. Albers, *Biopolymers 18*: 1889 (1979).
18. G.C. Kresheck and W.A. Hargraves, *J. Collloid Interface Sci. 48*: 481 (1974).
19. L.-E. Briggner and I. Wadsö, *J. Biochem. Biophys. Methods 22*: 101 (1991).
20. D.M. Bloor, J.F. Holzwarth, and E. Wyn-Jones, *Langmuir 11*: 2312 (1995).
21. D.M. Bloor, W.M.Z. Wan-Yunus, W.A. Wan-Badhi, Y. Li, J.F. Holzwarth, and E. Wyn-Jones. *Langmuir 11*: 3395 (1995).
22. F. Tiberg, *Ph. D. Thesis*, Lund University, Sweden, 1994.
23. B. Lindman and H. Wennerström, *Top. Curr. Chem., 87*:1 (1980).
24. D.H. Everett, *Pure Appl.Chem. 31*: 577 (1972).
25. (a) B. Andersson and G. Olofsson, *J. Chem. Soc., Faraday Trans.1, 84*:4087 (1988); (b) G. Olofsson and G. Wang, unpublished results.
26. J.E. Desnoyers, G. Caron, R. DeLisi, D. Roberts, A. Roux and G. Perron, *J. Phys. Chem. 87*:1397 (1983).
27. I. Johnson, G. Olofsson, and B. Jönsson, *J. Chem. Soc. Faraday Trans. 1. 83*: 3331 (1987).
28. P. Stenius, S. Backlund, and P. Ekwall, in *Thermodynamic and Transport Properties of Organic Salts* (P. Franzosini and M. Sanensi, eds), IUPAC Chemical Data Series No. 28, Pergamon, Oxford, 1980, pp 295-319.
29. G. Wang, *Ph.D. thesis*, Lund University, Lund, Sweden, 1997.
30. G. Olofsson and G. Wang, *Pure Appl. Chem. 66*: 527 (1994).
31. (a) J. van Stam, M. Almgren, and C. Lindblad, *Prog. Colloid. Polym. Sci. 84*: 13 (1991); (b) J. van Stam, W. Brown, J. Fundin, M. Almgren, and C. Lindblad, in *Colloid–Polymer Interactions,*(P.L. Dubin and P. Tong, eds.) ACS Symp. Ser. 532 1993, pp. 195-215.
32. G. Olofsson, *J. Phys. Chem. 89*: 1473 (1985).
33. B. Andersson and G. Olofsson, *J. Solution Chem. 18*: 1019 (1989).
34. E.A.G. Aniansson, S.N. Wall, M. Almgren, H. Hoffmann, I. Kielmann, W. Ulbricht, R. Zana, J. Lang, and C. Tondre, *J. Phys. Chem. 80*: 905 (1976).
35. I. Johnson, G. Olofsson, M. Landgren and B. Jönsson, *J. Chem. Soc. Faraday Trans.1, 85*: 4211 (1989).
36. B. Jönsson and H. Wennerström, *J. Phys. Chem. 91*: 338 (1987).
37. M. Landgren, *Ph.D. Thesis*, Lund University, Lund, Sweden, 1990.
38. B. Jönsson, M. Landgren, and G. Olofsson, *in Solubilization in Surfactant Aggregates,* (S.D. Christian and J.F. Scamehorn eds.), Marcel Dekker, New York, 1995, pp. 115-141.
39. M.N. Jones, *J. Colloid Interface Sci. 23*: 36 (1967).
40. W.A. Wan-Badhi, W.M.Z Wan-Yunus, D.M. Bloor, D.G. Hall, and E. Wyn-Jones, *J. Chem. Soc. Faraday Trans. 89*: 2737 (1993).
41. J. François, J. Dayantis, and J. Sabbadin, *Eur. Polym. J. 21*: 165 (1985).
42. N.W. Fadnavis and J.B.F.N. Engberts, *J. Am. Chem. Soc. 106*: 2636 (1984).
43. N.W. Fadnavis, H.-J. van den Berg, and J.B.F.N. Engberts, *J. Org. Chem. 50*: 48 (1985).
44. J.C. Brackman, N.M. van Os, and J B.F.N. Engberts, *Langmuir 4*: 1266 (1988).
45. J.C. Brackman and J.B.F.N. Engberts, *Chem. Soc. Rev.* (1993):85.

46. R. Zana, J. Lang, and P. Lianos, in *Microdomains in Polymer Solutions* (P. Dubin, ed.), Plenum, New York, 1985, pp. 357-368.
47. H. Lange, *Kolloid-Z. u. Z. Polymere 243*: 101 (1971).
48. B. Cabane and R. Duplessix, *J. Physique 43*: 1529 (1982).
49. K. Thuresson, B. Nyström, G. Wang, and B. Lindman, *Langmuir 11*:3730 (1995).
50. K. Thuresson, O. Söderman, P. Hansson and G. Wang, *J. Phys. Chem. 100*: 4909 (1996).
51. K. Persson, G. Wang, and G. Olofsson, *J. Chem. Soc. Faraday Trans. 90*: 3555 (1994).
52. J. Kevelam, J.F.L. van Breemen, W. Blokzijl, and J.B.F.N. Engberts, *Langmuir 12*: 4709 (1996).
53. J. Skerjanc, K. Kogej, and G. Vesnaver, *J. Phys. Chem. 92*: 6382 (1988).
54. J. Skerjanc and K. Kogej, *J. Phys. Chem. 93*: 7913 (1989).
55. J.E. Desnoyers, *Pure Appl. Chem. 54*: 1469 (1982).
56. G. Perron, J. Francoeur, J.E. Desnoyers, and J.C.T. Kwak, *Can. J. Chem. 65*: 990 (1987).
57. J.M. Sturtevant, *Ann. Rev. Phys. Chem. 38*: 4463 (1987).
58. V.K. Filippov, *Dokl. Akad. Nauk. SSSR; Phys. Chem. Sect. (Engl. Trans.) 242*: 802 (1978).
59. V.K. Filippov and G.G. Chernik, *Thermochim. Acta 101*: 65 (1986).
60. G.G. Chernik, *J. Colloid Interface Sci. 141*: 400 (1991).
61. J. Armstrong, B. Chowdhry, R. O'Brien, A. Beezer, J. Mitchell, and S. Leharne, *J. Phys. Chem. 99*: 590 (1995).
62. H. G. Schild and D. A.Tirrell, *J. Phys. Chem. 94*: 4352 (1990).
63. L. Robitaille, N. Turcotte, S. Fortin, and G. Charlet, *Macromolecules 24*: 2413 (1991).
64. E. Hecht and H. Hoffmann, *Langmuir 10*: 86 (1994).
65. E. Hecht, K. Mortensen, M. Gradzielski, and H. Hoffmann, *J. Phys. Chem. 99*:4866 (1995).
66. N.M. Mitchard, A.E. Beezer, J.C. Mitchell, J.K. Armstrong, B.Z. Chowdhry, S. Leharne, and G. Buckton, *J. Phys. Chem. 96*: 9507 (1992).
67. J. Armstrong, B. Chowdhry, J. Mitchell, A. Beezer, and S. Leharne, *J. Phys.Chem. 100*: 1738 (1996).
68. A. Carlsson, G. Karlström, and B. Lindman, *Colloids Surf. 47*:147 (1990).
69. B. Nyström, H. Walderhaug, and F.K. Hansen, *Langmuir 11*: 750 (1995).
70. K. Lindell and G. Wang, *J. Phys. Chem.*, submitted.
71. A. Carlsson, G. Karlström, and B. Lindman, *Langmuir 2*: 536 (1986).
72. L. Piculell and B. Lindman, *Adv. Colloid Interface Sci. 41*: 149 (1992).

9

Interactions of Polymers and Nonionic Surfactants

SHUJI SAITO Consultant, Nigawa-Takamaru 1-12-15, Takarazuka 665, Japan

DAN F. ANGHEL Institute of Physical Chemistry, Bucharest, Romania

I. INTRODUCTION

Interactions between polymers and surfactants in aqueous system are attracting growing interest in practice and in academic research. Upon addition of anionic surfactants solutions of neutral polymers show significant characteristic features. Such features are not normally displayed by cationic surfactants [1-3]. On the other hand, the pronounced interaction of cationic surfactants with anionic polyelectrolytes has been the object of detailed studies. In the case of nonionic surfactants, ever since the early stages of the study of polymer-surfactant interactions it has been demonstrated that interactions between nonionic surfactants and polymers are weak or non-existent [4]. Accordingly nonionic surfactants were simply neglected in their relation to polymers.

The differences between anionic, cationic, and nonionic surfactants are a very important clue for understanding the binding mechanism to polymers, or complex formation. Thus, studies on cationic and nonionic surfactants seemed to be of interest only in this context. Recently, interest in nonionic polymer-cationic surfactant systems, and particularly nonionic surfactant systems has resurfaced. This is because in fact interactions do exist when the polymer is moderately hydrophobic, and especially in the concentrated region the study of phase behavior has attracted considerable interest. In addition, hydrophobically modified polymers have a much stronger interaction with surfactants, including nonionic surfactants, and present a promising field for academic study as well as for application research [3].

Polyethylene oxide has a strong affinity toward polymeric carboxylic acids through a sequence of hydrogen bonds. Ethoxylated nonionic surfactants are likewise bound to polyacids and exhibit distinct behavior as micellar aggregates grow on the polymer. This emphasizes that in polymer-surfactant systems hydrophilic as well as hydrophobic parts participate together for stronger complex formation.

Several extensive and specific-topic reviews of polymer-surfactant interactions have appeared in recent years [1-7], but such reviews have scarcely addressed the case of nonionic surfactants. A review by one of the present authors was published in 1987 [8]. Following by more than a decade, the present work presents a substantially updated or rewritten review of the topic.

In this chapter, polymer-nonionic surfactant interactions in general are treated, but the specific focus is on systems of polyacids and ethoxy-type surfactants in aqueous solution. Adsorption of nonionic surfactants on solid acids and silica is excluded, although this process is closely connected with the above-mentioned topic. Nonionic surfactants are increasingly employed in treatments of native proteins as biologically mild processors and they are drawing new research interest. This review covers only some physico-chemical properties of interaction between typical water-soluble proteins and nonionic surfactants. We will start with a review of the role of surfactant headgroups in polymer-surfactant interactions, as discussed from the viewpoint of one of the present authors.

II. THE HEADGROUP PROBLEM IN SURFACTANTS

A. Background

The role of the headgroup has always been a central concern in the study of polymer-surfactant interactions. From the fact that anionic surfactants exhibit strong interaction or binding with neutral synthetic polymers such as polyvinylalcohol (PVA), polyethylene oxide (PEO), and polyvinylpyrrolidone (PVP), it seems that hydrophobic attraction will be at work in the interaction, as no conspicuous electrostatic forces operate between them. However, cationic and nonionic surfactants interact only slightly or not at all with these polymers. They interact markedly with more hydrophobic polymers like polypropylene oxide (PPO) or partially acetylated polyvinylalcohol (PVA-Ac), but they are also much weaker compared with anionic surfactants. Thus, the role of the surfactant headgroup has long perplexed researchers, and still intrigues [1-12].

In a comparison of headgroup charge in the interaction with polymers like PVA, PVP or PEO, it was argued that in case of sodium dodecylsulfate (NaDS) ion-dipole attractions would be in effect, however, in cationic surfactants the large size of headgroups such as trimethylammonium or pyridinium would negate or weaken such an attraction. This possibility was ruled out in early 1950s because dodecylammonium chloride (DACl), which has a similar headgroup size as that of NaDS, was found not to interact with these polymers. The ion-dipole attraction is screened by salt addition, but conversely the binding of NaDS to PEO or PVP was promoted by NaCl addition [1,2]. DS anions with alkali metal ions but also with NH_4^+, $(CH_3)_4N^+$, or $(C_2H_5)_4N^+$ as counterions were interactive with PVP or PVA-Ac [13]. Anionic surfactants with phosphate as headgroups ($R-PO_4Na$) interact with nonionic polymers including PEO, but their effect was considerably weaker in comparison with the corresponding sulfates [14]. Sodium dodecyl-$(EO)_n$-sulfates were hampered from interaction with PVP when the PEO chains were longer than about 6 EOs [15].

Complex formation between polyacrylic acid (PAA) and NaDS was suggested from viscosity measurements [16] and further confirmed by fluorescence methods [17]. This indicates that DS anions are bound to PAA and to partially dissociated PAA against long-range electrostatic repulsion, though weak in this case, and therefore a slight positive charge on the neutral polymer can not explain the favorable binding of anionic surfactants. Even if negative charge was conferred on PVP in alkaline media, tetradecylpyridinium bromide did not exhibit a strong cooperative affinity to it as is usually the case for anionic surfactants [12].

Studies on the simple salt effect on PVA-Ac [18] and PEO [19] in water revealed that anions were more influential on the solution behavior of the polymers. Such phenomena are not limited to a specific polymer. The interaction of ionic surfactants with nonionic polymers may therefore be considered as a kind of salt effect on the polymers, reinforced by the presence of a large hydrophobic moiety. The surfactant ion with its hydrophobic moiety is attracted more closely or bound to the polymers even in the form of micelles. It enhances its characteristic effect as a salt on the polymer at lower concentration than the ordinary salts [8,13,18,20-23]. Since the hydrophobic hydration water of organic ions is influenced by the counterions, we will first discuss the systematic exploration of the hydrophobic effect of the organic ions and medium effect of counterions on the interaction of surfactants with nonionic polymers. Of course the functions of cation and anion are inseparable, but we will start with the role of the counterion.

B.   The Role of the Counterion in Ionic Surfactants

The effect of a series of monoalkylammonium salts ($R-NH_3^+$ X^-) with various counterions X on the cloud point of 30.0 mol % acetylated PVA-Ac or on the reduced viscosity of PVP was studied [18,20]. The results were compared with the effects of NH_4X and tetraalkylammonium salts (R_4N^+ X^-), whose solution properties are known in terms of water structure around the anions and hydrophobic cations [20,24,25]. In a similar way, series of alkylsulfates and alkylcarboxylates with counterions including R_4N^+ and guanidinium$^+$ were investigated [13]. The cloud point of the moderately hydrophobic PVA-Ac was found to be more sensitive than viscosity as a qualitative measure of interaction on addition of those organic salts.

Figure 1 shows the cloud point of a PVA-Ac solution on addition of various inorganic ammonium salts and short chain $R-NH_3X$ salts in the dilute range. Whereas NH_4Cl and NH_4Br differ considerably in changing the cloud point, very short alkyl chain chloride salts have little effect. As the chainlength increases, the effect of $R-NH_3Cl$ becomes more pronounced. The steep rise at higher concentration can be ascribed to cooperative binding of the cations on the polymer. The medium effect of the counterion is limited.

For longer alkyl chain lengths, and with different anionic counterions following the lyotropic series, the $R-NH_3^+$ salts exhibited quite a divergent behavior with PVP [18,20], and the appearance of the interaction was entirely different from that of anionic surfactants. For example, whereas addition of anionic surfactants generally raised the viscosity of a solution of nonionic polymer like PVP or PVA-Ac markably, Figure 2 shows that addition of cationic surfactant dodecylammonium with F^-, Ac^- (acetate), Cl^-, and NO_3^- as counterions hardly changed the viscosity, but with SCN^- or I^- counterions the

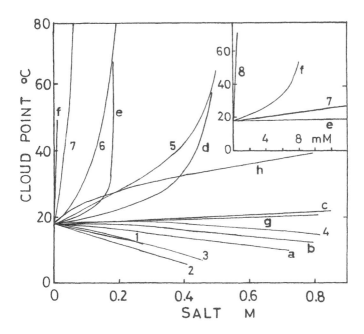

FIG. 1 Change of cloud point of 0.21% PVA-Ac (30.0 mol % acetate content) solution by addition of ammonium and n-alkylammonium halides. (a) NH_4Cl and methyl-NH_3Cl; (b) ethyl-NH_3Cl; (c) butyl-NH_3Cl; (d) hexyl-NH_3Cl; (e) octyl-NH_3Cl; (f) dodecyl-NH_3Cl; (g) NH_4Br; (h) NH_4I. The curve for NH_4SCN is a little higher than the one for (h) NH_4I[20]. The same by addition of sodium n-alkylcarboxylates and sodium n-alkylsulfates: (1) Na formate; (2) Na acetate; (3) Na butyrate; (4) Na methyl-SO_4; (5) Na butyl-SO_4; (6) Na hexyl-SO_4; (7) Na octyl-SO_4; (8) Na dodecyl-SO_4. From ref. 18, reprinted with permission of John Wiley & Sons, Inc.

viscosity initially decreases in the low concentration region and increases strongly at higher surfactant concentration region above the cmc [20]. Notably, octylammonium thiocyanate precipitated PVP and PVA-Ac [23]. Water-insoluble polyvinylacetate was solubilized by dodecylammonium thiocyanate (DASCN) but not by DACl or by dodecylpyridinium thiocyanate [22]. From dye solubilization studies also interaction was inferred between long-chain R-NH_3^+ and PVP, PEO or PVA-Ac in the presence of SCN^- counterions [21]. Enhanced binding of dodecylammonium$^+$ to these polymers by SCN^- or I^- was confirmed by conductivity and more directly by equilibrium dialysis measurements [26,27]. This behavior may be due to concomitant binding of SCN^- or I^- and surfactant cation to the polymer initially as ion pair for inducing further binding of the surfactant. But there is no evidence that these anions alone bind

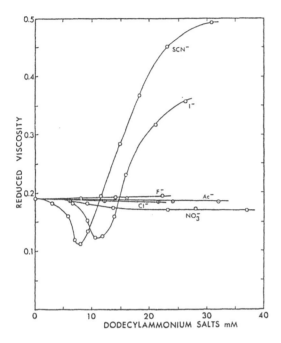

FIG. 2 Effect of dodecylammonium salts of various counterions on the reduced viscosity of 0.17% PVP solution at 25°C. From ref. 20 with permission.

directly and explicitly to PVP.

In dodecyltrimethylammonium or dodecylpyridinium salts, strong counterion-induced surfactant binding was observed with PVA-Ac, but not with PVP though this polymer interacted weakly with hexadecylpyridinium thiocyanate [21]. It seems that some mutual steric factor in the hydrophobic parts of both cosolutes plays a role in this kind of interaction.

This counterion effect may be interpreted as follows. In aqueous $R_4N^+X^-$ ($X^- = F^-$, Cl^-, Br^-, and I^-) solutions, the activity coefficients for the fluorides are very high, but those for the iodides fall below the values predicted by the Debye-Hückel law. This is explained by the concept of "structural salting-out and salting-in" derived from incompatibilities of water structure around cation and anion. R_4N^+ with clathrate or cage-like hydration is self-salting-out by the most strongly electrostrictive F^-, but self-salting-in by a very strongly water-structure breaking I^- ion: the I^- ion makes water available for the self-salting-in of the hydrophobic cation [24]. Thus, it is considered that, in contrast to electrostrictive anions such as F^-, Ac^- or Cl^-, strongly water-structure-breaking anions such as SCN⁻ and I⁻ tend to induce self-assembly of hydrophobic cations. As an example, the cmcs for DAF, DACl, DAI, and DASCN, respectively, are 13, 11, 7.0, and 5.7 mM at 30°C [21]. When a polymer is present, structure breaking counterions make the hydrophobic parts associate by forming ion pairs. Structure breaking counterions seem to favor hydrophobic contact between polymer and DA⁺ Anionic counterions such as ethylsulfate, butylsulfate, or butyrate, whose hydrophobic groups are attracted to the hydrophobic part of the polymer, do not induce a binding effect for surfactant cations [20,21].

Likewise, in case of the interaction between alkylsulfates⁻ and PVP or PVA-Ac a similar viscosity behavior as observed in DASCN is observed when changing the counterion from Na⁺ to guanidinium, a strongly water-structure breaking cation (cmcs for Na and guanidinium octylsulfates at 30°C are 135 and 61 mM, respectively), though the effect was not so pronounced as in DASCN, and it did not appear with R_4N^+ as counterion [13]. Therefore, the same mechanism holds with both cationic and anionic surfactants. A detailed discussion is given in refs. 13 and 20.

These same phenomena may be related to the fact that SCN⁻ and ClO_4^-, another strongly water-structure-breaking anion, stabilized the α-helix of some basic polyamino acids such as poly-L-lysine, poly-L-arginine, and also a random copolymer of L-lysine and L-phenylalanine [28-32]. As an aside, HLB values of lysine, arginine, and phenylalanine are, respectively, 6.3, 3.1, and 11.0 [33].

C.   The Role of Headgroup Charge and Nonionic Surfactants

Figure 1 also shows the cloud point of PVA-Ac solution on addition of sodium n-alkylsulfates or alkylcarboxylates. The alkylsulfates raise the CP as the alkyl chainlength increases from methyl to octyl. The alkylcarboxylates from formate to butyrate on the other hand cause nearly the same extent of cloud point depression. This is due to the very strong electrostrictive effect of -COO⁻ that offsets the hydrophobic contribution of the short alkyl chains. The difference between sulfate and carboxylate headgroups of short chain alkyl anions in the cloud point change is indeed remarkable, taking into account that NaDS and sodium laurate both have a strong interaction with PVA [34]. Thus, in addition to the counterion effect, the hydrophobicity around the alkyl portion near the ionic headgroups may be different dependent on the sign and nature of ionic heads, even though SO_4^- and NH_3^+ are regarded mild in affecting the water structure. Moreover, the size of headgroup, steric and polar factors such as hydrogen bonding or hydration in both components, etc. [2] make a contribution to the interaction. When the salts with longer alkyl chains accumulate locally on the polymer, and when they form micellar assemblies there, the distinction in the interactions is demonstrated even more strikingly.

It is noted from Figure 1 that the steep rise in the cloud point of PVA-Ac takes place far below the respective cmc, especially in the case of RSO_4Na (cmcs for Na hexylsulfate, Na octylsulfate, and octylammoniumchloride are, respectively, 1.2, 0.13, and 0.25 M). The effect of RNH_3Cl on the cloud point is lower than that of RSO_4Na with the same R, the difference being equivalent to that of a difference of a few methylene groups. The cmcs of RNH_3Cl or $R(CH_3)_3NCl$ are higher than that of $NaR\text{-}SO_4$ of the same chain length[35]. In hydrophilic polymers like PVP or PEO, the hydration sheath around thehydrophobic parts is perturbed to some extent by the neighboring hydrophilic parts. Nevertheless, NaDS is bound by PVP, whereas cetylammonium chloride interacts only slightly [34a], and DACl not at all (Figure 2). In binding to PVP and PEO, the hydrophobic parts of $R\text{-}NH_3^+$ or of other cationic surfactants seem less hydrophobic than the corresponding $R\text{-}SO_4^-$ and this apparent difference is much larger than with PVA-Ac. These cases concern the interaction between polymers and hydrophobic parts of micellar surface, but the problem of anion vs. cation in surfactants persists even below the concentration of micellar binding, and when micellar binding takes place on polymer, the phenomenon appears more pronounced. NMR studies have shown that some methylene units adjacent to the anionic headgroup of NaDS are involved in the binding to PEO [36,37].

The influence of ionic head charge is not limited to surfactant ions but is also evident in the interactions between PVP and dye ions [38] and small organic ions [39-41]. Phenylalanine at pH 10 and naphthoate, for example, were bound but this amino acid at pH 2 and naphthylamine hydrochloride were not [39,40]. Even for inorganic salts, anions usually cause a variety of effects on the bulk water structure and determine properties at air-water and hydrophobic interfaces, originating from the opposite orientation of water molecules around anions compared to cations, and from a wider range of sizes [42-44]. The least hydrated anions are most strongly adsorbed at the interfaces [45].

From these facts we may speculate that in polymer-anionic surfactant systems the long-alkyl chain anions dominate around the hydrophobic parts of polymers over the strongly hydrated cationic counterions such as Na^+ or K^+. In the cationic surfactant systems the attraction of long-chain cations to polymers is smaller than that of the corresponding anions and moreover reduced by strongly hydrated anionic counterions such as F^- or Cl^-. Hydrophobic attraction of the long-chain cations to the polymer can be increased by the presence of the least hydrated and strongly water-structure breaking counterions like SCN^- or I^-. On the other hand, in anionic surfactants their binding is also affected by the structure breaking guanidinium counterion, but not so strongly as is the case for cationic surfactants with SCN⁻ or I⁻ [13].

It may be deduced that two-point (hydrophobic and polar) binding of surfactants to neutral polymers is a requirement for stronger interaction, and the ionic head-counterion effect (salt effect) is one of polar attractions. More hydrophobic cationic surfactants with much lower cmc's would be unfavorable for the interaction because they prefer to form their own micelles. Thus, even only slightly more hydrophobic polymers may possibly have some interaction with cationic surfactants by covering the headgroup influence. In this regard, nonionic surfactants would take a position in between anionic and cationic surfactants when we consider only the dilute concentration range relative to their low cmc's and neglect effects of their bulky headgroups or hydration. This is discussed further in Section IIIA of this chapter.

The interactions between n-dodecyldimethylamine oxide (DDAO) and PPO, polyvinylmethylether (PVME), or PEO were studied at different pHs. This surfactant is nonionic but semipolar at pH 7, but becomes cationic at pH 2-3 without a drastic change in headgroup size. From pH 7 to lower pH, the Gibbs energy of micelle formation of DDAO in the presence of PPO or PVME decreases, and the aggregation numbers of the micelles is reduced. This is due to the fact that at lower pH electrostatic repulsion on the micellar surface is diminished by binding to the polymers in comparison to the nonionic micelles. DDAO interacts with PPO and PVME, but not with PEO at any degree of protonation despite its small head [5a,46]. These findings may imply that the reason for the absence of interactions between nonionic surfactants and hydrophilic polymers such as PVP and PEO is not primarily their large headgroups but maybe the weak mutual attraction due to lack of negative charge, as hypothesized above.

III. INTERACTIONS OF POLYMERS AND NONIONIC SURFACTANTS

Nomenclature: in the following, monodisperse ethoxylated alkylether surfactants are denoted as C_mE_n and polydisperse surfactants as $(EO)_nC_m$ or PEO-alkylethers.   PEO-octylphenyl-, and PEO-nonylphenylethers are abbreviated as $(EO)_n$ OP and $(EO)_n$ NP, respectively.

A. Dilute Region

Typically hydrophilic polymers such as PVA, PEO, and PVP show no sign of interaction with ethoxylated nonionic surfactants by dye solubilization, viscosity [34], and surface tension [47] methods. No binding between PVP and Triton X-100 ((EO)$_{9-10}$OP) was found by the ultraviolet absorption method at the cmc [48], or between PVP and alkyldimethylphosphine oxide, a nonionic surfactant, by calorimetric measurement [49]. From gel permeation chromatography it was observed that PEO and PEO-NP had a weak interaction [50]. As described in Section II, bulkiness of hydrophilic headgroups is not the only reason for deterring the nonionic surfactants from interaction with those polymers. The surfactants do display some interaction with mildly hydrophobic nonionic polymers such as PVA-Ac and PPO [51], and PEO-dodecylether solutions solubilized water-insoluble PPO with a molecular weight of 2000 [34]. See also Section III.C.

Recently, the interaction between C$_{12}$E$_5$ and PEO of high molecular weight was studied by dynamic light scattering and fluorescence methods over a temperatures range up to the cloud point. Clusters of small surfactant micelles were formed within the PEO coil, leading to its expansion. The excluded volume effect resulting from the addition of high molecular weight polymer to the sytem drives formation of the surfactant micelle [135,136].

Interaction is also demonstrated, for example, by the rise of cloud point of mildly hydrophobic polymers upon addition of nonionic surfactants. Figure 3 shows such changes in cloud point of 30.0 mol % acetylated PVA-Ac solution [51]. Aqueous solutions of (EO)$_n$OP do not exhibit clouding in the temperature range studied, therefore the rise in cloud point of PVA-Ac or PPO, is attributed to binding of the surfactants to these polymers. On a per mol basis, the nonionic surfactants (cmc<1 mM) are far less effective than NaDS (cmc: 8 mM) in raising the cloud point, but those with longer PEO chains are more effective than DACl (cmc 11 mM, see Figure 1) and dodecyl-trimethylammonium chloride (cmc 20 mM). With increasing surfactant concentration, the cloud point rise due to the addition of nonionic surfactant decreases, and is even exceeded by the cationic surfactant. The crossing of the curves is ascribed to the fact that at higher temperature the nonionic surfactants become less hydrophilic, while the cationic surfactant concentration is above the cmc. From Figure 3 it is noted that the nonionic surfactants with higher cmc are more effective in raising the cloud point despite the longer PEO headgroups. This increase may be related to the number of EO groups on PVA-Ac rather than to surfactant binding. Addition of nonionic surfactant to PVA-Ac also results in a greater solubilization ability for the oil-soluble dye Yellow OB [51]. This increase is due to micellar surfactant binding on PVA-Ac. No such increase is observed with PVA or PVP. PVA-Ac with 68 mol % acetate groups can be solubilized in an (EO)$_{15}$OP solution. Possibly, the weak hydrophobic attraction between the acetate groups in PVA-Ac is overcome by surfactant binding.

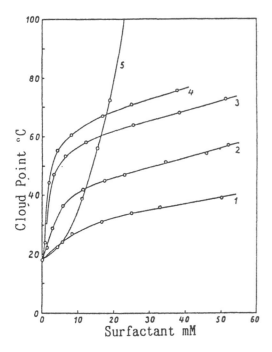

FIG. 3 Change of cloud point of 0.21% PVA-Ac (30.0 mol % acetate content) solution by addition of surfactants. 1: $(EO)_9OP$; 2: $(EO)_{15}OP$; 3: $(EO)_{30}OP$; 4: $(EO)_{50}OP$; 5: dodecyltrimethylammonium chloride. From ref. 51 with permission.

Binding of $(EO)_{14}NP$ and $(EO)_{20}NP$ to hydroxyethylcellulose (HEC) was detected by light scattering [52]. In this work, the apparent molecular weight of HEC was found to increase with increasing surfactant concentration by preferential adsorption of the micelles around the polymer chain. At higher concentrations, when the solvent compositions in the neighborhood of the polymer and in the bulk are nearly the same, the apparent molecular weight of the HEC-surfactant complexe becomes about 4 times of that of the original HEC.

Interaction between PPO and n-octyl β-D-thioglucopyranoside (OTG), a nonionic surfactant, was studied by measurement of the cloud point, Krafft temperature, and cmc, and by microcalorimetry [53]. Contrary to the case of NaDS, the cmc of OTG (8-9 mM) was not changed by addition of PPO. The enthalpy of complex formation with PPO was found to be $+4.3$ kJ mol^{-1}, whereas that of the micellization was $+4.5$ kJ. Since the Gibbs energy change of micellization is unchanged by the addition of PPO, the endothermic enthalpy is compensated by a positive entropy change due to the release of water molecules from the polymer. OTG was bound to PPO only above the cmc. There was no evidence of interaction between PEO and OTG. Furthermore, aggregation numbers of OTG in one micelle in the absence and presence of PPO, measured using static fluorescence quenching, were 156±10 and 96±3, respectively. These results suggest that, although PPO is soluble in apolar solvents, this polymer is probably located in the outer region of the OTG micelles, similar to the complex of PEO-NaDS [5].

Interaction of hydroxypropylcellulose (HPC) and OTG or n-octyl ß-D-glucopyranoside (OG) was studied by a fluorescence technique [54]. Pyrene-labeled HPC was employed alone and also as probe added to HPC at the same concentration. In both kinds of experiments the interaction began at the same surfactant concentration, which coincided with the cmc (23 mM), similar to the case of PPO with OTG but unlike the systems PPO-NaDS (discussed above) and HPC-NaDS, in which the interaction started below the cmc [55]. From these fluorescence measurements, it was concluded that OTG or OG interact with pyrene-labeled HPC but not with unlabeled HPC. No interaction between HPC and $C_{12}E_8$ or dodecyl β-D-maltoside could be detected, the cloud point of HPC remaining constant at 39°C [56].

B. Concentrated Region

Phase behavior of polymer-surfactant-water systems is a new research area treated in Chapter 3 of this volume. Here we will mention only some studies involving nonoionic surfactants.

The cloud point of UCON, a linear random copolymer of ethylene and propylene oxides, was found to increase upon addition of $C_{12}E_8$ [57]. The two-phase region consists of an isotropic phase concentrated in both components and a dilute phase containing a small amount of surfactant and even less of UCON. This trend appeared even more pronounced at higher temperature. It is noted that both UCON and $C_{12}E_8$ are preferred solvents for each other compared to water.

The cloud point of a polymer in aqueous solution is not always raised by surfactant addition. The cloud point of ethylhydroxyethylcellulose (EHEC) was lowered by addition of a small amount of the more water-soluble $C_{12}E_8$, followed by an increase with increasing surfactant concentration, more markedly so at higher concentrations of both components [57].

At higher temperatures EHEC and $C_{12}E_8$ tend to separate, forming two isotropic phases, one concentrated in both components and another more dilute. This phenomenon could be attributed to an attractive interaction and a strong complex formation between EHEC and the surfactant. However, viscometry of dilute EHEC solutions at $25°C$ did not indicate an interaction with PEO-NP [65]. Conversely, the cloud point of the surfactant is also influenced by the polymer. In the interaction between EHEC and less water-soluble $C_{12}E_4$, the cloud point of the surfactant decreases upon a small addition of the polymer, and then increases by further addition [57]. The miscibility of PEO and C_4E_1 or C_8E_4 in water was explored at various temperatures. From the pattern of phase behavior it was indicated that with increase in molecular weight of PEO and in micelle-forming ability of the surfactant, the systems turned less miscible [58].

In the case of very hydrophilic polymers such as agarose and dextran, their interaction with surfactants was different from the above-mentioned cases. Cloud points of $C_{12}E_6$ and $C_{12}E_8$ were found to be lowered by addition of agarose at relatively high concentrations, leading to segregation into a surfactant-rich and a polymer-rich phase due to incompatibility of the two components, or repulsive interactions between the polymer coil and micelle. This phenomenon is called segregative phase separation, or depletion flocculation [59,60].

A similar cloud point decrease occurs in the interaction between EHEC and NaDS in its dilute region. The depression is especially remarkable in the presence of salts. In the salt-free case, the degree of binding of NaDS to the polymer below its cmc is small, and the ionic strength effect of the high concentration of free surfactant ions and counterions in solution suppresses the repulsion between the polymer chains charged with the surfactant ions [61,62]. This mechanism is different from that just mentioned above for nonionic surfactants.

C. Hydrophobically Modified Polymers

Hydrophobically modified water-soluble polymers (HMP) are materials with hydrophobic group(s) (*e.g.* C_8-C_{18}) either end-positioned or randomly substituted onto the hydrophilic backbone polymer chain. These compounds are known as associative thickeners with various applications [63]. In aqueous solutions of HMP, the hydrophobic sidegroups aggregate to form intra- and interchain micelle-like clusters. The interchain crosslinkages generate large three-dimensional networks. HMP produces viscous solutions at low and moderate concentrations, which undergo marked changes by addition of surfactants [3,63], or salts [64] when the HMP is a polyelectrolyte. Here, we will deal only with the interaction between HMP and nonionic surfactants.

In mixed solutions at a fixed concentration of a hydrophobically modified PEO-urethane block copolymer (R-NHCO-$(E_n$-DI$)_2$-E_n-OCONH-R; R: a long-chain alkyl, E_n: PEO, DI: hexamethylene or toluene diisocyanate) and various concentrations of PEO-NPs, a viscosity maximum appeared at a constant ratio of surfactant to hydrophobic end group concentration. The ratio increased with increasing surfactant hydrophobicity. The viscosity decrease after the maximum was steeper as the surfactant was more hydrophilic [65]. These observations are explained as follows. At low concentrations, the added surfactants associate with the hydrophobic clusters of polymers by a noncooperative process that increases the strength of the aggregates. A rearrangement of the polymer-surfactant mixed micelles may also increase the number of aggregates. Both effects raise the viscosity. At higher surfactant concentrations the mixed micelles will contain only one polymer hydrophobe, and the network structure is lost. Surface tension curves also indicate that at low concentrations the nonionic surfactants associate noncooperatively with the polymer and at high concentrations both bound complex and free micelles coexist. For NaDS, the surface tension isotherm exhibits noncooperative binding below a critical surfactant concentration (T_1 or cac, see Section IV.A.1.). Above this surfactant concentration cooperative binding is observed similar to systems such as PEO-NaDS [2,93] or PAA-$C_{12}E_8$ (see Figure 8 later in this chapter). This may be ascribed to the fact that the nonionic surfactants are hardly interactive with PEO, whereas NaDS is bound not only to the hydrophobic domains but also cooperatively to the PEO blocks of PEO-urethane, (Section III.A.). A different mechanism holds for the interaction of OG or OTG with unlabeled and fluorescently labeled, hydrophobically modified poly(N-isopropylacrylamides with a small content of N-R-acrylamides) (PNIPAM's) [66a,b]. The interaction is cooperative and takes place at a critical concentration (T_1) lower than the cmc. Fluorescence measurements with both pyrene and bis(1-pyrenylmethyl)ether (Dipyme) [66a] as probes and the data obtained with fluorescently labeled PNIPAM's [66b] support this conclusion. Mixed micelles are formed between the alkyl side groups substituted on the polymer chain and the surfactant molecules. At the onset of interaction and at a constant sidegroup substitution, the number of OTG molecules per alkyl side group (N) was found to increases from 6 to 12 when the side group was changed from C_{10} to C_{18}. At saturation, N had a constant value of ca. 30 irrespective of alkyl side group and hydrophobe content. The static fluorescence measurements demonstrated that among the components appearing in the system the free surfactant micelles were most fluid and the clusters of polymer alkyl side groups most rigid.

   Using fluorescence techniques, the ionic surfactants NaDS and hexadecyltrimethylammonium chloride, were found to associate noncooperatively with PNIPAM's. Whether this difference in binding mechanism between ionic and nonionic surfactants is due to the headgroup charge or length of hydrophobic tail is not yet clear [66b]. The N values were different for each surfactant.

Other issues addressed are the locus and efficiency of surfactant binding [67]. In aqueous solutions of poly(disodium maleate-hexadecyl vinylether) polysoap, cryo-transmission electron microscopy showed long thread-like clusters. They are interchain aggregates linked by the hexadecyl groups from different polymer chains. When $C_{10}E_6$ or $C_{12}E_6$ was added, they were bound to these microdomains and decomposed the structure at high concentration. $C_{10}E_6$- and $C_{12}E_6$-polysoap mixtures produced the same viscosity patterns in spite of the big difference in cmcs. This may suggest that the binding efficiency depends very little on the length of hydrophobic tail of the surfactants.

The association between hydrophobically modified poly(sodium acrylate-alkylacrylamide) (HMPA) and a series of $C_{12}E_n$ surfactants (n = 3, 4, 5, and 8) was systematically investigated [68-71]. Association produces major changes in both the rheological properties and phase behavior of the HMPA-nonionic surfactant mixtures. In the presence of $C_{12}E_8$, small mixed micelles are formed that act as crosslinkages between the polymer chains [68-70]. Surfactant binding to the polymer induces a pronounced decrease in the micellar mobility as evidenced by molecular self-diffusion measurements [68]. When the surfactants form giant vesicles, as in the case of $C_{12}E_4$, a small amount of HMPA dramatically increases the stability of micellar and lamellar phases. The structure may consist of crosslinkages between the surfactant vesicles and the alky side groups of the polymers. A sharp increase in viscosity and gelation was observed at a certain temperature [69-71]. The phenomenon was reversible and correlated well with the transition from micelles to bilayers observed in binary systems of water and surfactants. The thermal gelation temperature could be adjusted over a wide temperature range by using surfactant mixtures [70,71].

While HEC hardly binds $C_{12}E_8$, its hydrophobically modified counterpart (HMHEC) with less than 1 weight % alkyl groups (C12-C18) has a strong interaction with the nonionic surfactant. The association behavior of HMHEC and its interaction with $C_{12}E_8$ were studied by static fluorescence with pyrene as probe, surface tension and viscosity measurements [72]. The hydrophobic groups of HMHEC associate with each other to form micelle-like clusters above a critical polymer concentration of 500 ppm. In dilute polymer solutions the surfactants modify the polymer structure while in concentrated solutions they enhance polymer-polymer association.

D. Proteins

Nonionic surfactants are widely employed as mild processors in the extraction of native proteins from lipids and biomembranes. Physico-chemical studies on the interaction between proteins and nonionic surfactants are relatively rare compared with the number of studies on ionic surfactants [33]. We will report on some recent studies where typical water-soluble proteins are mentioned.

Like hydrophilic synthetic polymers, water-soluble proteins rarely interact with nonionic surfactants. From equilibrium dialysis measurements at pH 9.2 and ionic strength of 0.1, it was found that the saturation binding of Triton X-100 to bovine serum albumin (BSA) (4:1 on a molar basis) was less than that of Na deoxycholate (15:1), and far less than NaDS (160:1, free from buffer and salt). No binding of Triton X-100 occurred to ovalbumin [73]. Different from NaDS, the binding of Na deoxycholate or Triton X-100 did not bring about denaturation and no change was observed in the optical rotation.

Binding of $C_{12}E_6$ to BSA or lysozyme was determined by the surface tension method, in which the free nonionic surfactant alone was assumed to contribute to the surface activity of the mixed systems [74]. In the binding isotherm for the surfactant on BSA at 25°C, pH 5.6, and without buffer, the maximum number of mols of nonionic surfactant bound per mol of protein was found to approach a constant value of about 4 with increasing protein concentration, and this agreed with the result obtained by the dialysis technique with Triton X-100 [73]. The dependence of the amount of saturation binding of nonionic surfactants on protein concentration was more pronounced in the lysozyme systems than in the BSA systems. This was attributed to the tendency of lysozyme to aggregate in solution.

Binding isotherms for BSA-Triton X-100 and BSA-$C_{12}E_6$ systems followed the Scatchard equation assuming uniformity of binding sites. Binding isotherms for the lysozyme systems, however, obeyed the Freundlich equation, which is based on the heterogeneity of the binding sites. The stronger tendency of self-association of lysozyme in solution may be responsible for the heterogeneity of the sites [74]. Binding isotherms for these nonionic surfactants to BSA did not indicate cooperativity [73]. At pH 7.0 BSA had 2 cooperative binding sites for Triton X series [75,76].

Interactions of two unfolded lysozymes, reduced and carboxamido-methylated, with $C_{10}E_6$, $C_{12}E_6$ or $C_{12}E_8$ at pH 2.5 were studied by equilibrium dialysis, surface tension, circular dichroism and fluorescence [77,78]. The binding isotherm exhibits two regions: in the dilute surfactant region below a free concentration of 6.2 x 10^{-5} M for $C_{12}E_6$, (cmc: 8 x 10^{-5} M, 25°C) the binding was very small (0.04 mol surfactant per aminoacid residue), whereas above this level the amount of binding increased abruptlyand greatly (the mol ratio: 0.5-0.7 at the maximum). In the latter region, different from the native lysozyme, aggregation of the unfolded protein itself did not occur appreciably. In the native lysozyme [74], as the measurements were limited to the dilute region, no cooperative binding was observed. Probably, the binding of surfactant takes place in an all-or-none manner at a critical concentration of free surfactant (T_1) which is lower than the cmc. The binding isotherms for these three surfactants superimpose when the free surfactant concentration is normalized to the respective cmc. The binding is hydrophobic, and no interaction with the PEO moiety is involved. According to circular dichroism spectra, the main-chain conformation of the reduced lysozyme becomes more or less similar to that of the native protein as the amount of bound surfactant increases. This interaction induced no appreciable change in tertiary structure of these proteins.

OG, which has a high cmc (23.4 mM) is bound to various water-soluble globular proteins such as ribonuclease, lysozyme, α-chymotrypsin, pepsin, deoxyribonuclease, ovalbumin, BSA, bovine catalase, fibrinogen, or *aspergillus niger* catalase [79]. The binding isotherms for these proteins, determined by equilibrium dialysis, also exhibit a steep rise at a free OG concentration below its cmc, and could be fitted to the Hill equation with the positive coefficients indicating cooperativity. The enthalpy of binding, measured directly, was found to be very small, hence the binding was attributed to a hydrophobic process. A model was proposed in which OG molecules are bound around the protein forming a prolate or oblate ellipsoid for lower or higher molecular weight protein, respectively. No evidence of denaturation was found.

IV. INTERACTIONS BETWEEN POLYMERIC ACIDS AND ETHOXYLATED NONIONIC SURFACTANTS

A. Experimental Results

1. Complex formation

As elaborated in Section II, the role of ionic surfactant headgroup in the interaction with nonionic polymers is still controversial. Therefore, the reactivity of hydrophilic headgroup of ethoxylate-type nonionic surfactants with proton-donor polymers deserves attention.

For instance, not only PEO but also PVP and many other types of water-soluble nonionic polymers bind small protic substances such as carboxylic acids or phenols. The amount of binding has been studied by various methods [7,80]. As a result of binding, the polymers in solution contract. Likewise, they form complexes with polycarboxylic acids in water both through hydrogen bonding and hydrophobic effects. Above about pH 5 no complex is formed between PAA and PEO, and below pH 3 there is an insoluble complex with a

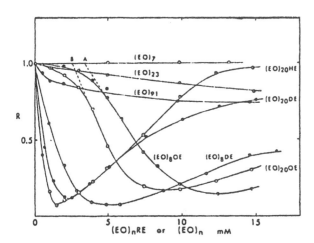

FIG. 4 Reduced viscosity ratio R of 24 mM PAA solutions in the presence of (EO)$_n$RE (OE: octyl-, DE: dodecyl-, HE: hexadecyl ether), and PEO relative to the PAA solution at the initial pH 3.25 and 25 °C. From ref. 89 with permission.

stoichiometric composition with respect to proton donor and acceptor [81]. When the ligand of either polymer was too bulky, the composition deviated from unity mol ratio [82]. In nonaqueous media, complex formation is suppressed [83,84]. This kind of polymer-polymer complex formation has been used for template or matrix polymerization, in which polymer radicals grow along the template polymer [85,86].

The equilibrium of complex formation between proton-donor and acceptor polymers depends on the molecular weight of both components [81,87]. When the molecular weight of the polymeric acid is high, polyacrylic acid (PAA) forms a stable complex with PEO when the PEO degree of polymerizaton is above 200, while for polymethacrylic acid (PMA), which is more hydrophobic than PAA, the PEO degree of polymerization has to be 40 or less [81,99]. In contrast to PAA-PPO systems, clouding of PMA-PPO solutions in appropriate conditions does not occur even at 90°C [88]. These results emphasize the importance of hydrophobic attraction in interpolymer complexation.

The same applies to the interactions of polymeric acids with ethoxylated nonionic surfactants [89,90]. Figure 4 shows viscosity change of a PAA solution at pH 3.25 upon addition of various concentrations of PEO or PEO-alkylethers. In this pH range no precipitation appears [89]. Above pH 5.2, or a degree of NaOH neutralization of 0.20, the PAA interaction with (EO)$_{20}$dodecyl disappears almost entirely, as discussed above for the interaction between PAA

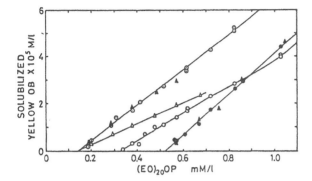

FIG. 5 Yellow OB solubilization in $(EO)_{20}OP$ solution in the presence of PAA without (circle) and with (triangle) 0.6 mM HCl at 20 °C. ●,▲: surfactant alone and with HCl; ○,△: surfactant + 1.16 mM PAA without and with HCl; dotted circles and triangles: surfactant + 11.6 mM PAA without and with HCl, respectively. From ref. 91 with permission.

and PEO. The minimum in the viscosity curve indicates that at high PAA/surfactant ratios the complex contracts while at low ratios it expands and/or crosslinks with other polymer chains. With PEO, however, the reduced viscosity of the PAA solution is affected much less. As a result of fixation of PAA carboxyls with EOs, the pH of solution rises, but this pH increase also is much higher with alkylethoxylate surfactants than with PEO [89]. Thus, in the nonionic surfactants, the PAA-EO interaction is remarkably reinforced by the presence of the hydrophobic moiety. In other words, two-point binding of surfactant is the basis of a strong interaction as described in Section II.

The polyacid-alkylethoxylate systems also have solubilization ability [91]. Figure 5 shows the results of Yellow OB solubilization in PAA-$(EO)_{20}OP$ solutions. In the mixed systems of fixed PAA and varied surfactant concentrations, the dye was solubilized above a certain surfactant concentration, which was below the cmc and dependent on the PAA concentration. At lower PAA concentration, and higher degree of dissociation of PAA (higher pH), there is less hydrogen bonding with the surfactant EO groups. However, by addition of enough HCl to suppress dissociation of PAA in dilute solution but not so much as to yield complex precipitation, *i.e.*, at a pH of about 3.2, the surfactant concentration at the start of dye solubilization is independent of PAA concentration, as long as PAA concentration is not much higher than that of the surfactant [91]. At this concentration the surfactants form a micelle-like assembly on the PAA chain. Following similar cases in polymer-ionic surfactant interaction as described below, this concentration is thus denoted as cmc_P, cac (critical aggregation concentration [92]), or T_1 [93].

In the interaction of PVP, PEO [2], or poly(1,3-dioxalane) with NaDS [94], surface tension, conductance, dialysis, solubilization, viscosity, and fluorescence measurements indicate the existence of three regions, and hence two transition points as a function of NaDS concentration. There is no measurable binding in the dilute range (region I), cooperative binding in region II, and finally in region III both free micelle and complex coexist. In these systems too the transition point between the regions I and II, denoted as T_1, is not strongly dependent on polymer concentration and temperature. At the transition point between the regions II and III, denoted T_2, the binding attains saturation.

The reduced viscosity of solutions of fixed PAA and varied nonionic surfactant concentration also indicate a critical concentration, T_1, as the transition between slow and steep viscosity declines in the very low surfactant concentration region, suggesting that at T_1 the cooperative binding of the surfactants on PAA chains begins [91]. This transition can also be determined by other methods, as shown later.

In addition to the mode of surfactant organization on the polymer chain in the interaction of polymeric acid and nonionic surfactant, two factors should be taken into consideration: binding of the surfactant as a whole molecule, and the degree of hydrogen bonding in the hydrophilic moiety. The change of the reduced viscosity is a hydrogen bond-sensitive phenomenon and not directly related to the binding as a whole molecule, as will be shown next.

2. Effect of PEO moiety of nonionic surfactants

a. Polyacrylic acid

Figure 6 presents change of reduced viscosity of a PAA solution upon addition of $(EO)_n$dodecyls. Longer PEO moieties, result in larger viscosity decreases at lower surfactant concentrations, with a simultaneous narrowing of the valley in the curves. We recall that binding of PEO to PAA is strengthened with increasing PEO molecular weight of PEO [81]. At high $(EO)_{50}$dodecyl concentrations the reduced viscosity is higher than that of the original PAA solution. The long PEO chains protruding from the PAA may contribute to this increased viscosity. In Figure 6, when the concentration of the nonionic surfactant is expressed in terms of EO units, the concentrations of EO at the reduced viscosity minima in this series are similar and about 1.6 times of the PAA monomer unit concentration.. We will come back to this observation at the end of Section IV.A.2. The pH rise accompanying complex formation reaches a plateau at high surfactant concentration, suggesting that hydrogen bonding reaches saturation. This concentration coincides with the viscosity

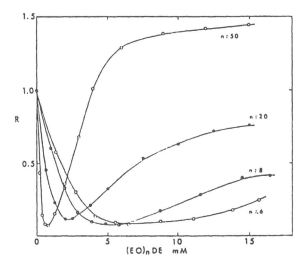

FIG. 6 Reduced viscosity ratio R of 24 mM PAA solutions in the presence of $(EO)_n$dodecylether to the PAA solution at the initial pH 3.25 and 25 °C. From ref. 89 with permission.

minimum [89].

As shown in Figure 7, T_1 for the PAA-$(EO)_{10}$OP system is independent of temperature between 4 and 38°C [95]. T_1 was determined for a series of $(EO)_n$OP and $(EO)_n$NP (Table1) and of $(EO)_n$alkylethers (Table 2) with PAA at various temperatures, using the Yellow OB solubilization method [96]. As shown later in Section IV.C., cmc and T_1 determined with other solubilizates at various temperatures are identical within experimental error [97]. The $(EO)_n$alkylethers employed were of commercial grade but with narrowly polydisperse PEO chains. The T_1 values are always lower than their respective cmcs, and are lower for the more hydrophobic moieties. Surprisingly, when the hydrophobic moieties are the same, T_1s are independent of both temperature and a wide range of EO numbers, whereas their cmcs are markedly different and dependent on temperature. T_1 was determined also by surface tension, pyrene probe fluorescence and pH [98].

Figure 8 shows the surface tension for $C_{12}E_8$ alone and with PAA at pH 3.6. In the low concentration region, cmc and T_1 are the break points of the surface tension curves without and with PAA, respectively. In Figure 9 the the intensity ratio I_1/I_3 of the fluorescence spectrum or the polarity parameter of a pyrene probe in $C_{12}E_8$ solutions is plotted against surfactant concentration with and without PAA. Comparison of the two curves leads to the conclusion

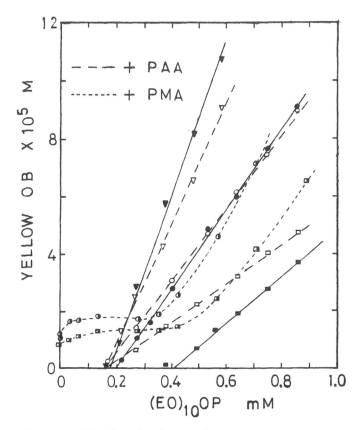

FIG. 7 Solubilization of Yellow OB in $(EO)_{10}OP$ solutions without (solid lines with filled marks) and with 13.5 mM of PAA (broken lines with open marks) and of PMA (dotted lines with half-filled marks) at 4°C (squares), 25°C (circles), and 38°C (inverted triangles) at the initial pH 3.5. From refs. 95 and 103 with permission.

that at high surfactant concentration the probe is located in a micelle-like environment in the complex. Once again, the break points at low concentration with and without PAA are T_1 and cmc, respectively. Upon addition of surfactant, the pH of PAA solution begins to rise at T_1. These values together with those derived from other methods are summarized in Table 3 and are in accord within experimental errors with the data for corresponding polydisperse surfactants in Table 2.

The fact that T_1 is unaffected by EO number but dependent on the hydrophobic moiety of the surfactant may be interpreted as follows.

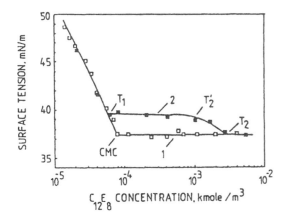

FIG. 8 Surface tension isotherms of $C_{12}E_8$ at 25 °C. Curve 1: without PAA; curve 2: with 10.2 mM PAA. Reprinted from ref. 98, *Colloids Surf. A.90*, with permission of Elsevier Science NL, Sara Burgerhartstraat 25, 1055 KV Amsterdam, The Netherlands.

TABLE 1 Cmc, T_1, and Precipitation Limit Concentrations (PLC) of Polyoxyethylene Octylphenyl and Nonylphenyl Ethers with 13.5 mM PAA at Various Temperatures

	Temp (°C)	cmc (mM)	T_1 (mM)	PLC (mM)
$(EO)_{10}OP$	4	0.40	0.15	0.15
	14	0.30	0.15	0.15
	25	0.20	0.15	0.15
	38	0.17	0.15	0.15
$(EO)_{20}OP$	4	0.79	0.15	0.15
	14	0.59	0.15	0.15
	25	0.45	0.15	0.15
$(EO)_{10}NP$	4	0.09	0.05	0.055
	14	0.07	0.05	0.05
	25	0.06	0.05	0.05
$(EO)_{20}NP$	4	0.22	0.05	0.05
	14	0.13	0.04	0.04
	25	0.11	0.04	0.04

Source: ref. 96.

TABLE 2 Cmc and T_1 values of Polyoxyethylene Alkyl Ethers with 13.5 mM PAA at Various Temperatures

	Temp (°C)	cmc (mM)	T_1
$(EO)_5C_{12}$	4	0.075	0.045
	14	0.065	0.05
	25	0.055	0.045
$(EO)_{10}C_{12}$	4	0.17	0.045
	14	0.12	0.05
	25	0.085	0.045
	38	0.075	0.05
$(EO)_{15}C_{12}$	4	0.175	0.05
	14	0.165	0.05
	25	0.11	0.045
$(EO)_{10}C_{14}$	4	0.02	0.005
	14	0.01	0.005
	25	0.005	0.005
$(EO)_{20}C_{14}$	4	0.035	0.005
	14	0.025	0.005
	25	0.015	0.005

Source: ref. 96 .

Hydrophobic attraction between PAA and PEO and hydrogen bonding between carboxyl and EO groups stabilize the complex. In the interaction between PEO and PAA, a much longer PEO chain is needed than the EO length of most ethoxylate surfactants [81,99,100]. When attached to a hydrophobic moiety, even a very short EO chain is able to interact with PAA because of the hydrophobic stabilization. In such a case, the influence of the hydrophobic environment of the surfactant upon the hydrogen bonding of its EO chain with PAA can reach only a limited distance from the hydrophobic moiety. Supposedly, irrespective of the EO chain length, only a definite segment of the EO sequence close to the hydrophobic moiety is bonded to PAA. The end portion is free and pendent in solution. In a model of the complex depicted in Figure 10 (left, without solubilizate), a micellar aggregate is wrapped around by PAA chain with hydrogen bonding and hydrophobic attraction. In this case, if hydrophobic exposure of the packed micellar surface per surfactant molecule is the same independent of the EO chain length, the bound EO segment by PAA would be of the same length, and therefore, T_1 would be constant [98]. In this respect, "a definite segment of EO sequence" as above would be independent of the alkyl chain length when the hydrophobicity of the alkyl group is sufficiently high. Surfactants with longer hydrophobic moieties have lower cmcs and tend to aggregate on PAA at lower T_1s.

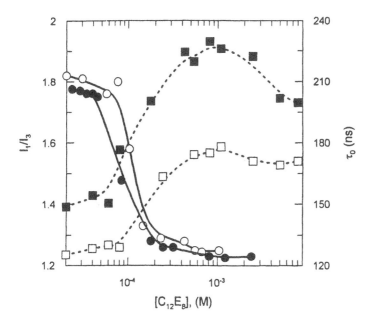

FIG. 9 Fluorescence intensity ratio (I_1 / I_3) (●,○) and the lifetime (■ □) of pyrene fluorescence vs. $C_{12}E_8$ concentration without (open symbols) and with (closed symbols) 10.2 mM PAA at 25 °C. From ref. 113 with permission.

As shown by the temperature-independence of T_1, the overall change in enthalpy in complex formation may be zero, but a long sequence of hydrogen bonds in the cooperative binding may be responsible for the stable complex formation. The change of standard Gibbs energy from ordinary micelles to polymer-bound micelles is given by $\Delta G_P° = RT\ln(T_1/\text{cmc})$ [2]. From Tables 1-3, T_1 is lower than cmc, and therefore, the complex with a longer PEO-chain surfactant is more stabler than the free micelle.

In Figure 8, above T_1 the surface tension of the surfactant solutions with PAA is almost constant until T_2', and then decreases and from T_2 follows the surface tension of the solutions in the absence of PAA. T_2 coincides with the reduced viscosity minimum point (T_V) of the PAA-surfactant solutions and also with the pH plateau point (Table 3). We can conclude that at T_2' micellar aggregates on PAA and single surfactants begin to coexist. Aggregates continue to form by hydrogen bonding (pH increasing, viscosity decreasing), and at T_2 the ordinary free micelles appear in solution and binding by hydrogen

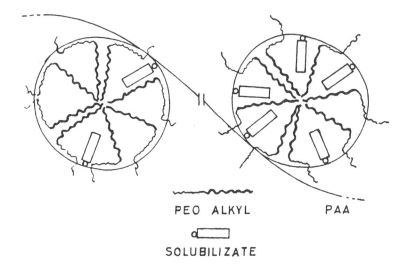

PEO   ALKYL   PAA

SOLUBILIZATE

FIG. 10 Models of a nonionic surfactant-PAA complex and solubilization. Right; a model at low temperature. Modified from ref. 97.

TABLE 3 Cmc for $C_{12}E_6$ and $C_{12}E_8$ and Their Transition Points in the Interaction with 10.2 mM PAA at 25 °C

Method	$C_{12}E_6$				$C_{12}E_8$			
	cmc, $Mx10^5$	T_1 $Mx10^5$	T_2' $Mx10^4$	T_2 $Mx10^3$	cmc, $Mx10^5$	T_1 $Mx10^5$	T_2' $Mx10^4$	T_2 $Mx10^3$
Surface tension	7.0	5.0	10.0	3.0	8.0	6.3	4	2.8
Fluorescence	6.8	5.2	9.5		8.0	5.9	9.4	
Viscosity		5.1		3.0 *		6.2		2.8*
pH		8.0		3.1		8.0		2.7
Yellow OB solubilization	6.5	4.9			6.8	6.1		

Source: reprinted from ref. 98, *Colloids Surf. A90*, 89 (1994), with permission of Elsevier Science NL, Sara Burgerhartstraat 25, 1055KV Amsterdam, the Netherlands.

bonding ends. Surfactant binding increases further by hydrophobic association alone (pH constant, viscosity increasing). As will be described in Section IV.A.4., the surfactant aggregation number of the aggregates is constant and a little smaller than that of free micelles over a wide range of concentration beyond T_V and T_2. Therefore, above T_V and T_2 aggregates are formed on the PAA chain in accordance with the "beads on a string" or "necklace" model. At higher surfactant concentration each complex may be reorganized without changing the total number of hydrogen bonds with PAA and offering more PAA chain for further hydrophobic binding until saturation. Recently, it has been proposed that the string of complexes takes an oblate shape above T_V and T_2 [137]. The continued surfactant binding to PAA (and also to carboxy vinyl polymer, CVP, see next section) beyond T_2 and T_V shows that in these systems the nature of T_2 is different from that of the binding saturation in nonionic polymer-NaDS systems. The situation seems similar to systems of polyelectrolyte and ionic surfactants of opposite charge.

At T_V, as the free surfactant concentration is very close to the cmc, T_V - cmc is considered to be the amount of bound surfactant. As described in Section IV.A.3., in some systems T_2 by surface tension was a little higher than T_V [101], but the free surfactant concentration at T_V is taken as approximately as equal to the cmc. As the cmcs for PEO-dodecyls are very low, the ratio of EO units to carboxyl groups in the complexes, estimated at the reduced viscosity minima are about 1.8 and 2.2, respectively, for $C_{12}E_6$ and $C_{12}E_8$, and about 1.6 for polydisperse $(EO)_n$dodecyls (Figure 6). These values suggest that at T_V, at most about half of the bound surfactant EOs do not participate in hydrogen bonding with PAA but are dangling and/or looping from the string. Recent T_1, T_2 and T_V determinations for $C_{12}E_8$ at various PAA concentrations show that T_1 is almost independent of PAA concentration [138], as observed previously in other systems using dye solubilization [91]. With increasing PAA concentration, both T_2 and T_V rise, T_2 more strongly than T_V [138].

The surface layer of the PAA-$C_{12}E_8$ system was studied by ellipsometry and surface tension at pH 3 [102]. The results indicated that surface adsorption of PAA with the surfactant occurred only just above T_1 in the amount of 0.36 g/mL with a thickness of 15-22 nm, which is of the same order of magnitude as the diameter of a PAA coil.

The lifetime (τ_o) of a fluorescence probe in $C_{12}E_8$ solution without quencher was found to rise abruptly above a concentration coinciding with the cmc (Figure 9) [113], as the micelle protects the solubilized probe against the quenching by oxygen dissolved in the water. In the presence of PAA the lifetime increases further above T_1 (Table 3), suggesting that PAA shields the micelle by wrapping around it as shown in Figure 10. At high surfactant concentrations the lifetime in the PAA-added solutions decreases. This is attributed to the appearance of free micelles. The polymer-surfactant interaction tends to slow down the quenching process, as previously noted in complexes of the anionic polysaccharide hyaluronan with decyl- or dodecyltrimethylammonium bromides [114].

b. Polymethacrylic acid

In PMA-(E0)$_{10}$OP solutions, T_1 could not be determined as unambiguously with the viscosity and Yellow OB solubilization methods [91] as in PAA solution (Figure 7), since PMA is more hydrophobic and its aqueous solution has some dye solubility even without surfactant [103]. Dye solubilization, and probably surfactant binding by PMA takes place in two steps, one at low, and one at high surfactant concentration. In the dilute region, the surfactants are considered to be bound noncooperatively by PMA. In the concentrated region, they form micellar aggregates. The transition from the low binding plateau to the steep slope is taken to be T_1. Table 4 summarizes cmcs and T_1s of various inhomogeneous surfactants with PMA at 4 and 25°C determined by solubilization (see also Section IV.C.). Two-step binding isotherms have been observed in the interaction between PMA and cationic surfactant [104].

In contrast to PAA, with PMA in some cases T_1 is much higher than the cmc and strongly dependent on the EO number rather than the hydrophobic moiety of the surfactants. When the hydrophobic moiety is the same, the surfactant with a larger EO number has a lower T_1. A PMA chain in water has hydrophobic microdomains [92,105] which may serve to incorporate

TABLE 4 Cmc and T_1 of Surfactants with 13.5 mM PMA Determined with Various Solubilizates at 4 and 25°C

Temp (°C)	(EO)$_{10}$OP cmc	T_1	(EO)$_{20}$OP cmc	T_1	(EO)$_{15}$C$_{12}$ cmc	T_1	(EO)$_{10}$C$_{14}$ cmc	T_1	(EO)$_{20}$C$_{14}$ cmc	T_1
	mM		mM		mM		mM		mM	
				Yellow OB						
4	0.40	0.45	0.79	0.3	0.175	0.7	0.02	0.47	0.035	~0.4
25	0.20	0.35	0.45	0.2	0.11	0.4	0.005	0.4	0.015	0.3
				Naphthyl Red						
4			0.78	0.3					0.02	
25			0.44	0.2					0.01	
				Sudan I						
25									0.02	0.3

Source: ref 103.

solubilizate without surfactant (see Section IV.C.). This may be related to the noncooperative surfactant binding in the dilute region and thus to higher T_1. At T_1 the mol ratio of EO for $(EO)_{20}OP$ per carboxyl of PMA is 0.4 at 25°C, suggesting a fairly crowded binding compared with the most insoluble state as shown later in Table 5. In the dilute region, the microdomains seem to have a stronger tendency to isolated binding of nonionic surfactant rather than to micellar aggregation.

Interactions between PAA or PMA with PEO having a small hydrophobic anchor (phenyl, t-butylphenyl, naphthyl, n-octyl, etc.) or PEO-hydroquinone-PEO were studied by means of viscosity and potentiometric titration, and similar results as mentioned above were obtained [100,106-108]. The curve of specific viscosity of the polyacid solution $vs.$ concentration of the PEO-derivative exhibits a minimum. The pH continues to rise slightly beyond the minimum. Thus, these non-micelle-forming ethoxylates also behave like nonionic surfactants with the polyacids, and it is to be noted that the viscosity rise at high concentration occurs even without micellar aggregation. PEO-hydroquinone-PEO was bound to the polyacids but less strongly than the PEO-phenylethers of similar molecular weights [100,106,107].

c. Carboxy vinyl polymer (CVP)

CVP (Carbomer, Carbopole) is a water-dispersible PAA loosely interlinked through polyalcohols. By neutralization with alkali an aqueous dispersion becomes transparent and more viscous, forming a gel at high concentration. CVP is widely used as a thickener for stabilizing industrial emulsions. When a translucent CVP aqueous dispersion was mixed with a nonionic surfactant at pH 4, it became more turbid and flocculated [109]. Due to its network structure the CVP coil is considered to be more compressed than that of PAA at the same pH. At a given pH the carboxyls inside the coil are probably less dissociated, and thus, the dispersion solution is flocculated upon mixing with the nonionic surfactant. The reduced viscosity curve of a dilute CVP dispersion solution with nonionic surfactant concentration has a similar valley shape, as observed for PAA in Figure 4.

In Figure 11 the amount of $(EO)_{10}OP$ or $(EO)_{30}OP$ bound to CVP in the flocculated state is plotted against the free surfactant concentration. Crosses on the isotherms mark the binding at T_V, $i.e.$ at the minimum in reduced viscosity, at which the free surfactant concentration approximates the cmc. In the isotherms, beyond the cmc of the free solution and T_V, binding continues to increase until saturation, and the once shrunken CVP chain loosens up intramolecularly and then disperses. This type of isotherm suggests a cooperative binding mode. In comparing the amounts of binding of $(EO)_{10}OP$ and $(EO)_{30}OP$ to CVP, although the latter surfactant is stronger

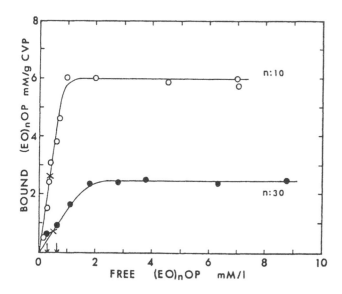

FIG. 11 Binding of (EO)$_n$OP (n=10 and 30) per g of CVP as a function of surfactant concentration in 0.05% CVP at 25°C. The cross marks binding at the reduced viscosity minimum. Arrow indicates cmc (0.32 mM and 0.65 mM for n=10 and 30, respectively) of free surfactant solution. From ref. 109 with permission.

in changing the reduced viscosity, the amount of binding is greater in the former shorter chain surfactant with a lower cmc even below saturation. This is because (EO)$_{10}$OP occupies less space per molecule on the polymer chain. Since observations such as the viscosity decrease and pH shift of the mixed solution are related to the length of PEO moiety, *i.e.*, the number of hydrogen bonds, and not to the binding as a whole molecule, the implication of the CVP binding results is important.

The ratio of EO units to acid units of CVP (1 g of CVP contains approximately 13 meq. acid units) at the reduced viscosity minimum is around 2, and around 5 at saturation binding for both surfactants. The results imply that in the most insoluble shrunken state about 50% of all the surfactant EOs in the complex, and in the saturation state only 20%, is used for hydrogen bonding with acid units, on the assumption that all carboxyls of the polymer are available. The remaining EOs are and especially will form loops and/or free tails pendent from the polyacid chain in the water phase.

3. Effect of hydrophobic moiety of nonionic surfactants

In a series of $(EO)_nR$ with the same n but different alkyl chains R, with longer R the effect of the surfactant on the reduced viscosity of a PAA solution becomes more pronounced in both magnitude and range of the viscosity decrease in the viscosity vs. concentration curve (Figure 4). In addition, the position of the minimum (T_V) shifts to a lower concentration. As described earlier, T_V - cmc is approximately equal to the amount of surfactant binding at T_V. At the viscosity minima for the PAA-$(EO)_{20}$octyl, -$(EO)_{20}$dodecyl and -$(EO)_{20}$hexadecyl systems(cmc 8 mM for the octyl, and below 0.1 mM for the other two), their (T_V - cmc) values are almost identical. In the same way, T_V - cmc for the PAA-$(EO)_8$octyl and -$(EO)_8$dodecyl systems (cmc 6 and 0.08 mM, respectively) coincide. A recent study on the PAA-$C_{10}E_8$, -$C_{12}E_8$ and -$C_{14}E_8$ systems by surface tension and viscosity measurements showed a similar trend [101].

Thus, at the viscosity minima, the degrees of binding for surfactants with the same PEO moieties were nearly equal irrespective of their alkyl moieties. For both the PAA-$(EO)_{20}$alkyls and -$(EO)_8$alkyls systems, the mol ratios of EO to carboxyl are about 1.6 (but with considerable error ranges). For the PAA-C_nE_8 series the ratios are about 1.7. This also illuminates the dominant contribution of hydrogen bonding to complex stabilization. The stronger hydrophobic attraction by longer alkyl chains leads to shrinking of the complex, i.e., to a lower minimum of the viscosity curve in Figure 4. This is in contrast to the identical viscosity minima in the series of PAA-$C_{12}E_6$ and -$C_{12}E_8$ [98], and PAA-$(EO)_n$dodecyls shown in Figure 6.

The plots of ln cmc and $\ln T_1$ for $C_{10}E_8$, -$C_{12}E_8$ and -$C_{14}E_8$ against the number of carbon atoms in the alkyl moiety were linear and of the same slope [101]. In applying the relation lncmc or $\ln T_1 = -n\omega/kT$ + constant, in which ω is the Gibbs energy change on transferring one methylene unit from micellar to aqueous environment, ω was found to be - 1.17 kT for both free micelle and complex [101], supporting the model shown in Figure 10. This value is comparable with other data for transfer of a methylene unit from hydrocarbon to water [4,110]. In PVP-sodium alkylsulfates systems the value for ω was - 1.1 kT for both micelle and complex [111], and it was concluded that the surfactants on the polymer are bound as micelles.

4. Aggregation number of surfactant in the complex

Dynamic fluorescence quenching [112] using the pyrene dimethylbenzo-phenone probe-quencher pair was used to determine the effect of

PAA and temperature upon the aggregation number of nonionic surfactants over a wide range of surfactant concentration [113]. The average aggregation numbers for free micelles of $C_{12}E_6$ and $C_{12}E_8$ (N_M) are 105 and 83, and those for the complexes (N_C) with 10.2 mM PAA below T_2 are 88 and 73, respectively, at 25°C. They change little in the surfactant concentration range studied. Thus, both N_M and N_C are larger for $C_{12}E_6$ than for $C_{12}E_8$. N_C is smaller than N_M but this trend does not hold in all polymer-surfactant systems [3,112]. The effect of temperature on the aggregation numbers at low concentration below the cloud point is larger for the $C_{12}E_6$-containing systems, and the aggregates begin to grow at 30°C. Around 45°C, the values for both N_M and N_C are about twice those at 25°C.

B.   Effect of Additives

1.   Acid (pH lowering)

As expected from the discussion on PAA-PEO interaction, an increase in pH leads to decomposition of the complex, while lowering the pH promotes hydrogen bonding between polymeric acid and surfactant. Low pH leads to coiling up and agglomeration of the strings of complexes, eventually resulting in precipitation depending on the ratio PAA/surfactant [115].

a. Polyacrylic acid
     Figure 12 shows the change in solubility of $(EO)_{10}OP$ from 0.53 mM as a function of PAA concentration at pH 2 and various temperatures. At high PAA concentration the supernatant surfactant concentration becomes constant. This concentration, below which no precipitation occurs is called the precipitation limit surfactant concentration (PLC), and is found to be independent of temperature. The PLCs for $(EO)_{10}OP$, $(EO)_{20}OP$, $(EO)_{10}NP$, and $(EO)_{20}NP$ with PAA are included in Table 1. For the same hydrophobic moieties, they are independent of both temperature and PEO chainlength, and coincide with the respective T_1 values. It implies that change from pH 3.5 to 2, or from a soluble to an insoluble state, is insignificant in determining T_1 [95]. The molecular weight of PAA did not affect the precipitation except for the very low molecular weights [116].
     The mol ratio of surfactant precipitated to the lowest PAA concentration required to reach the PLC gives composition of the most insoluble complex, provided that all the PAA added is included in the precipitate at this critical polymer concentration. Table 5 presents the ratios multiplied by the EO number of the respective surfactants at various temperatures [96]. These unit mol ratios are temperature-dependent but at the same temperature they are

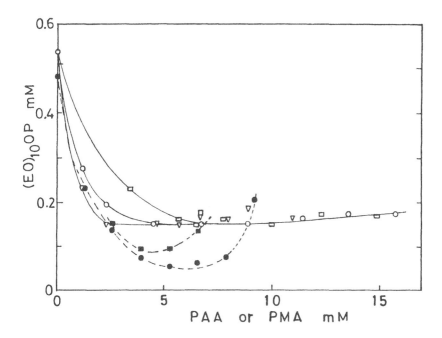

FIG. 12 Concentration of supernatant $(EO)_{10}OP$ in the precipitated system from 0.53 mM $(EO)_{10}OP$, plotted against PAA concentration (solid lines with open marks), and also from 0.485 mM $(EO)_{10}OP$ plotted against PMA concentration (broken lines with filled marks) at 4°C,(squares), 25°C (circles), and 38°C (inverted triangles) at pH 2. The difference from the initial surfactant concentration equals the amount of surfactant precipitated. From refs. 95 and 103 with permission.

almost equal to each other irrespective of EO number and hydrophobic moiety of the surfactants. This underlines the significance of hydrogen bonding in complex formation. At room temperature, the ratios are nearly 1, suggesting complexes with hydrogen-bonded structure at pH 2. At lower temperature, the EOs of the surfactants, probably in the terminal portion of the PEO moiety, tend to be surrounded by water molecules rather than by carboxyls, and therefore, the complex is less compact and more soluble. However, as can be seen in Figure 12 at high PAA content the overall composition of the complex becomes temperature-independent, as is also the case for T_1 and the PLC.

As noted above, at pH 2 the compositions of the complexes may actually be close to those of the soluble complexes at pH 3.5. This contention may be supported by data in Figure 13 for the reduced viscosity (broken lines) of PAA-$(EO)_{10}OP$, $-(EO)_{20}OP$, and $-(EO)_{20}NP$ systems at pH 3.5, compared to precipitation data at pH 2. In this figure, the minimum region of the viscosity

TABLE 5 Composition of Nonionic Surfactant Complexes with PAA or PMA
Estimated from Precipitation, at pH 2 and Various Temperatures.

T (°C)	Polyacid	mol unit ratio EO(C-PLC)/Polyacid*			
		$(EO)_{10}OP$	$(EO)_{20}OP$	$(EO)_{10}NP$	$(EO)_{20}NP$
4	PAA	0.6	0.9	0.8	0.7
	PMA	0.9	0.8	0.9	0.8
25	PAA	1	0.9	1.2	1.1
	PMA	0.9	0.8	1.2	1.1
38	PAA	1.8			

* C = initial surfactant concentration (M), EO = number of EO's per surfactant,
  PLC = Precipitation Limit Surfactant concentration (M),
  Polyacid = lowest concentration (M) to get PLC.
Source: refs 96 and 103.

curve almost covers the peak region of precipitation for each of the systems.
The supernatant surfactant concentrations corresponding to the viscosity
minimum of the soluble state, (T_V' values for $(EO)_{10}OP$ and $(EO)_{20}NP$ are 1.4
and 1.0 mM), are 0.45 and 0.1 mM, respectively, equal to the cmc of the
respective surfactant (Table 1). In other words, when the supernatant solution
of a system at pH 2 happens to be equal to the cmc, the total surfactant
concentration is equal to T_V at pH 3.5. On the other hand, in the soluble
complex at pH 3.5, since the free surfactant concentration at T_V is estimated to
be the cmc, as found in the PAA-$C_{12}E_6$ and-$C_{12}E_8$, and the -$C_{10}E_8$, -C_{12} E_8 and
-$C_{14}E_8$ systems [101] (see Sections IV.A.2. and 3.), the amount of surfactant
bound is given by T_V -cmc, i.e., 0.95 and 0.9 mM, for $(EO)_{20}OP$ and $(EO)_{20}NP$,
respectively, and coincides with the amounts of precipitation of the respective
surfactants at T_V.   The unit mol ratios of both complexes at T_V are
approximately 1.4 at pHs 3.5 and 2.

These results mean that upon lowering the pH the complex becomes more
compact and the EO chains are bound to the PAA without much affecting the
surfactant composition of each complex as a whole, although hydrogen bonding
in the insoluble complex is perhaps enhanced to some extent compared with the
soluble state, especially at the end portion of the EO moiety.

b. Polymethacrylic acid

In the case of PMA [103], the precipitation curves for $(EO)_{10}OP$ and
$(EO)_{20}OP$ at pH 2 are upwardly concave with respect to the abscissa (Figure 12,
broken lines), and the supernatant surfactant concentration at the lowest point
is denoted as the PLC.   Contrary to the case of PAA, the PLC in PMA is not

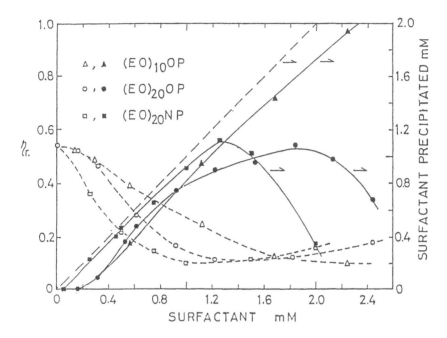

FIG. 13 Reduced viscosity of 13.5 mM PAA solutions with surfactants at pH 3.5 (broken lines, left ordinate) and the extent of surfactant precipitated at pH 2 (solid lines, right ordinate) plotted against surfactant concentration added at 25°C. The 45° line from the origin indicates the initial surfactant concentration before precipitation. From ref. 96 with permission.

constant but depends on the EO number and temperature, and is lower than T_1. The PLC may correspond to the starting point for single molecule binding, as described in Section IV.A.2., although it cannot be clearly determined by the solubilization method (Figure 7, and also Section IV.C.).

Table 5 also shows the compositions (as mol ratios) of the complexes of PMA with $(EO)_{10}OP$, $(EO)_{20}OP$, $(EO)_{10}NP$ and $(EO)_{20}NP$ calculated from the minima in the precipitation curves. The stoichiometric ratios are close to unity, indicating that hydrogen bonding plays the major role in the complex formation, even though PMA coils are more compact than PAA and hence carboxyls may be less available. The composition of the PMA complexes in the most insoluble state is similar to that of PAA complexes.

The difference in the precipitation curves for PAA and PMA (Figure 12) at high polymer concentration is remarkable. It may be explained that, in PAA, some of the chains anchor the micellar aggregates while others are free, whereas in the more hydrophobic PMA, the hydrophobic microdomains of

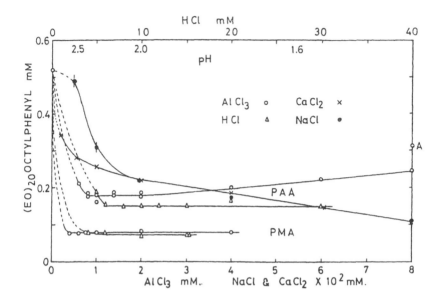

FIG. 14 Concentration of supernatant (EO)$_{20}$OP in the precipitated system from 0.52 mM solution in the presence of 15.6 mM PAA or 14.8 mM PMA by addition of HCl, or pH lowering (upper scale), NaCl, CaCl$_2$ (lower scale x 10^2) and AlCl$_3$ (lower scale) at various temperatures. Point A on the right-side ordinate indicates supernatant surfactant concentration at 20 mM AlCl$_3$ in the PAA system. No precipitation appeared in the dotted portion. From ref. 95 with permission.

additional PMA dissociate the aggregates into single molecules on its chains, leaving no free polymer chains. In general, the precipitation/dissolution phenomena in systems of polyacids and nonionic surfactants have something in common with those in systems of polyelectrolytes and ionic surfactants of opposite charge [2,4].

2. Salts

Neutral inorganic salts also were found to be effective for decreasing the solubility of the complex [115]. Among the salts with the same anions (chloride), Al^{3+} was the strongest precipitant followed by Ca^{2+} and Na^{+}. As shown in Figure 14, the amount of surfactant precipitated by salt addition was less than that precipitated by lowering the pH. The PLC in the PAA system with PMA-(EO)$_{20}$OP is almost the same for the addition of HCl or AlCl$_3$. The amount of precipitation reaches a maximum with $AlCl_3$, but with NaCl and $CaCl_2$ it increases steadily with salt concentration and no PLC seems to exist [95].

Neutral salts, especially of polyvalent cations, make the strings of complexes shrink, agglomerate and at the same time inhibit complex formation by blocking the carboxyls of polyacid. Moreover, salts may alter the solution behavior of both polymer and surfactant by their medium effects depending on the salt concentration. Thus, complex precipitation by salts is governed by the balance of these effects. In the case of $AlCl_3$ addition, because of its weak dissociation and strong blocking even in the dilute region, a maximum emerges in the amount of precipitation. Supposedly, the composition of the complex precipitated by Al^{3+} may be not much different from that in the solution state. With PMA no maximum was found by $AlCl_3$ addition perhaps due to the strongly bound surfactant is not replaced by Al^{3+}. With NaCl or $CaCl_2$ addition at high concentrations, the salting-out effect dominates over the blocking effect.

Anions exert their influence on the complex at high concentration through modification of the solvent properties following the lyotropic series: SCN^-, NO_3^- and Cl^- decreased the precipitation power in this order. The associated structure of liquid water is perturbed by various cosolutes, and the water-structure-promoting effect of hydrophobic solutes is weakened by water-structure-breaking substances such as urea, guanidinium$^+$, and some inorganic anions. Among the anions, SCN^- is strongest in this role and thus lowers the effect of inorganic cations most markedly (Section II).

A polymeric acid condensate of polyphenols such as tannin, and polyalcohols are employed for removal of nonionic surfactants in wastewater together with polyvalent cations in the neutral pH range [117]. The surfactant binding mechanism and the salt effect may be similar to the case of the PAA complex.

A PAA-$(EO)_9OP$ complex was precipitated by addition of a series of divalent bola-form cations, α,ω-diamine n-alkane dihydrochlorides ($ClNH_3$-R-NH_3Cl)}, and their effects were compared with that of Ca^{2+} [118]. The precipitation effects by the propane, butane, and hexane diamine cations and Ca^{2+} are almost the same, and only the octane salt is a slightly more efficient. The methylene group adjacent to an ionic headgroup is believed to be less hydrophobic than those farther removed [119], therefore the hydrophobicity of the cations NH_3^+-R-NH_3^+ should be less than that of the corresponding monovalent R-NH_3^+ ions. Moreover, monovalent long-alkyl chain cations like octylammonium and hexylammonium were more effective than thee divalent long alkylchain diammonium ions and even than Ca^{2+} in precipitating the complex. This implies that agglomeration of complexes by the hydrophobic effect is dominant over the purely electrostatic destabilization of the complexes by cations, and involves a complicated interplay of both functions.

3. Ionic surfactants

In practice, ionic surfactants are often mixed with nonionic surfactants. Because they interact with both nonionic surfactant and polymeric acid, their influence on complex formation are complicated [120]. Ionic surfactants were found to inhibit complex formation, and especially at low pH they prevent precipitation of the complex. Ionic surfactants replace nonionic surfactants fully from binding to CVP (Section IV.A.2.c.). Cationic surfactants require a higher amount to dissolve the precipitate than their anionic counterparts. This difference may be ascribed to the the fact that a fraction of the added cationic is used in electrostatic binding to PAA itself, preceding the stage of hydrophobic binding common with both ionic surfactants.

4. Polymers

a. Nonionic polymers

In principle, the binding mechanisms of ethoxylate-type nonionic surfactants and proton-acceptor nonionic polymers to a polymeric acid are considered to be the same. Thus, in the ternary systems of polymeric acid-nonionic surfactant-nonionic polymer, both nonionic solutes compete for the same carboxylic sites on the chain of polymeric acid. Figure 4 shows that the PEO-alkylethers overwhelm the free PEO of even much higher molecular weights in the reaction with PAA. Addition of PEO or PPO diminishes the precipitation of the PAA-$(EO)_{20}$OP complex at pH 2 only to a small extent. However, addition of PVP prevents $(EO)_{20}$OP fully from coprecipitating with PAA [121,122], and instead of the PAA-$(EO)_{20}$OP complex, a PAA-PVP complex separated, which is dispersed with further addition. PPO interacts with nonionic surfactants (Section III.A.), but there is no direct interaction between PVP and nonionic surfactant [34,48,51]. Addition of PVP also removes the nonionic surfactant from the CVP-$(EO)_{20}$OP complex, producing instead an insoluble complex CVP-PVP [120].

In PMA-$(EO)_{20}$OP systems, the effect of PVP addition was found to be small. Since PMA is more tightly coiled than PAA, PVP may not interfere with the complex so extensively as to alter the amount of precipitation greatly. However, it may partially replace the surfactant to form a water-dispersible or soluble ternary complex of PMA-PVP-nonionic surfactant mainly through hydrogen bonding. A three-component complex PAA-PVP-PEO appeared in the ternary system [123].

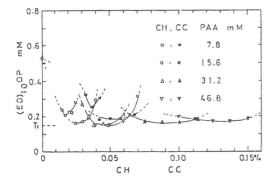

FIG. 15 Supernatant concentration of $(EO)_{10}OP$ precipitated from its 0.52 mM solution containing a fixed amount of PAA (7.8, 15.6, 31.2, and 46.8 mM) vs. concentration of chitosan hydrochlordie (CH) or cationized cellulose derivative (CC) added at 25 °C. The broken parts of the curves represent turbid systems. T_1 is the critical surfactant concentration for binding to PAA. Reprinted from ref. 124, *Colloids Surf. 62*, 57 (1992), with permission of Elsevier Science-NL, Sara Burgerhartstraat 25, 1055 KV Amsterdam, The Netherlands

b. Cationic polymers

The solution complexes PAA-$(EO)_{10}OP$, -$(EO)_{20}OP$,-$(EO)_{10}NP$ and - $(EO)_{20}NP$ precipitate with cationic polymers [124]. Figure 15 shows the supernatant concentration of $(EO)_{10}OP$ precipitated from its initial 0.52 mM solution with a fixed amount of PAA by addition of a cationic polymer, chitosan hydrochloride (CH) or a cationic cellulose derivative (CC) called hydroxyethylcellulose-2-hydroxypropyltrimethylammonium chloride ether. Although $(EO)_{14}NP$ and $(EO)_{20}NP$ interact with HEC [52] (section III.A.), their interaction with CC may be rather weak because the cationic charges make the whole polymer chain more hydrophilic. CH is considered not to interact with nonionic surfactants. Therefore, the precipitation of surfactants is a result of agglomeration of PAA and cationic polymers. As illustrated in Figure 15, the amount of $(EO)_{P,O}OP$ in the supernatant increases with increasing cationic polymer concentration. The lowest supernatant surfactant concentration is again denoted as the PLC. This PLC is constant, except for very low PAA concentrations. The same holds for other surfactants with PAA and added cationic polymer. The PLC values are independent of the size of the PEO moiety of the surfactants and of temperature. They coincide with the precipitation concentrations at low pH, and hence the PLC can be considered identical to T_1 (Table 1).

C. Solubilization Properties of Complexes

In this section we will discuss specific solubilization properties of some of the polymeric acid-nonionic surfactant complexes discussed earlier in Section IV.A.2 [91].

a. Polyacrylic acid

In systems of a fixed PAA and varied $(EO)_{10}OP$ concentration (Figure 7), solubilization of Yellow OB begins at a surfactant concentration (T_1) lower than the cmc. In the dilute region above T_1 solubilization increases linearly. As the surfactant concentration increases, the solubilization curves for the complex and micellar solutions are fairly close and may even cross. At high concentration, where the curves are parallel, free micelles participate increasingly in the solubilization. In Figure 7, at low concentration the difference in solubilization is due mainly to the shift of critical aggregation concentration and also to some change in the surfactant aggregation number due to the complexation with PAA.

At lower temperature the solubilization of Yellow OB in the PAA-$(EO)_{10}OP$ complex solution decreases, as illustrated in Figure 7, but this is not always so. In PAA-$(EO)_{20}C_{14}$ systems (Figure 16), the amount of solubilized Yellow OB decreases when the temperature decreases from 25 to 14°C but at 4°C the trend reversed dramatically [96]. A similar reverse with solubilization increasing at low temperature is observed in the series of PEO-alkylethers, and in PAA-$(EO)_{30}C_{16}$ solutions where Yellow OB solubilization increases strikingly when the temperature decreases from 25 to 4°C. This apparent anomaly become less pronounced with a further increase in the alkyl-chainlength from $(EO)_{20}C_{18}$ to $(EO)_{20}C_{22}$, and disappears with shorter PEO chains [97]. Such phenomena can be expressed in terms of the solubilization of a solubilizate by a micelle (Sm) or by a complex (Sc) as the mol ratio of solubilizate to surfactant, representing the slope of the linear part of solubilization curve just above the cmc or T_1, respectively, as seen in Figures 7 and 16. Table 6 summarizes Sm and Sc for series of PEO-alkylaryl and -alkyl ethers for Yellow OB at 4, 14, 25 and 38°C. When Sc/Sm is about 1, the dye does not appear to be affected by the presence of PAA, even though its contact with PAA is observed in spectral properties [91].

This unusual temperature effect was found only for the oil-soluble dyes Yellow OB (1-*o*-tolylazo-2-naphthylamine), Yellow AB (1-phenylazo-2-naphthylamine) and Naphthyl Red (1-phenylazo-4-naphthylamine, an isomer of Yellow AB), all having a common phenylazo-naphthylamine structure. The phenomenon was not present for phenyl-1-naphthylamine, Sudan I (1-phenylazo-2-naphthol), and dimethylaminoazobenzene, etc. [97].

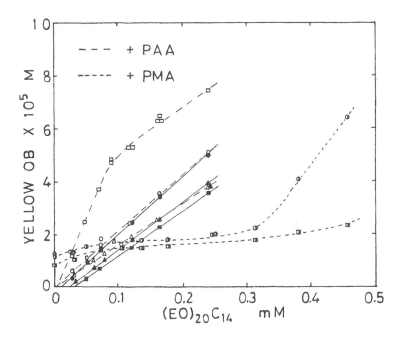

FIG. 16. Solubilization of Yellow OB in $(EO)_{20}C_{14}$ solutions alone (solid lines with filled marks), with 13.5 mM PAA (broken lines with open marks), and 13.5 mM PMA (dotted lines with half-filled marks) at 4 °C (squares), 14 °C (triangles), and 25°C (circles). From refs. 96 and 103 with permission.

Supposedly, in the complex the amino groups of the solubilizates and EOs of the surfactants compete for the PAA carboxyls. When the temperature decreases, the EOs in the terminal portion of the EO chain tend to be more hydrated and free, as depicted on the right hand side of Figure 10, and the surfactants with longer PEO moieties are more easily displaced. Thus, the balance moves in favor of the amines resulting in an abnormal solubilization increase. A stronger interaction between PAA and surfactants with longer alkyl and shorter PEO moieties may favor the surfactant EOs over the amino groups of the dye, hence, the solubilizates may behave less anomalously, as noted in Table 6 and also in the PMA cases shown below. The results suggest the existence of a complicated interplay of polymer, surfactant (PEO and hydrophobic moieties), solubilizate and water.in the complex, which moreover depends on temperature.

Although this reverse temperature effect was not observed with the PAA complexes of $(EO)_{10}OP$ and $(EO)_{20}OP$ (Figure 7), and with $(EO)_{10}NP$ and

TABLE 6 Solubilization of Yellow OB by Surfactant Micelle (Sm[1]) and by Complex with 13.5 mM PA (Sc[2]) at Various Temperatures

	4 °C			14 °C			25 °C			38 °C		
	Sm	Sc	Sc/Sm	Sm	Sc	Sc/Sm	Sm	Sc	Sc/Sm	Sm	Sc	Sc/Sm
$EO_{10}OP$	8.0	6.6	0.8	11.8	9.5	0.8	13.7	12.5	0.9	27	20	0.8
$EO_{20}OP$	4.4	5.4	1.2	9.8	9.5	1.0	12.0	12.7	1.1			
$EO_{10}NP$	8.3	7.0	0.8	12.2	9.3	0.8	18.0	12.4	0.7			
$EO_{20}NP$	6.7	7.1	1.1	9.6	11.1	1.2	13.5	14.5	1.1			
EO_5C_{12}	18	15	0.9	20	11	0.5	23	12	0.5			
$EO_{10}C_{12}$	11	14	1.3	14	13	0.9	18	16	0.9	24	21	0.9
$EO_{15}C_{12}$	9	18	2.0	15	14	0.9	18	16	0.9			
$EO_{10}C_{14}$	18	30	1.7	19	17	0.9	20	19	1.0			
$EO_{20}C_{14}$	17	57	3.3	18	16	0.9	23	22	1.0			
$EO_{10}C_{16}$	23	23	1.0	24	18	0.8	27	19	0.7			
$EO_{15}C_{16}$	23	43	1.9	26	21	0.8	26	21	0.8			
$EO_{30}C_{16}$	11	55	5.0	13	14	1.1	16	12	0.7			
$EO_{20}C_{18}$	28	35	1.3	28	28	1.0	32	28	0.9			
$EO_{30}C_{18}$	36	47	1.3				38	27	0.7			
$EO_{20}C_{22}$	17	19	1.1	18	19	1.1	31	23	0.7			
$EO_{30}C_{22}$	19	25	1.3	22	25	1.1	36	29	0.8			

[1] Mol ratio (Yellow OB/Surfactant)x10^2 above cmc
[2] Mol ratio (Yellow OB/surfactant)x10^2 above cac
Source: ref. 97.

$(EO)_{20}NP$ in the solubilization of Yellow OB and Yellow AB, it did occur in the case of Naphthyl Red in PAA-$(EO)_{20}OP$ (Figure 17). In the latter case, solubilization was higher not only at 4°C but also at 25°C than at 38°C. Thus, the anomaly is not restricted to low temperature, and no fundamental difference exists between PAA complexes of PEO-alkyl and PEO-alkylaryl type surfactants. A phenyl group in surfactant seems to have a significant influence on the competitive behavior of the solubilizate in the complex.

As seen from Figure 17, the (Sc/Sm) values for Naphthyl Red in PAA-$(EO)_{20}OP$ at 4, 25, and 38°C are about 35, 4.3, and 1.5, respectively. Likewise, those in PAA- $(EO)_{20}C_{14}$ are 17, 3.5, and 1.4, respectively. The (Sc/Sm) values for Naphthyl Red at 4 °C are extremely high compared with those for Yellow OB as shown in Table 6. It suggests that one surfactant molecule in a micellar aggregate alone can not solubilize so many dye molecules without a contribution from PAA. The mol ratio of Naphthyl Red bound at saturation per carboxyl of PAA without surfactant is 0.03.

Such a synergistic solubilization may be called the polymer-binding solubilization and is different from that due simply to shift of micellar aggregation from cmc to T_1 as mentioned earlier. It is also different from cases such as the solubilization of Yellow OB or Orange OT in PVP-Na alkylsulfates [2,125]. In the latter systems, the polymers wrapping around surfactant micelles probably act as a kind of solvent for solubilizates in the complex [126,127].

Despite the reverse temperature effect, cmc and T_1 are unchanged within experimental error by the solubilization of Yellow OB and Yellow AB in $(EO)_{20}OP$ solution without and with PAA except for one observation, i.e. the case of Naphthyl Red which has a lower T_1 (0.07 and 0.10 mM at 4 and 25°C, respectively, cf. Table 1). Naphthyl Red has a very high Sc in this complex at lower temperature and may mediate increasing surfactant binding to PAA, and vice versa. A further investigation may be warranted.

A similar solubilization increase with accompanying promotion of surfactant binding (lowering of T_1) was reported in the solubilization of Oil Yellow OB in solutions of alkyltrimethylammonium bromides in the presence of polyvinylsulfate. This phenomenon was called concerted cooperative binding [128]. A detailed discussion of these ionic systems can be found in chapter 11 of this book.

Solubilization saturation of Naphthyl Red has not been reported in PAA-$(EO)_{20}OP$. In PAA-$(EO)_{20}C_{14}$ systems, solubilization saturation of Yellow OB was found only at 0.1 mM surfactant at 4°C (Figure 16), and at 0.14 mM for Yellow AB [97]. As these concentrations (T_S) were much lower than the reduced viscosity minimum $(T_V, 1$ mM), this solubilization saturation is not related to the saturation of surfactant binding. It is likely that above T_S the competition of EOs for PAA overwhelms that of the dye amines.

b. Polymethacrylic acid

In PMA-$(EO)_{10}OP$ (Figure 7) and PMA-$(EO)_{20}C_{14}$ (Figure 16) systems, Yellow OB solubilization increases abruptly above a T_1 point which is higher than the corresponding cmc (Table 4). This is contrary to the observation in PAA-added systems. In addition, the anomalous increase of Yellow OB solubilization in the PAA-$(EO)_{20}C_{14}$ systems at low temperature was not observed in the PMA systems, perhaps because the surfactants are more strongly bound to PMA than to PAA, as already remarked in Section VI.A.2.

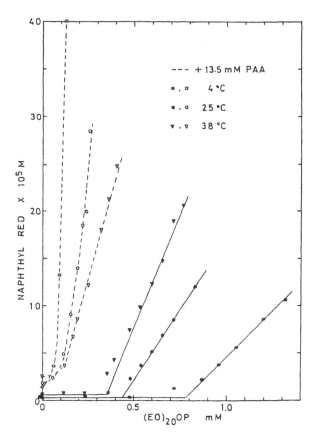

FIG. 17 Solubilization of Naphthyl Red by $(EO)_{20}OP$ solutions without (solid lines) and with 13.5 mM PAA (broken lines) at 4, 25, and 38°C. From ref. 97 with permission.

Compared with PAA, PMA in water dissolves higher amounts of oil-soluble dyes even without surfactant. The hydrophobic microdomains of the PMA chains [105,92] may be the loci for incorporating the solubilizates in dilute region. For some solubilizates the microdomains on PMA chains may cause different kinds of irregular solubilization behavior because the surfactants not only compete with the solubilizates for the microdomains but also change the solubilizability in them [103]. One example is solubilization of Sudan I in the PMA-$(EO)_{20}C_{14}$ system as depicted in Figure 18. It is likely that in the low concentration region the surfactants replace the solubilizates in the microdomains. The surfactant concentration at the minimum of the solubilization curve coincides with the T_1 value determined with Yellow OB (Table 4).

The complex including solubilized dye is precipitated almost intact by pH lowering or by salt addition. The complex PMA-$(EO)_{20}OP$, when separated, dried, and roughly ground, decomposes in polar solvents (ethanol, acetone, dioxane), but not in nonpolar ones (hexane, benzene, carbon tetrachloride), with the action of ethylether and chloroform in between these two groups. In water the decomposition rate of the complex was very slow because of swelling of the complex. Likewise, the solubilizate entrapped in the dried complex was extractable by the polar solvents group, but not by the others or by water [129].

D. Effect of Complex Formation on Colloids

1. Flocculation of colloids stabilized with nonionic surfactants by polymeric acids

Unlike the aqueous colloidal dispersions stabilized with ionic surfactants, those stabilized with nonionic surfactants alone were stable against the attack of salts at room temperature. They were, however, easily flocculated by polymeric acids at low pH or with neutral salts [130]. The flocculation arises as a result of complex formation between the polymeric acid and the surfactant on the surface of the dispersed particles. This mechanism is common in the flocculation by free PEO of lattices which have surface acid groups or are stabilized with PAA, or conversely of those stabilized with a surface PEO layer by polymeric acid [131].

For example, a latex stabilized with $(EO)_{20}$dodecyl is flocculated on addition of an appropriate concentration of polymeric acids at pH 2 or by salts. The efficiency of PAA as a flocculant varied with coadditives. HCl was most

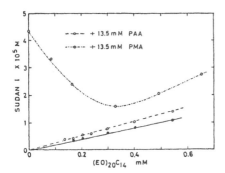

FIG. 18 Solubilization of Sudan I by $(EO)_{20}C_{14}$ solutions without (solid line) and with 13.5 mM PAA and PMA (broken lines) at $25°C$. From ref. 103 with permission.

effective followed by salts of polyvalent cations. In a proposed mechanism, the polyacid functions as a connecting bridge between particles coated with nonionic surfactants and acid, with salt compressing the bridge span leading to the coalescence of particles. CVP was found to flocculate the sol without any additive. Excessive polymer addition leads to restabilization of the dispersed system.

A temperature increase raises the flocculation efficiency, as the complex formation, or bridging in this case, is accelerated at high temperature. The temperature at which a colloidal dispersion stabilized with PEO-dodecyls flocculatesis is called the critical flocculation temperature (cft). The cft dropped significantly by addition of polymeric acid, but at high polymer concentration it rises again [132]. The similarity between the cloud point of the nonionic surfactant and cft of the micellar solution was also found to be valid for the cft of the stabilized dispersion [133].

2. Multiple emulsions

The Pluronic type surfactants PEO_X-PPO_Y-PEO_Z interact with PAA. The complexes formed can be used to prepare W/O/W emulsions [134]. The complex forms most efficiently at pH 2, resulting in a precipitate with a mol ratio of total EOs per carboxyl of almost 1. When more Pluronic is added to PAA at pH 2, the precipitate disperses into smaller particles. At pH 4, the particles also are smaller.

In preparing the W/O/W emulsions, a two-step emulsification procedure was followed. First, a complex of PAA-Pluronic was added to an oil containing a lipophilic surfactant such as Span 80, resulting in a primary W/O emulsion of which the oil phase contained the lipophilic surfactant. Then the emulsion was shaken with an external aqueous phase containing a hydrophilic surfactant such as Tween 80 or a Pluronic. The stability of W/O/W emulsions was found to depend on the particle size of PAA-Pluronic complex and on the lipophilic surfactant in the oil phase. Emulsion droplet size was also related to the complex size. The complex formed at pH 4 was not suitable for preparing stable multiple emulsions.

V. CONCLUDING REMARKS

The main characteristics of surfactants in aqueous solution are the tendencies to adsorb at interfaces and micelle formation. The latter is affected by cosolutes modifying the solvent medium, by small hydrophobic substances influencing the assembly of the surfactants, and most significantly by polymers or proteins through binding. When the polymers are linear, the binding of surfactants is not a two-dimensional adsorption process, but a one-dimensional molecular process in solution. When this process is indeed one of binding single surfactant molecules, the amount of binding is only small and the binding effect may not be difficult to determine. When the binding is cooperative along the polymer chain, the properties of both polymer and surfactant mutually affect the system properties and the solution state. The polymers gather and condense the surfactants, and surfactants aggregate locally on the polymer chain in a form and size different from the free micelles. This is the basic feature of polymer-surfactant interactions.

Systems of polymers with ionic surfactants have been studied extensively, and such systems have found various applications, as described in chapter 2 of this book. Although studies on interactions between nonionic surfactants and polymers were less numerous, research in this area has also progressed considerably in the last ten years, especially with respect to the role of surfactant structure. A number of distinctions of nonionic surfactant systems relative to ionic systems have been brought to light, particularly with respect to their interaction with polymeric acids. Still, many new and emerging problems and challenges are yet to be explored. Since the ethoxy-type surfactants occupy most of the nonionics market, their strong interaction with polymeric acids is of considerable interest and involves many applications. We hope that the present article will stimulate further research, and new experimental approaches.

ACKNOWLEDGMENT

The first author personally appreciates Momotani Juntenkan Ltd. for support of research on the subjects discussed in this chapter during his tenure with the company (1958-1991).

REFERENCES

1. E.D. Goddard, *Colloids Surf. 19*: 255, 301 (1986).
2. E.D. Goddard, in *Interactions of Surfactants with Polymers and Proteins* (E.D. Goddard and K.P. Ananthapadmanabhan, eds.), CRC Press, Boca Raton, 1993, Ch. 4.
3. B. Lindman and K. Thalberg, in *Interactions of Surfactants with Polymers and Proteins* (E.D. Goddard and K.P. Ananthapadmanabhan, eds.), CRC Press, Boca Raton, 1993, Ch. 5.
4. K. Hayakawa and J.C.T. Kwak, in *Cationic Surfactants, Physical Chemistry* (D.N. Rubingh and P.M. Holland, eds.), Marcel Dekker, New York, 1991, Ch. 5.
5. J.C. Brackman and J.B.F.N. Engberts, *Chem. Soc. Rev. 22*: 85 (1993).
5a. J.C. Brackman, *Ph.D. Thesis*, University of Groningen, 1990.
6. I.D. Robb, in *Anionic Surfactants. Physical Chemistry of Surfactant Action* (E.H. Lucassen-Reynders, ed.), Marcel Dekker, New York, 1981, p.109.
7. P. Molyneux, *Water-Soluble Synthetic Polymers: Properties and Behavior*, Vol. 2, CRC Press, Boca Raton, 1984.
8. S. Saito, in *Nonionic Surfactants. Physical Chemistry* (M.J. Schick, ed.), Marcel Dekker, New York, 1987, Ch. 15.
9. S. Saito and Y. Mizuta, *J. Colloid Interface Sci. 23*: 604 (1967).
10. F.M. Witte and J.B.F.N. Engberts, *Colloids Surf. 36*: 417 (1989).
11. P.L. Dubin, J.H. Gruber, J. Xia, and H. Zhang, *J. Colloid Interface Sci. 148*: 35 (1992).
12. K. Shirahama, K. Mukae, and H. Iseki, *Colloid Polym. Sci. 272*: 493 (1994).
13. S. Saito, T. Taniguchi, and K. Kitamura, *J. Colloid Interface Sci. 37*: 154 (1971).
14. J.C. Brackman and J.B.F.N. Engberts, *J. Colloid Interface Sci. 132*: 200 (1989).
15. S. Saito, *J. Colloid Sci. 15*: 283 (1960).
16. S. Saito, *Kolloid-Z. 137*: 93 (1954).
17. C. Maltesh and P. Somasundaran, *Colloids Surf. 69*: 167 (1992).
18. S. Saito, *J. Polym. Sci. A-1 7*: 1789 (1969).
19. M. Ataman, *Colloid Polym. Sci. 265*: 19 (1987).
20. S. Saito and K. Kitamura, *J. Colloid Interface Sci. 35*: 346 (1971).
21. S. Saito and M. Yukawa, *J. Colloid Interface Sci. 30*: 211 (1969).
22. S. Saito and M. Yukawa, *Kolloid-Z .Z. Polym. 234*: 1015 (1969).
23. S. Saito, *J. Polym. Sci. A-1 8*: 263 (1970).
24. W.-Y. Wen, S. Saito, and C.M. Lee, *J. Phys. Chem. 70*: 1244 (1966).
25. J.L. Kavanau, *Water and Solute-Water Interactions*, Holden-Day, San Fransisco, 1964.
26. S. Harada, K. Komatsu, and T. Nakagawa, *Nippon Kagaku Kaishi* (1974) 662. CA 81: 121224g (1974).
27. S. Harada, K. Komatsu, and T. Nakagawa, *Rep. Progr. Polym. Phys. Jpn. 17*: 101 (1974), 19: 17 (1976).
28. S. Makino and S. Sugai, *J. Polym. Sci. A-2 5*: 1013 (1967).
29. D. Puett, A. Cifferri, E. Bianchi, and J. Hermans, Jr., *J. Phys. Chem. 71*: 4126 (1967).

30. C. Ebert, G. Ebert, and W. Werner, *Colloid Polym. Sci. 251*: 504 (1973).
31. E. Peggin, A. Cosano, M. Jerbojevich, and G. Borin, *Biopolymers 11*: 633 (1972).
32. N. Murai, M. Miyazaki, and S. Sugai, *Nippon Kagaku Kaishi* (1976) 659, CA 85: 33544p (1976).
33. E. Dickinson, in *Interactions of Surfactants with Polymers and Proteins* (E.D. Goddard and K.P. Ananthapadmanabhan, eds.), CRC Press, Boca Raton, 1993, Ch. 7.
34. S. Saito, *Kolloid-Z. 137*: 98 (1954).
34a. S. Saito, *Kolloid-Z. 158*: 120 (1958).
35. K.P. Ananthapadmanabhan, *in Interactions of Surfactants with Polymers and Proteins* (E.D. Goddard and K.P. Ananthapadmanabhan, eds.), CRC Press, Boca Raton, 1993, Ch. 2.
36. B. Cabane, *J. Phys. Chem. 81*: 1639 (1977).
37. M.L. Smith and N. Muller, *J. Colloid Interface Sci. 52*: 507 (1975).
38. W. Scholtan, *Makromol. Chem. 11*: 131 (1953).
39. J. Eliassaf, F. Eriksson, and F.R. Eirich, *J. Polym. Sci. 47*: 193 (1960).
40. P. Molyneux and H.P. Frank, *J. Am. Chem. Soc. 83*: 3169, 3175 (1961).
41. P. Molyneux and M. Cornarakis-Lentzos, *Colloid Polym. Sci. 257*: 855 (1979).
42. J.E.B. Randles, *Disc. Faraday Soc. 24*: 194 (1957).
43. N.L. Jarvis and M.A. Scheiman, *J. Phys. Chem. 72*: 74 (1968).
44. W.J. Moore, *Physical Chemistry*,4th ed., Prentice-Hall, New Jersey, 1972, Ch. 10.
45. N.K. Adam, *The Physics and Chemistry of Surfaces*, 2nd ed., Oxford Univ., London, 1938, pp.134, 360.
46. J.C. Brackman and J.B.F.N. Engberts, *Langmuir 8*: 424 (1992).
47. M.J. Schwuger and H. Lange, *Proc.5th Int. Congr. Surface Activity 2*: 955 (1968).
48. F.A. Green, *J. Colloid Interface Sci. 35*: 481 (1971).
49. G.C. Kresheck and W.A. Hargraves, *J. Colloid Interface Sci. 83*: 1 (1981).
50. V. Szmerekova, P. Kralik, and D. Berek, *J. Chromatogr. 285*: 188 (1984).
51. S. Saito, *Kolloid-Z. Z. Polym. 226*: 10 (1968).
52. Y. Boscher, F. Lafuma, and C. Ouivoron, *Polym. Bull. 9*: 533 (1983).
53. J.C. Brackman, N.M. van Os, and J.B.F.N. Engberts, *Langmuir 4*: 1266 (1988).
54. F.M. Winnik, *Langmuir 6*: 522 (1990).
55. F.M. Winnik, M.A. Winnik, and S. Tazuke, *J. Phys. Chem. 91*: 594 (1987).
56. C.J. Drummond, S. Albers, and D.N. Furlong, *Colloids Surf. 62*: 75 (1992).
57. K.-W. Zhang, G. Karlström, and B. Lindman, *Colloids Surf. 67*: 147 (1992).
58. K.R. Wormuth, *Langmuir 7*: 1622 (1991).
59. L. Piculell and B. Lindman, *Adv. Colloid Interface Sci. 41*: 149 (1992).
60. M.H.G.M. Penders, S. Nilsson, L. Piculell, and B. Lindman, *J. Phys. Chem. 97*: 11332 (1993).
61. G. Karlström, A. Carlsson, and B. Lindman, *J. Phys. Chem. 94*: 5005 (1990).
62. A.A. Samii, B. Lindman, and G. Karlström, *Progr. Colloid Polymer. Sci. 82*: 280 (1990).
63. E.D. Goddard, in *Interactions of Surfactants with Polymers and Proteins* (E.D. Goddard and K.P. Ananthapadmanabhan, eds.), CRC Press, Boca Raton, 1993, Ch. 10.

64. K.T. Wang, I. Iliopoulos, and R. Audebert, *Polym. Bull. 20*: 577 (1988).
65. M. Huldén, *Colloids Surf. A 82*: 263 (1994).
66. F.M. Winnik, H. Ringsdorf, and J. Venzmer, *Langmuir 7*: (a) 905, (b) 912 (1991).
67. N. Kamenka, A. Kaplun, Y. Talmon, and R. Zana, *Langmuir 10*: 2960 (1994).
68. I. Iliopoulos and U. Olsson, *J. Phys. Chem. 98*: 1500 (1994).
69. A. Sarrazin-Cartalas, I. Iliopoulos, R. Audebert, and U. Olsson, *Langmuir 10*: 1421 (1994).
70. K. Loyen, I. Iliopoulos, U. Olsson, and R. Audebert, Progr. Colloid Polymer. Sci. 98: 42 (1995).
71. K. Loyen, I. Iliopoulos, R. Audebert, and U. Olsson, *Langmuir 11*: 1053 (1995).
72. K. Sivadasan and P. Somasundaran, *Colloids Surf. 49*: 229 (1990).
73. S. Makino, J.A. Reynolds, and C. Tanford, *J. Biol. Chem. 248*: 4926 (1973).
74. N. Nishikido, T. Takahara, H. Kobayashi, and M. Tanaka, *Bull. Chem. Soc. Jpn. 55*: 3085 (1982).
75. W.W. Sukow, H.E. Sandberg, E.A. Lewis, D.J. Eatough, and L.D. Hansen, *Biochemistry 19*: 912 (1980).
76. W.W. Sukow and J. Bailey, *Physiol. Chem. Phys. 13*: 455 (1981).
77. E. Tujii and H. Maeda, *Colloid Polym. Sci. 270*: 894 (1992).
78. H. Nishiyama and H. Maeda, *Biophys. Chem. 44*: 199 (1992).
79. J. Cordoba, M.D. Reboiras, and M.N. Jones, *Int. J. Biol.Macromol. 10*: 270 (1988).
80. M. Inoue and T. Otsu, *J. Polym. Sci. Polym. Chem. Ed. 14*: 1933, 1939 (1976).
81. T. Ikawa, K. Abe, K. Honda, and E. Tsuchida, *J. Polym. Sci. Polym. Chem. Ed. 13*: 1505 (1975).
82. G.D. Jaycox, R. Sinta, and J. Smid, *J. Polym. Sci Polym. Chem. Ed. 20*: 162 (1982).
83. H. Ohno, K. Abe, and E. Tsuchida, *Makromol. Chem. 179*: 755 (1978).
84. K. Abe, H. Ohno, A. Nii, and E. Tsuchida, *Makromol. Chem. 179*: 2043 (1978).
85. T. Bartels, Y.Y. Tan, and G. Challa, *J. Polym. Sci. Polym. Chem. Ed. 15*: 34 (1977).
86. D.W. Koestsier, Y.Y. Tan, and G. Challa, *J. Polym. Sci. Polym. Chem. Ed. 18*: 1932 (1980).
87. E. Kokufuta, A. Yokota, and I. Nakamura, *Polymer 24*: 1031 (1983).
88. S. Saito, *Colloids Surf. 19*: 351 (1986).
89. S. Saito and T. Taniguchi, *J. Colloid Interaface Sci. 44*: 114 (1973).
90. K.B. Musabekov, K.Zh. Abdiev, and S.B. Aidarova, *Kolloid Zh. 46*: 376 (1984). CA 101: 39121u (1984).
91. S. Saito, *Colloid Polym. Sci. 257*: 266 (1979).
92. D.Y. Chu and J.K. Thomas, *J. Am. Chem. Soc. 108*: 6270 (1986).
93. M.N. Jones, *J. Colloid Interface Sci. 23*: 36 (1967).
94. A. Benkhira, E. Franta, and J. François, *J. Colloid Interface Sci. 164*: 428 (1994).
95. S. Saito, *J. Am. Oil Chem. Soc. 66*: 987 (1989).
96. S. Saito, *Rev. Roumaine Chim. 35*: 821 (1990).
97. Saito, *J. Colloid Interface Sci. 158*: 77 (1993).
98. D.F. Anghel, S. Saito, A. Iovescu, and A. Baran, *Colloids Surf. A 90*: 89 (1994).

99. D. Eagland, N.J. Crowther, and C.J. Butler, *Eur. Polym. J. 30*: 767, (1994).
100. V.Yu. Baranovsky, S. Shenkov, I. Rashkov, and G. Borisov, *Eur. Polym. J. 27*: 643 (1991).
101. D.F. Anghel, S. Saito, A. Iovescu, and A. Baran, to be published.
102. C. Maloney and K. Huber, *J. Colloid Interface Sci. 164*: 463 (1994).
103. S. Saito, *J. Colloid Interface Sci. 165*: 505 (1994).
104. J.J. Kiefer, P. Somasundaran, and K.P. Ananthapadmanabhan, *Langmuir 9*: 1187 (1993).
105. G. Barone, V. Crescenzi, A. Liquori, and F. Quadrifoglio, *J. Phys. Chem. 71*: 2341 (1967).
106. V.Yu. Baranovsky, S. Shenkov, I. Rashkov, and G. Borisov, *Eur. Polym. J. 28*: 475 (1992).
107. V.Yu. Baranovsky, S. Shenkov, and G. Borisov, *Eur. Polym. J. 29*: 1137 (1993).
108. V.Yu. Baranovsky and S. Shenkov, *J. Polym. Sci. A. Polym. Chem. Ed. 34*: 163 (1996).
109. S. Saito and T. Taniguchi, *J. Am. Oil Chem. Soc. 50*: 276 (1973).
110. C. Tanford, *The Hydrophobic Effect: Formation of Micelles and Biological Membranes*, 2nd Ed., John-Wiley and Sons, New York, 1980, Ch. 3 and 7.
111. H. Arai, M. Murata, and K. Shinoda, *J. Colloid Interface Sci. 37*: 223 (1971).
112. F.M. Winnik, in *Interactions of Surfactants with Polymers and Proteins* (E.D. Goddard and K.P. Ananthapadmanabhan, eds.), CRC Press, Boca Raton, 1993, Ch. 9.
113. M. Vasilescu, D.F. Anghel, M. Almgren, P. Hansson, and S. Saito, *Langmuir 13*: 6951 (1997).
114. K.Thalberg, J. van Stam, C. Lindblad, M. Almgren, and B. Lindman, *J. Phys. Chem. 95*: 8975 (1991).
115. S. Saito, T. Taniguchi, and H. Matsuyama, *Colloid Polym. Sci. 254*: 882 (1976).
116. S. Saito, *Tenside 14*: 113 (1977).
117. K. Hagihara and S. Murakami, *Water Treatment Tech. (Japan) 14*: 581 (1973).
118. S. Saito, *Tenside 17*: 84 (1980).
119. C.V. Krishnan and H.L. Friedman, *J. Phys. Chem. 74*: 3900 (1970).
120. S. Saito, *Tenside 18*: 117 (1981).
121. S. Saito, *Colloid Polym. Sci. 260*: 613 (1982).
122. S. Saito, *Tenside 20*: 88 (1983).
123. S.K. Chatterjee, A. Malhota, and L.S. Pachauri, *Angew. Makromol. Chem. 116*: 99 (1983).
124. S. Saito and Y. Matsui, *Colloids Surf. 62*: 57 (1992).
125. H. Lange, *Kolloid-Z. Z. Polym. 243*: 101 (1971).
126. S. Saito, *Kolloid-Z. 154*: 19 (1957).
127. S. Saito, *J. Colloid Interface Sci. 24*: 227 (1967).
128. K. Hayakawa, T. Fukutome, and I. Satake, *Langmuir 6*: 1495 (1990).
129. S. Saito and Y. Matsui, *J. Colloid Interface Sci. 67*: 483 (1978).
130. S. Saito and M. Fujiwara, *Colloid Polym. Sci. 255*: 1122 (1977).
131. R. Evans and D.H. Napper, *Nature (London) 246*: 34 (1973).
132. L. Thompson and A. McEwen, *J. Colloid Interface Sci. 93*: 329 (1983).

133. D.B. Hough and L. Thompson, in *Nonionic Surfactants. Physical Chemistry* (M.J. Schick, ed.), Marcel Dekker, New York, 1987, Ch. 11.
134. M.L. Cole and T.L. Whateley, *J. Colloid Interface Sci. 175*: 281 (1995).
135. E. Feitosa, W. Brown, and P. Hansson, *Macromolecules 29*: 2169 (1996).
136. E. Feitosa, W. Brown, M. Vasilescu, and M. Swanson-Vethamuthu, *Macromolecules 29*: 6837 (1996).
137. V. Raicu, A. Baran, D.F. Anghel, S. Saito, A. Iovescu, and C. Radoi, *Progr. Colloid Polym. Sci.,* in press.
138. D.F. Anghel, S. Saito, A. Baran, and A. Iovescu, to be published.

10

Polyelectrolyte-Surfactant Interactions: Polymer Hydrophobicity, Surfactant Aggregation Number, and Microstructure of the Systems

RAOUL ZANA Institut C. Sadron (CNRS-ULP), 6, rue Boussingault, 67000 Strasbourg, France

I. INTRODUCTION

The reasons for the current interest in systems containing both polymers and surfactants are presented in several chapters of this volume. This permits us to go directly to the topic of the present chapter, that is the characterization of the surfactant aggregates present in polymer/surfactant systems, mainly by the *surfactant aggregation number*, N (number of surfactants making up an aggregate), and the effect of the *polymer hydrophobicity* on the value of N. This quantity is of special importance because its value permits one to evaluate the surfactant aggregate size and to obtain information on the *microstructure* of the systems. This chapter mainly concerns systems where the interaction between polymers and surfactants involves a binding of the surfactant onto the polymer.

Few experimental methods give access to the surfactant aggregation number in polymer/surfactant systems. Thus, conventional methods that are sensitive to the density of particles present in the system under investigation yield information on the polymer/surfactant complexes and not on the surfactant aggregates bound to the polymer. Light, neutron or X-ray scattering techniques usually measure a complex quantity which results from the convolution of the effects of size and shape, and of interaction between polymer-bound surfactant aggregates and of the polymers connecting these aggregates. The only methods which appear capable of measuring fairly accurately aggregation numbers of polymer-bound surfactant aggregates are the so-called fluorescence probing methods [1-4]. In these methods a fluorescent molecule (referred to as probe, thus fluorescence probing) is introduced in the system under investigation. The static and dynamic fluorescence properties of the probe solubilized in the surfactant aggregates are then used to obtain information on the aggregates.

This chapter focuses on charged polymers (polyelectrolytes) of varying hydrophobicity. Nonionic polymers are dealt with in Chapter 7 of this volume. A very large number of polyelectrolyte/surfactant systems have been investigated, differing by the nature of the polymer and surfactant [5,6]. A quick glance at the reported studies shows that the two most important parameters governing various aspects of polyelectrolyte/surfactant interactions are the nature (hydrophobicity) of the polyelectrolyte and the sign of its electrical charge. In general, no binding takes place when polyelectrolyte and surfactant have like charges. However if the polymer is sufficiently hydrophobic, hydrophobic interactions can overcome the electrostatic repulsions, resulting in the binding of surfactants on polyions of like charge. In fact the polymer hydrophobicity determines both the qualitative and quantitative aspects of the interaction: nature of the interaction (cooperative or non-cooperative), surfactant binding constant, size of polymer-bound surfactant aggregates and internal properties of the aggregates. Unfortunately, the polymer hydrophobicity is a parameter which is not easily quantified, making comparisons between different polymers difficult. However, systematic investigations have been performed on a series of polyelectrolytes of tunable hydrophobicity, namely a series of alternated copolymers of maleic acid and alkylvinylether, referred to as PSX where X is the number of carbon atoms in the alkyl group and n the degree of polymerization, first introduced by U. P. Strauss [7]. The hydrophobicity of these copolymers can be easily modified by changing the degree of neutralization, α, of the maleic acid moieties and/or the ether group carbon number, X. This review is largely based on the results reported for these copolymers. PSX polyelectrolytes of increasing hydrophobicity are discussed successively. In each instance studies for other polyelectrolytes are referred to, showing the similarities and differences with PSX.

$$(-CH-CH-CH_2-CH-)_n$$

with substituents: COOH, COOH, and O–C_XH_{2X+1}

(PSX)

This chapter is organized as follows. Section II examines various fluorescence probing methods for the determination of surfactant aggregation numbers, with particular emphasis on the conditions under which these methods yield valid results. Section III reviews the results for hydrophilic polyelectrolyte/surfactant systems. Hydrophobic polyelectrolyte/surfactant systems are reviewed in Section IV. Section V deals with polymers of moderate hydrophobicity and proteins. Section VI reviews polymer/surfactant interactions in systems of special interest.

Although this review mainly deals with surfactant aggregation numbers other properties of the bound surfactant aggregates are considered, for instance microviscosity and lifetime of fluorescent probes solubilized in the aggregates. The microstructure of polyelectrolyte/surfactant systems is also discussed.

II. FLUORESCENCE METHODS FOR THE DETERMINATION OF SURFACTANT AGGREGATION NUMBERS IN SURFACTANT-CONTAINING SYSTEMS

Surfactants are generally not fluorescent and their study by means of fluorescence probing requires the introduction of a fluorescent probe in the surfactant solution. This probe is added to the system and, in the ideal situation, is completely solubilized in the micelles. In the case of polymer/surfactant interaction studies the probe can be covalently attached to the polymer (it is then referred to as a label or tag). In both instances, the changes of the probe static and dynamic fluorescence properties with respect to those in pure solvents can be used to obtain information on the micelles [1-4,8]. Note that labelled systems may behave differently from systems with free probes [9-11]. Winnik and Regismond show in Chapter 7 of this volume how fluorescence probing can be used to obtain information on the micropolarity and microviscosity of surfactant aggregates.   This section focuses on using fluorescence probing to determine surfactant aggregation numbers. Note that a few surfactant molecules contain fluorescent groups, phenyl or phenylene, for instance.   Some specific aspects of their micellar solutions can be studied without addition of probe but not their aggregation number.

   The fluorescence decay of probes solubilized in micellar surfactant solutions in the presence of quenchers (time-resolved fluorescence quenching, TRFQ) was first investigated in 1974-75 [12,13] and the equations describing this decay in a fairly complex situation, i.e. with quenchers only partially solubilized in the micelles, were derived. These equations contained all the necessary elements for the application of TRFQ to the determination of micelle aggregation numbers and dynamic characteristics of the systems. In 1978 Turro and Yekta [14] showed that steady-state fluorescence quenching (SSFQ) measurements, performed with conventional spectrofluorometers, could also be used to obtain micelle aggregation numbers.   From then on fluorescence probing studies, including TRFQ and SSFQ became increasingly common. The theory underlying these methods developed in parallel, and showed the great wealth of information that TRFQ measurements were capable of providing. It also showed that analysis of the data can be very complex if the probe and quencher are not properly selected. In this section we recall the basic assumptions involved in fluorescence quenching in micellar systems, show under which conditions the methods can be used to obtain surfactant aggregation numbers, and also call for caution when using these methods.

A. Micellar Systems with Probes and Quenchers

Consider a surfactant solution of molar concentration C, containing micelles made of N surfactants, i. e., at a molar micelle concentration $[M] = (C - cmc)/N$ (cmc = surfactant critical micelle concentration), and a probe P and a quencher Q, at concentrations $[P]$ and $[Q]$. Probe and quencher are selected to be as little soluble as possible in the intermicellar solution. If such is not the case the amounts of micelle-solubilized quencher must be measured and the data analysis must take into account the fluorescence originating from the probe in the aqueous phase. At the outset it is assumed that $[P]/[M] \ll 1$ and that the average quencher occupancy number $R = [Q]_m/[M]$, $[Q]_m$ being the concentration of quencher solubilized in the micelles, can vary from zero (no added quencher) to say 0.5 to 3. Large values of R must be avoided as the accumulation of quencher in a micelle may significantly affect its aggregation number. Obviously, the smaller N, the smaller the value of R capable of inducing a change of N (the relevant parameter is the quencher mole fraction in the micelle, see below) and the smaller the maximum R value which can be used. The following assumptions are made:
(i)   the probe fluorescence decay is single exponential in the absence of quencher;
(ii) there is no limit to the quencher solubility in the micelles;
(iii) the quencher distribution among micelles is of the Poisson type, namely the probability P_i to find a micelle containing i quenchers is given by:

$$P_i = R^i e^{-R}/i. \tag{1}$$

A Poisson distribution of quenchers is obtained if one assumes that the quenching rate constant $k_{q,i}$ in a micelle containing i quenchers and one probe can be written as [1-4,12-14]:

$$k_{q,i} = i \, k_q \tag{2}$$

where k_q is the quenching rate constant in micelles containing one quencher. This assumption is consistent with that made in the treatment of the kinetics of surfactant exchange between micelles and bulk phase [15]. The Poisson distribution is further discussed below.

B. Time-Resolved Fluorescence Quenching in Micellar Systems: Determination of Micelle Aggregation Numbers

Under the above assumptions the fluorescence intensity I(t) at time t following a flash illumination of the micellar solution is given by [1-4,12,13,16-20]:

$$I(t) = I(0)exp\{-A_2t - A_3[1 - exp(-A_4t)]\} \tag{3}$$

where $I(0)$ is the fluorescence intensity at time zero and the A_is are constants independent of time but whose expressions depend on the systems investigated and on the properties of the probe and quencher (referred to as reactants below). The plots of $logI(t)$ vs t show a nearly exponential decay at short times ($t \ll 1/A_2$) and become linear at long times ($t \gg 1/A_4$).

In the absence of quencher the fluorescence decay is given by Eq. (4), where τ is the probe lifetime in the micellar environment.

$$I(t) = I(0)exp(-t/\tau) \tag{4}$$

1.   Immobile reactants in (nearly) monodisperse micelles

Immobile is used to indicate that the reactant residence times in the micelles are much longer than the probe lifetime, τ, that is, the reactant distribution is *frozen* on the fluorescence time scale. In this situation we have [16-20]:

$$A_2 = 1/\tau \qquad A_3 = R \qquad A_4 = k_q \tag{5}$$

and the surfactant aggregation number N is given by:

$$N = A_3(C - cmc)/[Q]_m \tag{6}$$

In such systems the micelle aggregation numbers are obtained fairly easily from the fits of equations (3) and (4) to the decay curves in the presence and in the absence of quencher which yield the values of the A_is and τ. The main advantage of the TRFQ method lies in its insensitivity to intermicellar interactions and to micellar shape. The aggregation numbers can be determined at finite concentration and in the presence of additives, including polymers. Nevertheless, the reactants must be selected such that the product τk_q (probe lifetime times unit intramicellar quenching rate constant) be larger than 1 and the decay curve (plot of $I(t)$ vs t) must be recorded over a time long enough in order that the linear variation of $logI(t)$ with t at long times be well developed and yield $A_2 = \tau^{-1}$. A_2 and τ are obtained from separate experiments in the presence and in the absence of quencher, respectively. In practice, the decay curves must be recorded over 4 to 6 probe lifetimes. This procedure greatly reduces the errors in the A_is.   The unit intramicellar quenching rate constant k_q is related to the viscosity of the medium in which the reactants move (often referred to as micelle microviscosity) and to the aggregation number, decreasing when either of these quantities increases [1-4,21-23].

According to the thermodynamics of micellar solutions, very low polydispersity is usually found in systems of spherical (or spheroidal) micelles [24]. For these relatively monodisperse micelles, it can be safely assumed that a single rate constant governs the quenching of the probe. The quality of the fit of equation (3) to the decay curve in the short time part of the decay, and the value of the aggregation number for a given surfactant reveal whether the assumption of nearly monodisperse micelles is valid.

2. Immobile reactants in polydisperse micelles

Polydisperse micelles are usually not spherical. In most instances they are elongated and the unit intramicellar rate constant k_q becomes a decreasing function of the micelle aggregation number (length) [1,4,21-23]. The system is then characterized by a distribution of quenching rate constants and Eq. (3) no longer provides a good fit to the decay curves, particularly at short times. Nevertheless, if the experiments are performed under conditions such that the linear decay of log I(t) at long times is reached, a quenching-average aggregation number N_q can be determined from the equation [25,26]:

$$N_q = \{(C - cmc)/ [Q]_m\}[logI(0)/I_\infty(0)] \tag{7}$$

where $I_\infty(0)$ is the intensity of the linear part of the decay at time zero. N_q now depends on $[Q]_m$ and is given by [25, 26]:

$$N_q = N_w - (1/2)\sigma^2\eta + (1/6)\Lambda\eta^2 + ... \tag{8}$$

where $\eta = [Q]_m/(C - cmc)$. N_w is the weight average aggregation number, σ^2 is the variance and Λ the skewedness of the distribution of aggregation numbers. Thus from the plot of N_q vs $[Q]_m$ at a given surfactant concentration C one can obtain N_w, σ and Λ. Almgren *et al.* reported how $I_\infty(0)$ can be determined [2,27].

Siemiarczuk *et al.* [28] have recently proposed a new method for determining the size distribution in polydisperse micellar systems. The decay data are analyzed to yield the distribution of the probe lifetimes in the presence of a quencher. The size distribution is obtained from the lifetime distribution.

3. Mobile reactants in monodisperse micelles

Mobile is used to indicate that the reactant distribution among micelles can change on the fluorescence time scale, i.e., a time comparable to the probe lifetime τ. The simplest cause of reactant distribution changes is that associated to a fast reactant (usually the quencher) exit from the micelles and reassociation to other micelles. This corresponds to short reactant residence times, T_R, in micelles. The value of T_R is determined by the reactant hydrophobicity. Fluorescence probes, such as pyrene and other polycyclic aromatic molecules, usually fulfill the condition $T_R \gg \tau$ [29]. This is not always the case for alkylpyridinium ions which are often used as quenchers of such probes. Nevertheless, even when using reactants that fulfill this condition, processes intrinsic to the micellar systems under investigation may result in fast reactant redistribution between micelles. Fusion (coagulation) and fission (fragmentation) of micelles constitute such processes [30,31]. The fluorescence decay curves corresponding to this situation still look biphasic but the long time part of the decay is now characterized by a decay rate constant $A_2 > 1/\tau$ because there is a finite probability for a micelle containing an excited probe to be visited by a quencher during the probe lifetime [1-4,16-19]. The experimental situation corresponding to mobile reactants can in principle be distinguished from that for immobile reactants by the increased slope of the long time part of the decay. Almgren *et al.* [32] showed that provided $R < 2$, Eq. (3) still holds to a very good approximation but the fitting parameters now are given by:

$$A_2 = (1/\tau) + k_q \langle R \rangle_s \qquad A_3 = R(1 - \langle R \rangle_s/R)^2 \qquad A_4 = k_q/(1 - \langle R \rangle_s/R) \qquad (9)$$

where $\langle R \rangle_s$ is the mean number of quenchers in micelles which also contain an excited probe during the long time, stationary-state (linear) part of the decay characterized by the decay rate constant A_2. The different types of redistribution mechanisms give rise to qualitatively different variations of $\langle R \rangle_s$ with [Q]. The dependence of $\langle R \rangle_s/R$ on the value of the ratio k_e/k_q, k_e being the pseudo first-order rate constant for redistribution has been numerically calculated [32]. The ratio $\langle R \rangle_s/R$ can then be obtained by an iterative procedure.

   One of the main difficulties in analyzing TRFQ data is that the decay curves for systems with immobile reactants in polydisperse micelles may look similar to those for systems with mobile reactants in monodisperse micelles, when the micelles are large. Indeed, the product $\tau \cdot k_q$ may then be close to 1 and the linear part of the decay becomes experimentally difficult to reach. The decay curve analysis then yields an apparent value of A_2 larger than τ^{-1}, even for immobile reactants. This emphasizes the necessity to record decay curves over a time interval as long as possible.

4. Mobile reactants in polydisperse micelles

This very complex situation, still investigated from the theoretical viewpoint [4], is rarely encountered in polymer/surfactant systems. Only approximate values of N can be obtained in such situations.

In summarizing this section, systems with immobile reactants can be distinguished from those with mobile reactants by checking whether the long time part of the decay curve yields the same slope in the absence and in the presence of quencher. Polydisperse systems with immobile reactants are characterized by experimental values of the aggregation number that decrease upon increasing quencher concentration, contrary to monodisperse systems.

5. Additional remarks

The above equations and conclusions also hold for systems where the monomeric probe in the ground state associates with a probe in the excited state, forming an excimer [20]. The best known case is that of pyrene. Ground state pyrene then acts as a quencher for excited state pyrene. In the decay equation (3) k_q is replaced by the unit rate constant for excimer formation. Excimer-forming probes at concentration such that $[P]/[M] \approx 1$ can be used to determine aggregation numbers. The main merit of this time-resolved excimer method is that fluorescent probes have a very low solubility in water. Its main drawback is that excimer formation is partly reversible, the more so the higher the temperature [20], contrary to quenching via charge or energy transfer. This introduces complications in the analysis of the data. Also, the measurements are then performed at fairly high probe concentrations where self-absorption may become a problem.

Surfactant aggregation numbers can be also obtained from the study of energy transfer between energy donor and acceptor (the naphthalene/pyrene pair, for instance) solubilized in the micelles. Both steady-state and time-resolved energy transfer methods have been used for studies of micellar solutions [33,34]. Time-resolved energy transfer appears capable of measuring also micelle radii [34]. The energy transfer methods have been applied to polymer/surfactant systems [35] but not for aggregation number measurements, yet.

The decay curves are generally recorded using a single photon counting apparatus in which the fluorescent probe is excited by means of short flashes of conventional or laser light. Such instruments are commercially available (EGG Princeton Applied Research (U.S.); Photochemical Research Associates (Canada); Applied Photophysics (U.K.)).

C.   Steady-State Fluorescence Quenching in Micellar Systems: Determination of Aggregation Numbers

Turro and Yekta [14] showed that equation (10) holds for the intramicellar quenching of a fluorescence probe:

$$\log(I_0/I_{[Q]}) = [Q]/[M] \tag{10}$$

where I_0 and $I_{[Q]}$ are the fluorescence intensities of the surfactant solution in the absence of quencher and in the presence of a quencher at concentration [Q], respectively. The micelle concentration [M] is obtained from the slope of the plot of $\log(I_0/I_{[Q]})$ vs [Q] and the surfactant aggregation number from Eq. (6).

The main advantage of SSFQ over TRFQ is that it only requires fluorescence intensity measurements at increasing quencher concentration, which are easily performed using a sensitive spectrofluorometer, while TRFQ measurements involve the recording of decay curves which necessitates relatively complex instrumentation. However, the use of SSFQ is much more restricted than TRFQ. Indeed, Eq. (10) holds only if following conditions are fulfilled:

(i)   the quencher distribution is of the Poisson type.

(ii)   the reactants are fully solubilized in the micelles with [P]/[M] << 1 and [Q]/[M] < 2-3.

(iii)   the reactants are immobile.

(iv)   the rate constant for intramicellar quenching k_q is much larger than the probe decay rate constant, $k = 1/\tau$: the smaller the value of the product $k_q \times \tau$ the larger the error on N.

(v)   the micelles are monodisperse.

Conditions (i) and (ii) are the same as for TRFQ. However, when using SSFQ one cannot know whether conditions (iii) and (iv) are met without performing TRFQ experiments on the system under investigation. Besides, condition (iv) may be seldom fulfilled for polymer/surfactant systems owing to the much higher microviscosity of polymer-bound surfactant aggregates, with respect to free surfactant micelles, resulting in low quenching rate constants (see below). Finally, the case of polydisperse micellar systems has not been dealt with theoretically in SSFQ. Expressions of $\log(I_0/I_{[Q]})$ have been derived for situations where conditions (iii) and (iv) are not fulfilled. However their use for the determination of aggregation numbers again implies fluorescence decay experiments. In view of the above, the use of SSFQ is not recommended, particularly for polymer/surfactant systems.

D. Quencher Distribution In Micellar Systems

A central assumption in both SSFQ and TRFQ is that the quencher distribution among micelles follows Poisson statistics (Eq. (1)). This assumption leads to Eqs. (3) and (10) which permit the determination of the aggregation number N. The Poisson distribution of solubilizates in micelles has been obtained by two differing approaches. The first one is based on kinetics and assumes that the association rate constant of a solubilizate to a micelle is independent of the micelle occupancy number whereas the exit rate constant is proportional to this number (Eq. (2)) [13,14,16-20]. The other approach is based on statistical mechanics [4]. It is of more general character and permitted the authors to deal with the case where the quencher interacts with the surfactant constituting the micelles. These interactions were shown to result in departures from Poisson statistics. This situation has been examined for the case where the quencher is a surfactant, using a mixed micellization approach [36]. The calculations resulted in Eq. (11) for the expression of the apparent aggregation number N_{app}, where β is the interaction parameter which appears in the theory of non-ideal mixed micelles and x the quencher mole fraction in the micelles:

$$N_{app} = N[1 + (1/2 - \beta)x] \qquad (11)$$

Most $|\beta|$ values are generally below 10 [37]. If the occupancy number R is kept around 1, then x = 0.015 for N = 60. The correcting term in Eq. (11) is about 0.15, larger than the overall experimental error in the measurements (5 to 10%). Care must therefore be exercised in selecting the quencher.

The validity of the Poisson distribution has been checked experimentally for several aromatic solutes [1, 4]. The theory confirmed that at quencher occupancy number R < 2-3, the Poisson distribution gives values of P_i (Eq. (1)) very close to those calculated with the binomial distribution [4]. The Poisson distribution thus appears firmly established.

E. Probes and Quenchers Used for Polymer-Surfactant Systems

The vast majority of fluorescence probing studies of polymer/surfactant systems use pyrene as probe owing to its favorable properties: extremely low water solubility (about 0.5 μM), long fluorescence lifetime (400 ns in hydrophobic media, 200 ns in water, and intermediate values in micellar solutions); the sensitivity of its emission and excitation spectra to the polarity of its environment; the formation of pyrene excimer; and the large number of compounds or ions able to quench its fluorescence [8,38].

Other much less used probes are the 1-pyrene-butyltrimethylammonium chloride and ruthenium(II)trisbipyridyl chloride (Rubipy). The most often used quenchers of pyrene fluorescence are benzophenone, dimethylbenzophenone (DMBP), dibutylaniline, and alkylpyridinium chloride surfactants with an alkyl chain with 12 to 16 carbon atoms. 9-methylanthracene can be used as quencher of the Rubipy fluorescence

F. Preparation of Surfactant Solutions for Fluorescence Probing Studies

Aromatic probes cannot be directly introduced in the system under investigation because their solubility in water is very low and they are used at very low concentration. Thus, in most instances an appropriate amount of a stock solution of probe in ethanol or tetrahydrofuran is injected into the solution while stirring. The amount of solvent thus introduced must be small (< 0.5 % of the volume of the solution) in order that its effect be completely negligible. The solution must then be stirred for several hours at 30-40°C.

The required amount of probe stock solution can also be introduced in a volumetric flask, spread over the flask wall and the solvent evaporated under a stream of air. The solution is then added and the probe solubilized by vortex stirring for a long time. This procedure may lead to results less reproducible than with the first method because solubilization of very hydrophobic probes may require days or weeks [39,40]. Winnik et al. [39] discussed artifacts that may arise from incompletely solubilized probes.

A last method consists in preparing a saturated aqueous solution of probe and using it as solvent for preparing the solutions to be investigated.

In systems with oppositely charged polymers and surfactants precipitation of a polymer/surfactant complex occurs readily. This is avoided by using dilute polymer and surfactant solutions, generally in the millimolar range of polymer repeat unit and sufficiently low [surfactant]/[repeat unit] ratios. Any local overconcentration may results in an irreversible precipitation of the complex. The full systems (polymer + surfactant + probe + quencher) can be studied as they are or after deoxygenation. Molecular oxygen is a strong quencher of the fluorescence of aromatic probes. Deoxygenation by successive freeze-pump-thaw cycles is usually recommended in TRFQ determinations of aggregation numbers. However, in systems with oppositely charged polymers and surfactants this procedure must be avoided as it often results in the irreversible precipitation of polymer/surfactant complexes.

III.   INTERACTIONS BETWEEN HYDROPHILIC POLYELECTROLYTES AND SURFACTANTS

In this review, hydrophilic polyelectrolytes are defined as polyelectrolytes whose aqueous solutions have a high viscosity reflecting their extended conformation. The viscosity may go through a minimum at low polyelectrolyte concentration. Also, a medium-sensitive fluorescence probe, such as pyrene, added to a solution of hydrophilic polyelectrolyte must report a fully water-like environment (values of the probe fluorescence lifetime and of the pyrene polarity ratio I_1/I_3 close to those in water [8,38]). PS1 and PS4 [41-43] and polyacrylic acid (PAA) [44] behave as typical polyelectrolytes at sufficiently high degree of neutralization α (above 0.1 for PS1 and PAA, and 0.4 for PS4). The hydrophilic character can be rapidly lost by modification of the polymer. Thus PS4 behaves like a hydrophilic polyelectrolyte at $\alpha > 0.4$ (high pH) but like a hydrophobic polyelectrolyte at $\alpha < 0.2$ (low pH) [41]. This is illustrated in Figure 1. Polymethacrylic acid shows a similar behavior but the hydrophobic/hydrophilic transition occurs at around $\alpha = 0.15$ (Figure 1).

A.   Hydrophilic Polyelectrolytes and Surfactants of Opposite Charge

A large number of studies have demonstrated the cooperativity in the binding of ionic surfactants to oppositely charged hydrophilic polyelectrolytes [5,6]. There is no binding at surfactant concentration, C, below a critical concentration C*. Then at C > C*, the surfactant binds cooperatively, that is the concentration of free surfactant, C_f, increases slowly with C whereas the concentration of bound surfactant, C_b, first increases very rapidly with C, then levels off. The surfactant binds to the polyelectrolyte in the form of aggregates. C* is referred to as the critical aggregation concentration (cac). The cac increases with the ionic strength of the system (owing to the screening of the electrostatic interactions between polymer and surfactant) and when the surfactant chain length is decreased [5,6,41,42]. The variation of log cac with the surfactant carbon number m for a homologous series of surfactants is linear, just like that of log cmc with m for the same surfactants. The values of the slopes of the two plots are rather close [42], a result which indicates that the free energies of transfer of a surfactant methylene group from the aqueous phase to a free micelle or a bound aggregate are nearly equal. Cac and cmc depend very similarly on the nature of the surfactant counterion [41].

1.   Aggregation numbers

There have been many studies aimed at the determination of surfactant aggregation numbers N in polyelectrolyte/surfactant systems. The results can be summarized as follows.

a.   N has been found to be independent of the concentration C_b of bound surfactant [43-48]. This behavior is illustrated by the results in Figure 2 for the binding of DTAC (dodecyltrimethylammonium chloride) to PS1 in the whole range of α values and PS4 at α ≥ 0.5, when PS4 behaves as a hydrophilic polyelectrolyte (Figure 1).   These results were obtained using TRFQ (pyrene/DMBP pair) and the partition of DMBP between aqueous phase and bound aggregates was taken into account [45]. The overall accuracy was rather poor, a consequence of the fairly low maximum polymer and surfactant concentrations used in the experiments in order to avoid precipitation.   The same difficulty is apparent in most studies of this type of systems.   The constancy of N indicates that the surfactant binding to polyions results in the formation of aggregates of constant composition and, thus, involving a constant

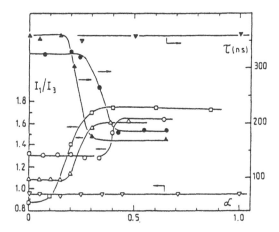

FIG. 1 Variations of the pyrene intensity ratio I_1/I_3 (□,O,Δ,∇) and lifetime τ (●,▲,▼) with the neutralization degree α for polymethacrylic acid (PMA, □); poly(maleic acid-*alt*-styrene) (PSS, O,●,), PS4 (Δ,▲) and PS10 (∇,▼) at 25 °C. Reproduced from ref. 41 with permission of the American Chemical Society.

length of polyion chain, in number increasing with C_b. Recall that anionic surfactant binding to nonionic polymers, such as PEO (polyethyleneoxide) or PVP (polyvinylpyrrolidone), is characterized by N values significantly increasing with the amount of bound surfactant, from rather low values to values close to that of the pure surfactant micelles, as if these polymers were characterized by a fixed number of binding sites [49].

b.   Figure 2 shows that N increases with α for PS4 but decreases for PS1. This difference has been attributed to the fact that the side methyl groups of PS1 are short and contribute very little to the formation of the mixed aggregates, the main interaction between PS1 and DTAC being of electrostatic nature [45]. On the contrary, with PS4, the side butyl chains interact hydrophobically with the surfactant alkyl chains and the aggregates are similar to mixed micelles of two surfactants, one with a dodecyl chain and the other with a butyl chain. The latter would form only small micelles and its mixed micellization with a longer chain surfactant results in micelles of aggregation numbers smaller than for the longer surfactant [45]. In their study of the PAA/TTAB (tetradecyltrimethylammonium bromide) system in the presence of 10 mM KBr by the pyrene excimer method [20] Kieffer et al.. [44] also observed an increase of N with α, even though the errors were rather large, as the PAA concentration used was about ten times lower than in the PS1 and PS4 study [45]. Kiefer et al. [44] showed that the measured aggregation numbers were much larger than expected from the analysis of the surfactant binding isotherms on the basis of Satake and Yang model [50] which has been widely

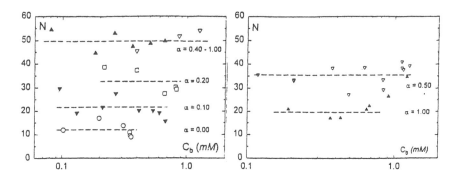

FIG. 2 Variation of the aggregation number of the polymer-bound surfactant aggregates in the PS1/DTAC system (left) at different neutralization degrees (α) of the polymer as indicated in the figure and in the PS4/DTAC system (right) at $\alpha = 0.5$ and 1.00, with the concentration of bound surfactant at 25°C. Reproduced from ref. 45 with permission of the American Chemical Society.

used in such studies. They concluded that conformational effects, not taken into account in this theory, were probably involved.

c.   The N values for polyelectrolyte-bound aggregates increased with the surfactant chain length as in polymer-free systems [46], but increased with temperature [51] and were independent of the nature of the surfactant counterion [46,47] and of the concentration of added salt [48], a series of results in contrast with those for pure (polyelectrolyte-free) surfactant solutions.

d.   The bound aggregates were shown to be relatively polydisperse with σ/N values ranging between 0.3 and 0.7 (σ = standard deviation of the distribution) [46,51,52].

e.   The most extensive determinations of N have been performed by Almgren et al. [46-48,51,52] for a variety of systems: NaPSS (sodium polystyrenesulfonate)/alkyltrimethylammonium bromides and chlorides [46]; NaPAA (sodium polyacrylate) or NaPSS/alkyltrimethylammonium bromides [47]; NaPAA/DTAB [48]; NaPVS (sodium polyvinylsulfate) or NaDxS (sodium dextransulfate)/DTAB [51]; and NaHy (sodium hyaluronate)/DeTAB and DTAB (decyl and dodecyl trimethylammonium bromides) [52]. These TRFQ studies used the pairs pyrene/DMBP and/or pyrene/DPyCl (dodecylpyridinium chloride) and non-deoxygenated systems. The results together with other ones [43-45,53] showed an important effect of the polyelectrolyte nature (charged group, pendant chain, and backbone) on the value of the aggregation number of the bound aggregates.   Thus, the N values in the PS1(α = 1)/DTAC [45], NaPAA/DTAB [47,48], NaHy/DTAB [52]; and Na-CMA(sodium carboxyamylose)/CTAB [53] were close to those for polymer-free aggregates. Note that these polyelectrolytes are all highly hydrophilic. On the contrary, the N values in the PS4(α = 1)/DTAC [45] and NaPSS/DTAB or CTAB [47] were found to be smaller than in the absence of polyelectrolyte. Both PS4(α = 1) and NaPSS behave like hydrophilic polyelectrolytes.   Nevertheless they already show an effect of the hydrophobicity arising from the presence of the butyl chains [45] and phenylene groups [54,55]. A partial penetration of the polymer chain in the micelles would result in lower N values [51].   Finally, the N values in the NaPVS/DTAB and DxS/DTAB systems were found to be larger than for pure DTAB micelles [51].   This result was supported by energy transfer experiments between pyrene and proflavine in the same systems [56].   It was interpreted as reflecting the specific nature of the sulfate group [51], similarly to the way the counterion nature can strongly affect micelle agregation numbers [57, 58].   Indeed, upon surfactant binding to the polyelectrolyte the surfactant counterions are expelled from the aggregate surface and replaced by polyelectrolyte charged groups (see below).

f.   In some polyelectrolyte/surfactant systems, such as the NaPAA/DTAB or CTAB systems, phase separation upon addition of salt yielded two liquid phases: a dilute phase and a concentrated one [47]. The values of N in the two phases differed only slightly, and the effect of the surfactant counterion and chain length were the same as in monophasic systems. However, the values of the pyrene lifetime in the two phases differed by as much as 15 % [47].

Three other studies relating to aggregation numbers in polyelectrolyte/surfactant systems are noteworthy. The first report of aggregation number of bound surfactant aggregates is that of Abuin and Scaiano [59] for the NaPSS/DTAB system. The crude estimate N = 7-10 was made on the basis of the steady state quenching of the xanthone triplet state by 1-methylnaphthalene. Chu and Thomas [60] reported values of 100-120 for the aggregation number of DeTAB aggregates bound to nearly fully neutralized polymethacrylic acid (pH 8). This value is much larger than that for free DeTAB micelles (N = 33). The experiments used TRFQ (pyrene/DPyCl, or pyrene excimer) and SSFQ. However the authors apparently did not take into account the concentration of free DeTAB. This would reduce the measured aggregation number to about 50-60, a value still larger than for free DeTAB micelles. Whereas all the above studies concerned polyanions and cationic surfactants, Fundin *et al.* [61] determined aggregation numbers in the polycation/anionic surfactant system: PCMA (polytrimethylammonioethyl-acrylate chloride)/SDS (sodium dodecylsulfate). The reported behavior is very different from that of polyanion/cationic surfactant systems, with very large values of N, close to and above 200 (as compared to 60 for free SDS micelles) and increasing with the salt content. Similarly, additions of DTAC (same charged group as PCMA) to SDS have been found to result in a more rapid increase of N than additions of SDS (same charged group as NaPVS or NaDxS) to DTAC [62].

2. Microviscosity in the bound surfactant aggregates

Several studies concluded that polyelectrolyte-bound surfactant aggregates are less fluid than free micelles of the same surfactant. Indeed, the intra-aggregate quenching rate constants were found to be systematically smaller in polymer-bound aggregates than in free micelles, even in cases where the former were smaller than the latter [44-46,52]. The microviscosity, η_i, of the bound aggregates relative to that of free DTAC micelles has been determined in the PS1/DTAC systems from the values of the intra-aggregate quenching rate constant, k_q, and also from fluorescence anisotropy measurements using the probe diphenylhexatriene [45]. The two methods yielded similar results, illustrated in Figure 3. η_i is seen to be independent of the DTAC concentration, a result supporting the conclusion that the aggregate composition is independent of the bound surfactant concentration. Also the relative values of η_i ranged between 2 and 12. These larger microviscosity values are responsible for the smaller quenching rate constants noted above. They would probably also result in smaller rate constants for all bimolecular reactions performed in polymer-bound aggregates which appear less attractive than free micelles for use as microreactors.

3. Pyrene fluorescence lifetimes

In all studies using aerated systems (that is in the presence of solubilized molecular oxygen, an efficient quencher of the pyrene fluorescence) the pyrene fluorescence lifetime, τ, was found to be larger in polyelectrolyte-bound aggregates than in free micelles [44-47,52,60]. The increase can be significant as illustrated by the results in Figure 4 for the PS1/DTAC system. The value of τ in free DTAC micelles is about 165 ns, but can reach 270 ns in PS1-bound aggregates, indicating that polymer-bound aggregates shield pyrene from oxygen quenching better than polymer-free micelles. The variations of τ are qualitatively similar to those of N for the same system and lead to the same

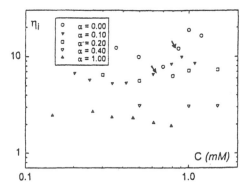

FIG. 3 Variation of the relative microviscosity of the polymer-bound surfactant aggregates with the surfactant concentration in PS1/DTAC systems at 25°C and different neutralization degrees. $\eta_i = 1$ corresponds to the free DTAC micelles. The values of η_i have been obtained from those of the quenching rate constant. The arrows correspond to the onset of precipitation of the polymer/surfactant complex. Reproduced from ref. 45 with permission of the American Chemical Society.

conclusion, namely that the number of bound micelles increases with the concentration of bound surfactant without changes of micellar characteristics such as size and composition [45,46]. Note that at $\alpha = 1$ the values of τ for additions of DTAB and DTAC are equal within the experimental error, over the whole range of concentration (Figure 4). This result indicates that the surfactant counterions are expelled from the surface of the bound aggregate, a conclusion supported by energy transfer experiments in similar polyanion/cationic surfactant systems [56]. It also leads to the conclusion that the polyelectrolyte chain must wrap fairly tightly around the bound aggregates in order for the surfactant charged groups to be neutralized by polymer charged groups.

The explanations of the larger τ values in polyelectrolyte-bound aggregates in terms of a larger hydrophobicity of these aggregates [44] or of the expulsion of the quenching counterions (Br⁻) from the aggregate surface do not hold [45]. Thalberg *et al.* [52] proposed that the increase of τ is due to the polymer shell around the bound surfactant aggregates which shields pyrene from oxygen quenching. It is likely however that the polymer also affects the quenching through the increased microviscosity which slows down probe and quencher motions in the aggregates [45].

The results in the preceding three sections strongly support a structural

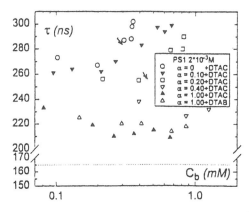

FIG. 4 Variation of the pyrene lifetime in the polymer-bound surfactant aggregates with the surfactant concentration in PS1/DTAC systems at 25°C and different neutralization degrees. The arrows indicate the onset of precipitation and the dotted line corresponds to the lifetime in free DTAC micelles. DTAB denotes dodecyl-trimethylammonium bromide. Reproduced from ref. 45 with permission of the American Chemical Society.

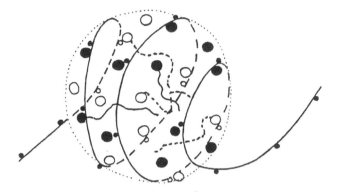

FIG. 5   Schematic representation of a surfactant aggregate bound to an oppositely charged hydrophilic polyelectrolyte: polymer charged groups, surfactant charged groups and polymer chain above (•, ● ——), and below (o, O——) the plane of the page, respectively. Only a few surfactant alkyl chains are represented, for sake of clarity.

model where the polyelectrolyte chain wraps around the bound surfactant aggregates with a large part of the surfactant charged groups in contact with polymer charged groups.   Two or three turns of the polyelectrolyte chain around a surfactant aggregate are sufficient to achieve this neutralization as represented schematically in Figure 5.   Part of the remaining polymer chain is used for complexing other micelles, the rest is free and connects the complexed micelles.

4.   Miscelleanous studies

This section reports results in less frequently studied systems of potential interest.   Uznanski *et al.* [63] synthesized polyionenes of the type $-[(CH_3)_2N^+(CH_2)_{10}]_n-$ with hydrophobic counterions such as dodecylsulfate or pyrenesulfonate and showed the presence of hydrophobic microdomains in these systems using photophysical methods.   TRFQ studies revealed that the aggregation number of the micelles of the nonionic surfactants $C_{12}E_8$ and $C_{12}E_6$ are increased in the presence of κ-carrageenan in solution or in the gel state [64]. This behavior was attributed to the existence of repulsive interactions between polymer and micelles.   Chu *et al.* [65,66] showed that in a polydiallyldimethylammonium chloride gel, SDS micelles are regularly arranged with a d-spacing of 3.7 nm.   Finally, Everaars *et al.* [67] reported the

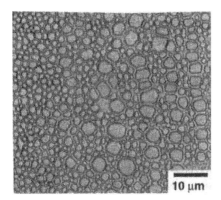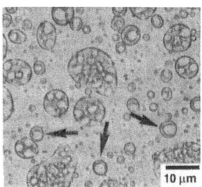

FIG. 6 Optical micrographs showing (left) densely packed vesicles in the hydrated DDDMAB/PAA film and (right) single vesicles surrounded by a layer of DDDMAB/PAA matrix. Some vesicles are indicated by arrows. Reproduced from ref. 67 with permission of the American Chemical Society.

formation of tissue-like structures (Figure 6) upon heating of an insoluble film of didodecyldimethylammonium bromide (DDDMAB) and polyacrylic acid covered with water.

B. Hydrophilic Polyelectrolytes and Surfactants of Like Charge

It is generally accepted that hydrophilic polyelectrolytes do not bind surfactants of like charge and few studies of such systems have been reported. However, the surfactant properties are affected by the polyelectrolyte in a similar manner as by a low molecular salt, i.e., the cmc should decrease and the aggregation number increase. This behavior was indeed observed with SDS in the presence of NaPAA or PS1(α = 1) [68]. However another study [53] reported that SDS binds to NaCMA (sodium carboxyamylose) as based on a decrease of cmc and aggregation number of SDS in the presence of 1.6 % NaCMA. This study used the pyrene excimer method to determine N, and SDS concentrations so close to the cac that the accuracy of the measurements was seriously impaired by the large error in the (C - cac) term which determines the value of N (see Eq. (6)). Besides, the micelles present in the system may have been induced by the relatively high pyrene concentration used. The addition of 2% NaCl apparently eliminated the interaction observed in the salt-free system. More measurements are required before a definite answer can be given concerning the existence of interaction in this system.

IV. INTERACTIONS BETWEEN HYDROPHOBIC
        POLYELECTROLYTES (POLYSOAPS) AND SURFACTANTS

Hydrophobic polyelectrolytes are defined as polyelectrolytes whose aqueous
solutions have a rather low viscosity reflecting their compact conformation,
which arises because the hydrophobic groups of the polyelectrolytes gather
together and form hydrophobic microdomains as in micellar solutions, thus the
name polysoaps or polyamphiphiles. Also, a medium-sensitive fluorescence
probe, such as pyrene, added to a solution of hydrophobic polyelectrolyte must
report a low polarity environment (values of the probe fluorescence lifetime τ
and of the pyrene polarity ratio I_1/I_3 close to those in surfactant micelles [8,38]).
In most instances the polysoap repeating unit is amphiphilic (PSX or
polyamines partially quaternized by alkylhalides [7]). The aqueous solutions of
PSX with $X \geq 8$ have been shown to contain hydrophobic microdomains at all
α [69,70]. Their behavior is illustrated by the results for PS10 in Figure 1 [41]:
both I_1/I_3 and τ have values independent of α and indicative of a hydrophobic
environment. At this stage it is recalled that surfactant binding to polysoaps is
very strong, starts at very low concentration and is non-cooperative: the
concentration of free surfactant increases monotonously with the total
surfactant concentration [42,69,70].

A.   Polysoaps in the Absence of Surfactant

This background is necessary for a better understanding of the behavior of
polysoap/surfactant systems. The number of studies of polysoaps in aqueous
solution is very large. For this reason only the salient features are presented.
        Fluorescence probing experiments confirmed that polysoaps of high
molecular weight have a zero cmc: microdomains exist at all concentration
[69,70]. They also permitted a rough estimate of the fraction β of PSX
repeating units forming microdomains, and of the average number N of
repeating units per microdomain [69-71]. Values for β and N are listed in
Table 1. They show an increase of β and N with the the alkyl chain carbon
number X, as expected. N was found to be independent of the PS16 molecular
weight, and increases with temperature, an effect attributed to the larger
conformational mobility of the polymer chain [71]. Values of N for PS6($\alpha = 1$)
were reported to be around 20 [72], then 44 in a later study [73]. These values
appear to be too large. For a copolymer poly(potassium maleate-*alt*-1-
octadecene) of molecular weight below 10,000, TRFQ with the 1-
pyrenebutyltrimethylammonium/dodecylpyridinium pair yielded N $\cong$ 24
indicating that a microdomain is constituted by one polymer chain [74]. For the

TABLE 1 Values of β (fraction of repeating units involved in microdomain formation) and of N (number of repeating units per microdomain) in PSX solutions at 25°C

Polysoap	α	$\beta^{(a)}$	$N^{(b,c)}$
PS16	0.00	1.00	90 (60)
	0.50	1.00	
PS12	1.00	0.66	92
	0.50	0.69	70
PS10	1.00	0.12	11 (14)
	0.50	0.20	23 (50)
	0.00	0.58	28 (80)
PS8	1.00		< 10
	0.50		13
PS6	1.00		very low
	0.50	0.12	< 10

(a) values from ref. 70; (b) values from ref. 69; (c) values in parenthesis from ref. 71.

same polysoap small angle neutron scattering, (SANS) revealed cylindrical micelles with $N \cong 200$, indicating that the aggregates contain several polymer chains. These apparently inconsistent results can be reconciled on the basis of the observation made in a study of aqueous solutions of PS16 ($\alpha = 1$) by transmission electron microscopy at cryogenic temperature [77]. The study showed the presence of very long worm-like micelles (Figure 7), made up by the end-to-end association of PS16 chains [77], whereas the values of N indicated that the microdomains probed by TRFQ were much smaller [69,71]. Other poly(potassium maleate-*alt*-1-alkene) with shorter alkene comonomers were investigated by SANS and the values of N were found to increase with the 1-alkene carbon number [76]. The polysoaps obtained by micellar polymerization of sodium undecenoate have also been reported to form one microdomain per polymer chain [78] by TRFQ. Engberts *et al.* [79,80] investigated the aggregation of poly(dimethyldiallylammonium-*co*-methyldodecyldiallylammonium) in aqueous solution. The polymers were probably of low molecular weight as they were characterized by finite cmc values. The aggregation numbers obtained by SSFQ were small, between 5 and 30, but the method of measurement may not have been fully appropriate. Finally Cochin *et al.* [81,82] showed that the N values for the microdomains in solutions of poly(hexadecyldimethyl(vinylbenzyl)ammonium chloride) were independent of concentration as expected for an intramolecular association and increased with temperature. Shorter homologues did not form microdomains. This last result was an important one. Indeed the comparison between PSX [69, 71] and the polysoaps investigated by Cochin *et al.* [81,82] indicates that the microdomain-forming ability is largely determined by the number of side alkyl chains per carbon atom in the polysoap chain, in agreement with recent theoretical calculations [83].

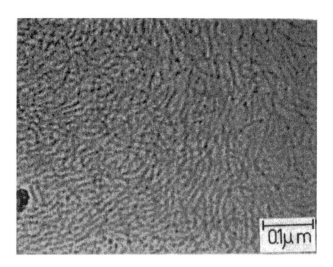

FIG. 7 Cryo-transmission electron micrograph of 0.40 wt % solution of PS16($\alpha = 1$) showing entangled thread-like micelles that are much longer (several microns) than expected on the basis of the PS16 molecular weight. Reproduced from ref. 77 with permission of the American Chemical Society.

Other important features of hydrophobic microdomains in polysoap solutions are their very high microviscosity [69,84-88], much larger than for polymer-bound surfactant aggregates, and also their low polarity [69,71,86-88]. Thus the microviscosity of PS16($\alpha = 1$) microdomains is about 30 times larger than that of free DTAC micelles [69]. In the PS16($\alpha = 1$) system nearly all repeating units are included in the microdomains [70]. The alkyl chain motions are then much hindered because the motion of any chain requires motion of part of the polysoap main chain and of some adjacent alkyl chains. As a result η_i values are high, and the cause of the very low quenching rate of pyrene by alkylpyridinium ions in these systems [69,71,81,82]. Determinations of aggregation numbers by SSFQ in pure polysoaps systems should thus be avoided. Reactant redistribution (Section II.B.3) was found to occur on the fluorescence time scale for polysoaps with a hexadecyl alkyl chain where successive microdomains are separated by very short distances [69,71,81]. This migration may involve reactant hopping from one domain to the next or reactant diffusion along the hydrophobic polysoap segments connecting successive microdomains [71].

B. Polysoaps and Surfactants of Opposite Charge

Measurements in PS12/DTAC and PS16/DTAC systems yielded surfactant aggregation numbers N that increase linearly with the concentration of bound surfactant as illustrated in Figure 8 [69]. This behavior is very different from that for hydrophilic polyelectrolytes shown in Figure 2. Another important point is that the N vs C_b plots extrapolate to the origin. This has been interpreted as indicating that the bound surfactants simply swell the polysoap hydrophobic microdomains preexisting the addition of surfactant, without creating new microdomains [69]. This non-cooperative binding of surfactants to polysoaps has been confirmed by titration calorimetry for the interaction between poly(sodium acrylate-co-n-alkylmethacrylate) or poly(acrylamide-co-n-alkylmethacrylate) and single tail surfactants, including DTAB [89].

The variations of other characteristics of polysoap/surfactant systems are also very different from those of hydrophilic polyelectrolyte/surfactant systems. Thus the microviscosity η_i of the PS16($\alpha = 1$)/DTAC aggregates relative to that of free DTAC micelles decreases as the concentration of bound surfactant

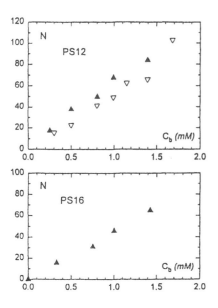

FIG. 8   Variation of the surfactant aggregation number of the polymer-bound surfactant aggregates in the PS12/DTAC and PS16/DTAC systems at 25°C and $\alpha = 0.50$ (∇) and 1.00 ($\bullet$), with the concentration of bound surfactant. Reproduced from ref. 69 with permission of the American Chemical Society.

increases but remains nearly constant for the PS10(α = 1)/DTAC systems (Figure 9) [69]. The binding of surfactant introduces some chains in the PS16 (α = 1) microdomains that can move independently from the polysoap alkyl chains, thereby increasing the overall fluidity. The situation is different for the PS10(α = 1)/DTAC system. Indeed in the absence of surfactant only a small fraction of repeating units form microdomains (Table 1) and the initial microviscosity is lower. Upon binding of surfactant (decreasing η_i), some free repeating units become incorporated in the growing microdomains (increasing η_i). These two opposite effects leave η_i nearly unchanged but fairly high (Figure 9). The decrease of η_i upon binding of surfactants by long-chain polysoaps explains the observed decrease of the pyrene fluorescence lifetime in aerated systems (Figure 10) as the quenching by molecular oxygen is then facilitated [69]. Also, the intra-aggregate quenching rate constant should be increased, again in agreement with the experimental observations. This facilitates the determination of N by TRFQ. In the PSX series, the binding of DTAC to hydrophilic homologs results in a decrease of the solution viscosity reflecting the polyelectrolyte coiling around the bound micelles [45]. To the contrary, the binding of DTAC to the hydrophobic homologues results in an increase of solution viscosity owing to the polysoap microdomain swelling by the surfactant [69]. Figure 11 shows a model of binding of a surfactant to an positely charged polysoap [69].

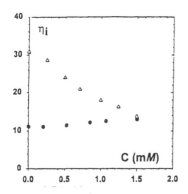 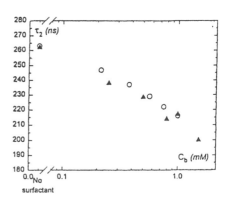

FIG. 9 (left) Variation of the microviscosity of the bound aggregates in PS16(α = 1)/DTAC (Δ) and PS10(α = 1)/DTAC ($\bullet$) at 25°C. Reproduced from ref. 69 with permission of the American Chemical Society.

FIG. 10 (right) Variation of the pyrene fluorescence lifetime with the concentration of bound surfactant in PS12(α = 1)/DTAC ($\blacktriangle$) and PS12(α = 1)/DTAB (O) systems. Reproduced from ref. 69 with permission of the American Chemical Society.

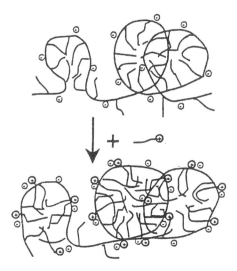

FIG. 11 Model for the binding of a cationic surfactant to an anionic polysoap which forms intramolecular hydrophobic microdomains. The polymer charged groups are located close to the surfactant charged groups replacing the surfactant counterions at the aggregate surface, while the surfactant alkyl chains swell the microdomains. Reproduced from ref. 69 with permission of the American Chemical Society.

C. Polysoaps and Surfactants of Like Charge

The interaction of SDS with polysoaps of like charge, the poly(potassium maleate-*alt*-1-alkene), with 1-alkene = 1-decene, 1-tetradecene and 1-octadecene, has been investigated by fluorescence intensity and anisotropy [88]. The results suggested that SDS binds to the 1-octadecene copolymer, resulting in a decrease of microviscosity similar to the observation with oppositely charged polysoaps and surfactants (see above) [69].

More direct evidence has been obtained for the interaction between SDS and PS16($\alpha = 1$) [90,91]. The viscoelasticity of pure PS16($\alpha = 1$) solutions disappeared and the solution viscosity decreased upon additions of SDS or of the nonionic surfactants $C_{10}E_6$ and $C_{12}E_6$ [90, 91]. Self-diffusion measurements showed that only a small fraction of the added SDS was bound to the polymer [91]. Cryo-transmission electron microscopy directly visualized the progressive break-up of the plurimolecular worm-like PS16 micelles into isolated microdomain-containing macromolecules, upon surfactant additions (Figure 12) [90,91]. The fact that a fairly small amount of bound SDS was sufficient for bringing about this break-up was interpreted as indicating that the

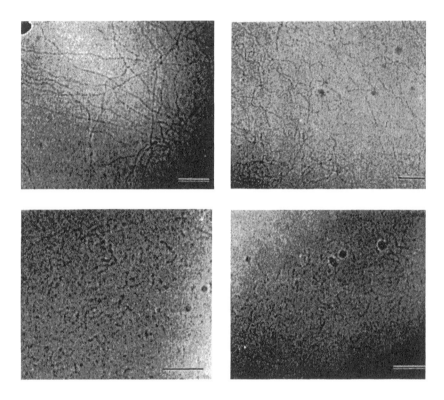

FIG. 12 Cryo-transmission electron micrographs of 9 mM (monomol/L) PS16(α = 1) solutions in the presence of increasing amounts of SDS expressed as P = [SDS]/[PS16]. Top left, P = 0.57: entangled worm-like micelles; Top right, P = 0.79: worm-like micelles coexisting with some shorter ones; Bottom left, P = 1.0 and bottom right, P = 1.14: short worm-like micelles coexisting with spheroidal micelles. The bar equals 100 nm in all micrographs. Reproduced from ref. 90 with permission of the American Chemical Society.

surfactant was preferentially bound at the junctions between PS16 chains, resulting in their break-up, as in the case of intermolecular association in associating polymer solutions (see Section VI.A) [90].

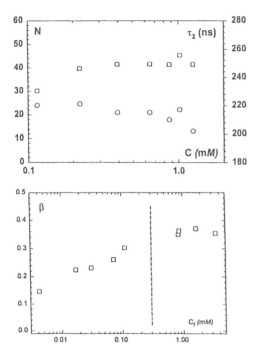

FIG. 13 PS4(α = 0.25)/DTAC systems. Bottom: Binding isotherm (the vertical line indicates the onset of precipitation). Top: Variation of the pyrene fluorescence lifetime (O) and aggregation number ($\square$) with the DTAC concentration. Reproduced from ref. 69 with permission of the American Chemical Society.

V. INTERACTIONS BETWEEN POLYELECTROLYTES OF INTERMEDIATE HYDROPHOBICITY AND SURFACTANTS

In this section we consider polyelectrolytes of intermediate hydrophobicity where the hydrophobic microdomains may be extremely small or not present but are readily induced. Typical polyelectrolytes of this type are those showing a polyelectrolyte/polysoap transition for instance when their degree of neutralization is increased (see below). Proteins constitute another type of such (hetero)polyelectrolytes. Most associating polymers, which generally contain a small fraction of monomers with a strongly hydrophobic group would fall in this category. These are dealt with separately in Section VI.

A. Polyelectrolytes

Figure 1 shows that PS4 goes from a polysoap conformation at low α to a polyelectrolyte conformation at $\alpha > 0.4$ on the basis of the values of the pyrene intensity ratio I_1/I_3 and lifetime τ. The interaction of PS4($\alpha = 0.25$) with DTAC has been investigated as the neutralization degree 0.25 falls in the polysoap/polyelectrolyte transition range of PS4. The binding isotherm (Figure 13, bottom) was found to be non-cooperative as for polysoaps, but the aggregation number of the bound surfactant and the fluorescence lifetime varied little with the surfactant concentration (Figure 13, top), as for hydrophilic polyelectrolytes. These results suggested the presence of a small number of microdomains prior to surfactant addition, thus a non-cooperative binding, and the formation of new microdomains upon surfactant binding. A binding model is presented in Figure 14 [69].

Polymethacrylic acid (PMA) and PSS (poly(maleic acid-*alt*-styrene)) also show a polysoap/polyelectrolyte transition upon increasing α (Figure 1). This is also the case of copolymers made of an ionic hydrophilic monomer and a hydrophobic one, as for instance poly(acrylic acid-*co*-ethylmethacrylate) upon increasing content of ethylmethacrylate [92].

Several studies have been concerned with the conformational transition of PMA upon increasing α but none reported on surfactant aggregation numbers

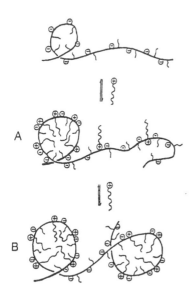

FIG. 14 Model of binding of a cationic surfactant to a weakly hydrophobic polyelectrolyte capable of forming very few and small microdomains (most repeating units are free). (A): swelling of a preexisting microdomain by the bound surfactant with incorporation of free repeating units; (B) generation of an additional microdomain upon further surfactant binding. Reproduced from ref. 69 with permission of the American Chemical Society.

in the transition range. At low pH PMA appears to be made of clusters capable of solubilizing hydrophobic molecules and comprising about 100 repeating units [93, 94]. A fluorescence probing study (I_1/I_3 measurements) of poly(acrylic acid-co-ethylmethacrylate) copolymers showed that the transition pH increased with the ethylmethacrylate content [92]. Additions of cationic surfactants such as DeTAB resulted in the formation of hydrophobic microdomains at DeTAB concentrations decreasing upon increasing ethylmethacrylate content. Unfortunately the effect of pH was not investigated and the binding isotherms were not determined. Nevertheless the measured DeTAB aggregation numbers appear to decrease upon increasing hydrophobicity of the copolymer, an effect expected on the basis of the results reviewed above.

B. Proteins

Protein/surfactant interactions have been much investigated and the topic was recently reviewed [95]. These interactions are more complex than in synthetic polyelectrolyte/surfactant systems owing to the heterogeneity of surfactant binding sites in proteins, and to the protein denaturation upon surfactant binding. Nevertheless the "necklace and bead" model used to represent the structure of many polymer/surfactant systems also appears to hold for protein/surfactant systems. This section reviews a few recent studies about two well investigated protein/surfactant systems which give some of the essential features of these systems. The proteins considered are bovine serum albumin (BSA), a representative of the globular proteins, gelatin (which is one of the most widely investigated proteins as far as interactions with surfactants are concerned), and its native, multiple helical form, collagen.

Starting with BSA, the reported studies agree that SDS binding is cooperative but disagree on whether binding involves one step [96] or two steps [97a]. The first binding step is almost not apparent when SDS is replaced by the cationic DTAB which exhibits lower binding than SDS [97b] (Figure 15). This first binding step has been attributed to a specific, non-cooperative binding involving sites of high affinity for the surfactant [98]. The second step is cooperative and corresponds to the formation of surfactant aggregates on the denatured BSA which then has an extended conformation somewhat similar to that of hydrophilic polyelectrolytes [98]. Denaturation results from surfactant binding [98]. Note that the nonionic HMEHEC (hydrophobically modified ethylhydroxyethylcellulose) also shows two-step binding of SDS with a non-cooperative first step followed by a cooperative second step [99]. Pyrene fluorescence probing (I_1/I_3) showed the existence of hydrophobic regions in pure BSA [98]. These regions are probably responsible for the non-cooperative

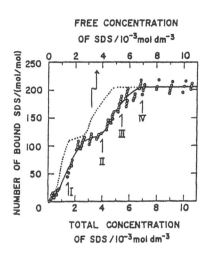

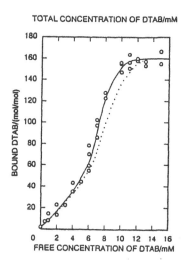

FIG. 15   Binding isotherms of SDS (left) and DTAB (right) to BSA, showing the two binding steps for SDS and only the cooperative step for DTAB. Reproduced from ref. 97 with permission of Academic Press.

binding step as in the case of moderately hydrophobic polyelectrolytes (see preceding paragraph). Fluorescence probing of the BSA/SDS system confirmed that the bound surfactant is in the form of micelle-like aggregates, with a lower polarity, a higher microviscosity and a longer pyrene lifetime than in free SDS micelles [98], a series of results similar to those for polyelectrolyte/surfactant systems.   The aggregation numbers determined using the pyrene excimer method are smaller than those of free SDS micelles, and increase with the surfactant concentration [98], a behavior similar to that for nonionic polymer/anionic surfactant [49] and for polysoap/oppositely charged surfactant systems (see above).   For instance the surfactant aggregation number in a 35 mM SDS solution (buffer 0.6 M) was 106, as compared to only 43 in the presence of 1 % BSA.   Similarly low aggregation numbers have been reported for the BSA/LiDS (lithium dodecylsulfate) system using small angle neutron scattering [100].

Gelatin can form extended networks, stabilized by hydrophobic interactions between gelatin chains.   As is the case for BSA, hydrophobic microdomains were detected with the pure protein and interactions were much stronger with anionic than with cationic surfactants [101-103].   The binding of SDS is not cooperative at low surfactant concentration but becomes cooperative at higher concentration [103].   A detailed NMR study has permitted Miller *et al.* [104] to determine the locus of binding of the dodecylsulfate aggregates. SSFQ with the Rubipy/9-MeA pair yielded aggregation numbers for the bound

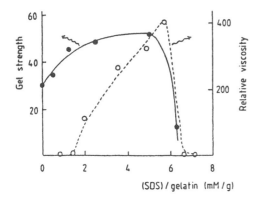

FIG. 16 Variation of the gel strength (in dyne/cm^2) and of the relative viscosity of the gelatin/SDS system with the SDS concentration. Reproduced from ref. 102 with permission of Springer-Verlag.

dodecylsulfate smaller than for free SDS micelles [105]. Using small angle neutron scattering, Cosgrove *et al.* [106] showed that at concentrations just above the cac the gelatin network is disrupted by SDS binding to the gelatin strands, resulting in a decrease of network mesh size. Gelatin diffusion studies in the presence of SDS confirmed the formation of a tighter network [107]. Another interesting result is the maximum in viscosity and gel strength vs SDS concentration plots for the gelatin/SDS system (Figure 16) [102], a result similar to that for many associating polymers discussed in Section VI.A.

Collagen is the native, highly helical form of gelatin. At low protein charge density (pH 2.5) the cooperativity of SDS binding to collagen is low [108] suggesting that the formation of SDS micelles is prevented by the low charge density and hydrophobicity of collagen. The cooperativity is recovered at pH 6.0 when collagen adopts a random coil conformation.

VI.　POLYMER-SURFACTANT SYSTEMS OF SPECIAL INTEREST

A.　Water-Soluble Associating Polymers (WSAP)

Most WSAPs are copolymers of a water-soluble monomer and of a very hydrophobic comonomer at low mole fraction, usually below 0.1. In aqueous solution the hydrophobic moieties self-associate intra- or intermolecularly. Intermolecular associations give rise to three-dimensional networks and to interesting rheological properties, thus the use of associating polymers as thickeners [109]. The increase of hydrophobic co-monomer mole fraction results in a maximum of viscosity which reflects the competition between intermolecular associations, predominant at low hydrophobe content, and intramolecular associations, which become predominant at high hydrophobe content. The associating polymer then becomes a polysoap. The self-association can be enhanced by addition of salt. Upon surfactant additions at constant polymer content, the solution viscosity first increases, goes through a maximum, then decreases, as in the case of protein/surfactant systems [110]. The model used to explain this behavior is shown in Figure 17.

Various techniques, particularly pyrene fluorescence spectroscopy, have been used to show the existence of hydrophobic microdomains in aqueous solutions of associating polymers [110-116]. These microdomains were found to be small with 10 or less hydrophobes per microdomain for both ionic [110]

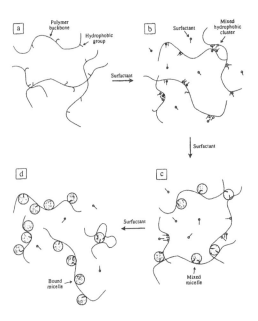

FIG. 17 Schematic representation of the binding of a surfactant to an associating polymer. (a) WSAP solution in the absence of surfactant: few contacts between hydrophobes; (b) same solution in the presence of a small amount of surfactant below system cac: non-cooperative binding results in the formation of small mixed aggregates and an increase in intermolecular contacts; (c) above system cac: cooperative binding results in larger mixed aggregates and strengthening of intermolecular contacts; (d) well above the cac: the hydrophobes are saturated and the system approaches phase separation, when polymer and surfactant are oppositely charged. Reproduced from ref. 110 with permission of the American Chemical Society.

and nonionic [111-114] WSAPs.

Associating polyelectrolytes interact with surfactants of like and opposite charge [110, 115-118] and also with nonionic surfactants [119]. For WSAPs capable of forming microdomains the first stage of binding is non-cooperative and involves individual surfactants. At higher surfactant concentration, the WSAP/surfactant systems exhibit a critical aggregation concentration (cac) above which the binding is cooperative [115]. Overall the behavior of WSAP/surfactant systems is close to that of protein/surfactant systems described in the previous section. An investigation of the binding stoichiometry revealed that the amount of bound surfactant per hydrophobe increases with the surfactant chain length and when the hydrophobe content decreases [116].

There have been few determinations of surfactant aggregation number in WSAP/surfactant systems. Magny *et al.* [110] reported on the aggregation numbers of DTAC bound to sodium polyacrylate hydrophobically modified by 1 to 3 mole % dodecyl or octadecyl groups, obtained using SSFQ and TRFQ (pyrene/dodecyl or hexadecyl pyridinium ion). The total number of surfactant ions and polymer side groups was only slightly larger than the aggregation number of the free DTAC micelles. Dualeh and Steiner [113] investigated the binding of SDS to a hydroxyethylcellulose hydrophobically modified by dodecyl groups. SSFQ with the Rubipy/9-MeA pair yielded surfactant aggregation numbers only slightly lower than those for free SDS micelles. The surfactant aggregation numbers increase with increasing SDS concentration at constant hydrophobe concentration, and are independent of the hydrophobe content at constant SDS concentration.

B. Hydrophobically End-capped Poly(ethyleneoxide)

Although these polymers are not charged they are reviewed briefly because their behavior is very similar to that of WSAPs. The PEO (polyethyleneoxide) moiety usually has a molecular weight of several thousands. The hydrophobic end-groups are directly attached to the PEO moiety (HMPEO) or through a urethane spacer group (HMUPEO). These polymers self-associate in water but the type of association, open or micellar, is still under discussion. The variations of the pyrene fluorescence intensity ratio I_1/I_3 [120] and of the excimer emission of dipyme [121] with the polymer concentration were indeed very pronounced but this reflected the partition of pyrene between water and the hydrophobic aggregates resulting from the end-group self-association, because the volume of the hydrophobic phase was very small in the systems investigated. Nevertheless it has been argued that the association of HMPEO is of the open type (no cmc) [122]. Microcalorimetry showed that the onset of association (cmc?) occurs at very low concentrations, 0.014 wt %, equivalent to a concentration of 3×10^{-5} M of dodecyl groups for $C_{12}EO_{200}C_{12}$ [123]. The authors concluded that "presumably the association is gradual, with the size of the aggregates varying with concentration." Note that the value of the cmc given above is only slightly lower than that for polyethyeleneoxide monododecyl ether ($C_{12}EO_x$) surfactants (cmc < 10^{-4} M). Also $C_{12}EO_{2x}C_{12}$ can be considered as dimers of $C_{12}EO_x$ surfactants, and dimeric surfactants always have cmcs much lower than their monomeric counterparts [124].

At higher concentration these polymers form rosette-like or flower-like micelles with the PEO chains looping in the aqueous solution. At still higher concentration an increasing number of polymer chains form bridges between micelles, giving rise to micelle clusters (microgels), as represented in Figure 18, and to microgel networks which possess interesting rheological properties [120-122,125,126]. This behavior explains the use of modified PEO as thickener.

The WSAP micelles have been characterized by fluorescence probing [120-122,125-131]. The polarity of the micellar interior was found to be somewhat lower than that of SDS micelles but comparable to that of $C_{12}EO_x$ surfactants [125]. However EPR measurements suggested the polarity to be substantially smaller than for SDS but larger than for $C_{12}EO_8$ and to be independent of concentration [127]. The micelle microviscosity was found to be much larger than for conventional surfactant micelles, by both fluorescence probing and EPR, and independent of the polymer concentration [121,125-127]. The reported values of the end-group aggregation numbers depend somewhat on the presence of the urethane spacer and on the alkyl chain carbon

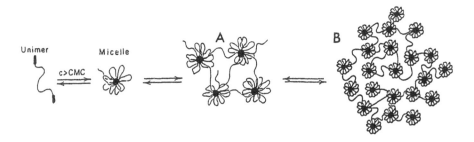

FIG. 18 Evolution of the structure in aqueous solutions of HMPEO or HMUPEO with increasing polymer concentration, showing the rosette-like or flower-like micelles, micelle cluster (A) and microgel network (B). The black dots represent the end-group aggregates. Reproduced from ref. 131, *Colloids Surf. A.112*: 239 (1996) with permission of Elsevier Science-NL, Sara Burgerhartstraat 25, 1055 KV Amsterdam, the Netherlands.

number but were always smaller than for micelles of conventional surfactants with the same alkyl chain [120,125-131]. For instance, Yekta *et al.* [126] using the time-resolved pyrene excimer method reported N values ranging from 18 to 28 for several hexadecyl chain-terminated HMUPEO whereas the aggregation number of similar conventional micelles would be at least 100. Alami *et al.* [129] reported concentration-independent N values around 16 and 30 for $C_{12}EO_{460}C_{12}$, when using the time-resolved pyrene excimer and the TRFQ (pyrene/DMBP pair) methods, respectively. This difference is much larger than the usual experimental error in such determinations and may reflect problems associated with fluorescent impurities present in the systems, and the high micelle microviscosity which reduces the quenching efficiency. Persson and Bales [127], using EPR, reported the concentration-independent value N = 31 ± 6 for the aggregation number of $C_{12}EO_{200}C_{12}$. Conventional micelles with a dodecyl chain have aggregation numbers of 60-90. The above studies assumed that all end-groups contribute to micelle formation. This may not be the case and the determined aggregation numbers should be considered as upper bound values.

In general, hydrophobically modified PEO interacts more strongly with surfactants than the parent PEO. Interaction has been reported to occur with anionic [127,132-134], cationic [133], and nonionic surfactants [127,135] whereas the parent PEO interacts only with anionic surfactants at room temperature. Also, binding appeared to be cooperative with both SDS and DTAB [133]. However, binding is stronger with SDS than with DTAB, as indicated by the binding isotherms (Figure 19 left) and the variations of viscosity (Figure 19 right) for a HMUPEO solution upon addition of SDS and DTAB [133]. The difference may reflect the fact that binding of SDS occurs at PEO ends as well as on the PEO chain whereas DTAB would bind only at the PEO ends [134]. Note that the binding isotherm of SDS to the polymer apparently shows two steps, as for proteins (Section V.B). The viscosity maximum is typical of associating polymers. It was also observed with the $C_{12}EO_{460}C_{12}/C_{12}EO_8$ system [135]. At high surfactant concentration the network of rosette-like micelles is completely broken up, and the hydrophobic end-groups are saturated with surfactant, in the same manner as in associative polymer/surfactant systems as depicted in Figure 18. The aggregation number of the mixed micelles formed by bound surfactants and end-groups were somewhat smaller than those of the pure surfactant micelle (case of the $C_{12}EO_{460}C_{12}/C_{12}EO_8$ system [135]). Figure 20 shows that the number of end-groups per micelle ($N_{endgroup.}$) increased nearly linearly while the number of surfactants per micelle ($N = N_{tot} - N_{endgroup}$) decreased with increasing $C_{12}EO_{460}C_{12}$ concentration (C_p). The decrease of N upon increasing C_p at constant surfactant concentration is similar to that reported for nonionic

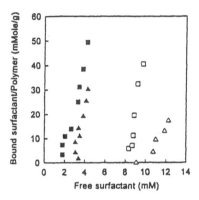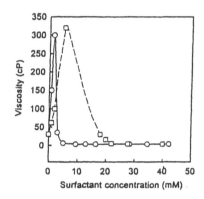

FIG. 19 Left: binding isotherms of SDS (■,▲) and DTAB (□,Δ) to PEO (▲,Δ) and to HMUPEO (■,□); Right: zero shear viscosities of a 1 wt % solution of HMUPEO in D₂O as a function of the concentration of added SDS (□) and DTAB (O). Reproduced from ref. 133 with permission of the American Chemical Society.

polymer/anionic surfactant systems [49]. In Figure 20, at a given polymer concentration (4 wt %, for instance), the surfactant aggregation number increases with the surfactant concentration [135], a behavior similar that for hydrophobic polysoaps as discussed in Section IV. The increase was not linear though, a difference which may reflect the fact that the microdomains are formed intermolecularly with HMPEOs and intramolecularly with polysoaps.

C. Polyelectrolyte-Surfactant Complexes in Organic Solvents and in the Solid State

Antonietti *et al.* [136,137] have observed that the polyelectrolyte/surfactant complexes that precipitate out when mixing aqueous solutions of NaPSS (poly(sodium styrenesulfonate)) and alkyltrimethylammonium bromides can be redissolved in organic solvents such as dimethylformamide, tetrahydrofuran, ethanol or isopropanol, where they behave like polyelectrolytes, *i.e.* exhibiting a decrease of the reduced viscosity upon increasing complex concentration. Cast films of some of these complexes showed highly ordered mesophases of the lamellar type [136,138-140], the most striking structure being an undulating lamellar phase, the "egg-carton phase" (Figure 21) with high undulation amplitude [138-140]. Such structures appear to be of general character since

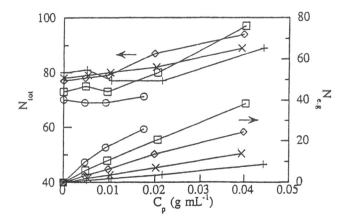

FIG. 20 Variation of the total average aggregation number and of the end-group aggregation number in the $C_{12}EO_{460}C_{12}/C_{12}EO_8$ system with the polymer concentration C_p in $C_{12}EO_8$ solutions at concentrations of (O) 2, (□) 5, (◊) 10, (×) 20, and (+) 40 mM. Reproduced from ref. 135 with permission of the American Chemical Society.

they were observed with several polyion/surfactant ion complexes: copolymers of 2-acrylamido-2-methyl-1-propanesulfonic acid with N-octadecylacrylamide, at different ratios of the two monomers, with CTAB [138], and poly-(dimethyldiallylammonium chloride)/natural lipid (II) [139,140].

Kabanov *et al.* [141,142] also investigated complexes of polycations or polyanions with the anionic surfactant AOT (sodium diethylhexylsulfo-succinate) in organic solvents in the presence of water. The complexation of polyelectrolytes of the ionomer type (partially sulfonated polystyrene) by AOT in *m*-xylene resulted in a partial disruption of the aggregates formed by the ionomer [141]. Poly(N-ethyl-4-vinylpyridinium bromide) was found to be soluble in octane and hexane in the presence of AOT [142]. In the presence of water, the polyelectrolyte was incorporated in the water pools formed by AOT where it adopted a rather compact conformation [142]. At low water content the authors proposed a comb-like model for the complex.

FIG. 21 Computer model of the "egg-carton phase." The lamellae consisting of the polyelectrolyte chains and of lipid head groups exhibit tetragonal thickness undulations. Reproduced from ref. 139 with permission of the American Chemical Society.

VII. CONCLUSIONS AND SUMMARY

The main characteristics of polyelectrolyte/surfactant interactions are summarized in Table 2. The similarities and differences between the different types of polymers according to their hydrophobicity are clearly seen. Moderately hydrophobic polyelectrolytes, proteins and associating polymers behave similarly in many respects, even though their structures are very different. Their capacity to give rise to small hydrophobic microdomains, either by themselves or assisted by the presence of surfactants, determines their behavior.

TABLE 2 Characteristics of the Various Polymer-Surfactant Systems Reviewed

Polymer	Nature of Binding and Binding model	Aggregation Number N	Aggregate Microviscosity
Hydrophilic Polyelectrolytes	Bind only oppositely charged surfactants cooperatively. *Necklace and beads model*	Independent of [bound surfactant]	Higher than for free micelles
Hydrophobic Polyelectrolytes	Bind all surfactants non-cooperatively. *Swelling of polysoap microdomains*	Increases linearly with [bound surfactant]	Higher than for free micelles
Moderately Hydrophobic Polyelectrolytes	Non-cooperative binding *Swelling of preexisting microdomains and creation of additional ones*	Nearly constant	Unknown
Proteins	Binding first non cooperative then cooperative, weaker for cationic than for anionic surfactants *Necklace and beads model after first step*	N smaller than for free surfactant micelles	Higher than for free micelles
Associating Polymers (including HMPEO)	Cooperative binding, perhaps in two steps for anionic surfactants *Mixed micellization*	N smaller than for free surfactant micelles, increases with [bound surfactant]	Higher than for free micelles
Non-Ionic Polymers (PEO, for instance)	Cooperative binding of anionic surfactants. Binds cationic surfactants only when sufficiently hydrophobic *Necklace and bead model*	N smaller than for free surfactant micelles, decreases with increasing polymer hydrophobicity, increases with [bound surfactant]	Probably higher than for free surfactant micelles

REFERENCES

1. R. Zana, in *Surfactant Solutions. New Methods of Investigation* (R. Zana, ed.) Marcel Dekker, New York, 1987, pp. 241-294 and references therein.
2. M. Almgren, *Adv. Colloid Interface Sci. 41*: 9 (1992) and references therein.
3. M.H. Gehlen, and F.C. De Schryver, *Chem. Rev. 93*: 199 (1993) and references therein.
4. A.V. Barzykin, and M. Tachiya, *Heter. Chem. Rev. 3*: 105 (1996) and references therein.
5. See for instance: *Interactions of Surfactants with Polymers and Proteins* (E.D. Goddard and K.P. Ananthapadmanabhan, eds.) CRC Press, Boca Raton (Florida, USA), 1993.
6. K. Hayakawa, and J.C.T. Kwak, in *Cationic Surfactants. Physical Chemistry.* (D. N. Rubingh, and P.M. Holland, eds.) Marcel Dekker, New York, 1991, pp.189-248 and references therein.
7. See U.P. Strauss in ref 5, pp. 277-294.
8. K. Kalyanasundaran, and J.K. Thomas, *J. Am. Chem. Soc. 99*: 2039 (1977).
9. P. Chandar, P. Somasundaran, and N.J. Turro, *Macromolecules 21*: 950 (1988).
10. F. Quina, E.A. Abuin, and E. Lissi, *Macromolecules 23*: 5173 (1990).
11. H. Morawetz, *Macromolecules 29*: 2689 (1996).
12. P. Infelta, M. Grätzel, and J.K. Thomas, *J. Phys. Chem. 78*: 190 (1974).
13. M. Tachiya, *Chem. Phys. Lett. 33*: 289 (1975).
14. N.J. Turro and A. Yekta, *J. Am. Chem. Soc. 100*: 5951 (1978).
15. E.A.G. Aniansson and S.N. Wall, *J. Phys. Chem. 78*: 1024 (1974); 79:857 (1995).
16. P. Infelta, *Chem. Phys. Lett. 61*:,88 (1979).
17. A. Yekta, M. Aikawa, and N.J. Turro, *Chem. Phys. Lett. 63*: 543 (1979).
18. M. Tachiya, in *Kinetics of Nonhomogeneous Processes* (G.R. Freeman, ed.) J. Wiley, New-York, 1987, pp. 575-650.
19. M. Almgren, in *Kinetics and Catalysis in Microheteregeneous Systems* (M. Grätzel, and K. Kalyanasundaram, eds.) Marcel. Dekker, New York, 1991, pp. 63-113.
20. S. Atik, N. Nam, and L.A. Singer, *Chem. Phys. Lett. 67*: 75 (1979).
21. P. Lianos, J. Lang, C. Strazielle, and R. Zana, *J. Phys. Chem. 89*: 1019 (1982).
22. M. van der Auweraer, and F. C. De Schryver, *Chem. Phys. 111*: 105 (1987).
23. M. Almgren, J. Alsins, E. Mukhtar, and J. van Stam, *J. Phys. Chem. 92*: 4479 (1988).
24. J.N. Israelachvili, D.J. Mitchell, and B.W. Ninham, *J. Chem. Soc. Faraday Trans. 2 77*: 601 (1976).
25. M. Almgren, and J.E. Löfroth, *J. Chem. Phys. 76*: 2734 (1982).
26. G.C. Warr, and F. Grieser, *J. Chem. Soc. Faraday Trans. 1 82*: 1813 (1986).
27. M. Almgren, J. Alsins, J. van Stam, and E. Mukhtar, *Prog. Colloid Polym. Sci. 76*: 68 (1988).
28. A. Siemiarczuk, W.R. Ware, and Y.S. Liu, *J. Phys. Chem. 97*: 8082 (1993).
29. M. Almgren, F. Grieser, and J.K. Thomas, *J. Am. Chem. Soc. 101*: 279 (1979).
30. R. Zana, and C. Weill, *J. Phys. Lett. 46*: L953 (1985).

31. A. Malliaris, J. Lang, and R. Zana, *J. Phys. Chem. 90*: 655 (1986).
32. M. Almgren, J.-E. Löfroth, and J. van Stam *J. Phys. Chem. 90*: 4431 (1986).
33. P. Koglin, D.J. Miller, J. Steinwändel, and M. Hauser, *J. Phys. Chem. 85*: 2363 (1981).
34. K.-J. Choi, L. A. Turkevich, and R.Loza, *J. Phys. Chem. 92*: 2248 (1988).
35. T. Itaya, H. Ochiai, K. Ueda, and A. Imamura, *Polymer 35*: 2004 (1994).
36. M. Almgren, P. Hansson, and K. Wang, *Langmuir, 12*: 3855 (1996).
37. P.M. Holland, in *Mixed Surfactant Systems* (P. M. Holland, and D. N. Rubingh, eds.) ACS Symp.Ser. 501, Washington D.C., 1992, pp. 31-44.
38. D.C. Dong, and M.A. Winnik, *Photochem. Photobiol., 35*: 17 (1982).
39. F.M. Winnik, M.A. Winnik, H. Ringsdorf, and J. Venzmer, *J. Phys. Chem. 95*: 2583 (1991).
40. A.V. Kabanov, I.R. Nazarova, I.V. Astafieva, E.V. Batrakova, V.Y. Alakhov, A.A. Yaroslavov, and V.A. Kabanov, *Macromolecules 28*: 303 (1995).
41. W. Binana-Limbélé, and R. Zana, *Macromolecules 20*: 1331 (1987).
42. M. Benrraou, R. Zana, R. Varoqui, and E. Pefferkorn, *J. Phys. Chem. 96*: 1468 (1992).
43. O. Anthony, *Ph.D thesis*, University L. Pasteur, Strasbourg, France, 1995.
44. J.J. Kiefer, P. Somasundaran, and K. P. Ananthapadmanabhan, in *Polymer Solutions, Blends, and Interfaces* (I. Noda, and D.N. Rubingh, eds.) Elsevier Science Publ., 1992, pp. 423-444.
45. O. Anthony, and R. Zana, *Langmuir 12*: 1967 (1996).
46. M. Almgren, P. Hansson, E. Mukhtar, and J. van Stam, *Langmuir 8*: 2405 (1992).
47. P. Hansson, and M. Almgren, *Langmuir 10*: 2115 (1994).
48. P. Hansson, and M. Almgren, *J. Phys. Chem 99*: 16684 (1995).
49. R. Zana, J. Lang, and P. Lianos, in *Microdomains in Polymer Solutions* (P. Dubin, ed.) Plenum Press, New York, 1985, pp. 357-368.
50. I. Satake, and J.T. Yang, *Biopolymers 15*: 2263 (1976).
51. P. Hansson, and M. Almgren, *J. Phys. Chem 99*: 16694 (1995).
52. K. Thalberg, J. van Stam, C. Lindblad, M. Almgren, and B. Lindman, *J. Phys. Chem. 95*: 9875 (1991).
53. Z. Zhen, and C.-H. Tung, *Polymer 33*: 812 (1992).
54. Z. Gao, J.C.T. Kwak, and R.E. Wasylishen, *J. Colloid Interface Sci. 126*: 371 (1988)
55. Z. Gao, R.E. Wasylishen, and J.C.T. Kwak, *J. Phys. Chem. 94*: 773 (1990).
56. K. Hayakawa, T. Nakano, and I. Satake, *Langmuir, 12*: 269 (1996).
57. J. Appell, G. Porte, A. Khatory, F. Kern, and S. J. Candau, *J. Phys. (France) II 2*: 1045 (1992).
58. R. Makhloufi, E. Hirsch, S. J. Candau, W. Binana-Limbélé, and R. Zana, *J. Phys. Chem. 93*: 8095 (1989).
59. E.A. Abuin, and J.C. Scaiano, *J. Am. Chem. Soc. 106*: 6274 (1984).
60. D.-Y. Chu, and J.K. Thomas, *J. Am. Chem. Soc. 108*: 6270 (1986).
61. J. Fundin, W. Brown, and M. Swanson Vethamuthu, *Macromolecules 29*: 1195 (1996).

62. A. Malliaris, W. Binana-Limbélé, and R. Zana, *J. Colloid Interface Sci. 110*: 114 (1986).
63. P. Uznanski, J. Pecherz, and M. Kryszewski, *Can. J. Chem. 73*: 2041 (1995).
64. L. Johansson, P. Hedberg, and J.-E. Löfroth, *J. Phys. Chem. 97*: 747 (1993).
65. F. Yeh, E. Sokolov, A. Kholkhlov, and B. Chu, *J. Am. Chem. Soc. 118*: 6615 (1996).
66. B. Chu, F. Yeh, and A. Kholkhlov, *Macromolecules 28*: 8447 (1995).
67. M. D. Everaars, A. C. Nieuwkerk, S. Denis, A. T. Marcelis, and E. J. Sudhölter, *Langmuir 12*: 4042 (1996).
68. W. Binana-Limbélé, and R. Zana, *Colloids Surf. 21*: 483 (1986).
69. O. Anthony, and R. Zana, *Langmuir 12*: 3590 (1996).
70. O. Anthony, and R. Zana, *Macromolecules 27*: 3885(1994).
71. W. Binana-Limbélé, and R. Zana, *Macromolecules 23*: 2731 (1990).
72. J.-L. Hsu, and U. P. Strauss, *J. Phys. Chem. 91*: 6328 (1987).
73. V. S. Zdanowicz, and U. P. Strauss, *Macromolecules 26*: 4770 (1993).
74. (a) D.-Y. Chu, and J. K. Thomas, *Macromolecules 20*: 133 (1987)
    (b).D.-Y. Chu, and J.K. Thomas, in *Polymers in Aqueous Media* (J.E. Glass, ed.), Advances in Chemistry Series N° 223 American Chemical Society, Washington D.C., 1989, pp. 325-341.
75. L. B. Shih, E. Y. Sheu, and S. H. Chen, *Macromolecules 21*: 1387 (1988).
76. L. B. Shih, D. H. Mauer, C. J. Verbrugge, C. F. Wu, S. L. Chang, and S. H. Chen, *Macromolecules 21*: 3225 (1988).
77. D. Cochin, F. Candau, R. Zana, and Y. Talmon, *Macromolecules 25*: 4220 (1992).
78. D.-Y. Chu, and J. K. Thomas, *Macromolecules 24*: 2212 (1991).
79. J. Kevelam, and J. B. F. N. Engberts, *J. Colloid Interface Sci. 178*: 87 (1996).
80. G. J. Wang, and J. B. F. N. Engberts, *Langmuir 12*: 652 (1996).
81. D. Cochin, F. Candau, and R. Zana, *Macromolecules 26*: 5755 (1993).
82. D. Cochin, F. Candau, and R. Zana, *Polym. Int. 30*:491 (1993).
83. M.S. Turner, and J. F. Joanny, *J. Phys. Chem. 97*: 4825 (1993).
84. M. Aizawa, T. Komatsu, and T. Nakagawa, *Bull. Chem. Soc. Jap. 50*: 3107 (1977).
85. M. Sisido, K. Akiyama, Y. Imanishi, and I. M. Klotz, *Macromolecules 17*:198 (1984).
86. T. Seo, S. Take, K. Miwa, K. Hamada, and T. Iijima, *Macromolecules 24*: 4255 (1991).
87. T. Seo, S. Take, T. Akimoto, K. Hamada, and T. Iijima, *Macromolecules 24*: 4801 (1991).
88. (a) M. J. McGlade, F. J. Randall, and N. Tcheurekdjian, *Macromolecules 20*: 1782 (1987)
    (b).M.J. McGlade, and J.L. Olufs, *Macromolecules 21*: 2346 (1988).
89. J. Kevelam, J.F.L. van Breemen, W. Blokzijl, and J.B.F.N. Engberts, *Langmuir 12*: 4709 (1996).
90. R. Zana, A. Kaplun, and Y. Talmon, *Langmuir 9*: 1948 (1993).
91. N. Kamenka, A. Kaplun, Y. Talmon, and R. Zana, *Langmuir 10*: 2960 (1994).

92. V.A. De Oliveira, M.J. Tiera, and M.G. Neumann, *Langmuir 12*: 607 (1996).
93. M.J. Snare, K.L. Tan, and F.E. Treloar, *J. Macromol. Sci.-Chem. A17*: 198 (1982).
94. D.Y. Chu, and J.K. Thomas, *J. Phys. Chem. 89*: 4065 (1985).
95. K.P. Ananthapadmanabhan, in ref. 5, p. 319.
96. A. Chen, D. Wu, and C.. Johnson, Jr., *J. Phys. Chem. 99*: 828 (1995).
97. (a) K. Takeda, M. Miura, and T. Takagi, *J. Colloid Interface Sci. 82*: 38 (1981); (b) K. Takeda, H. Sasaoka, K. Sasa, H. Hirai, K. Hachiya, and Y. Moriyama, *J. Colloid Interface Sci. 154*: 385 (1992).
98. N.J. Turro, X.-G. Lei, K. P. Ananthapadmanabhan, and M. Aronson, *Langmuir 11*: 2525 (1995).
99. (a) K. Thuresson, O. Söderman, P. Hansson, and B. Lindman, *J. Phys. Chem. 100*: 909 (1996)
    (b) K. Thuresson, and B. Lindman, *J. Phys. Chem.101*: 6460 (1997)
100. H. Guo, N.M. Zhao, S.H. Chen, and J. Texeira, *Biopolymers 29*: 335 (1990).
101. R. Wustneck, E. Buder, R. Wetzel, and H. Hermel, *Colloid Polym. Sci. 267*: 429 (1989).
102. M. Henriquez, E. Abuin, and E. Lissi, *Colloid Polym. Sci. 271*: 960 (1993).
103. H. Fruhner, and G. Kretzschmar, *Colloid Polym. Sci. 267*: 839 (1989).
104. D.D. Miller, W. Lenhart, B.J. Antalek, A.J. Williams, and J.H. Hewitt, *Langmuir 10*: 68 (1994).
105. T.H. Whitesides, and D.D. Miller, *Langmuir 10*: 2899 (1994).
106. T. Cosgrove, S.J. White, A. Zarbakhsh, R.H. Heenan, and A.M. Howe, *Langmuir 11*: 744 (1995).
107. P.C. Griffith, P. Stilbs, A.M. Howe, and T. Cosgrove, *Langmuir 12*: 2884 (1996).
108. M. Henriquez, E. Lissi, E. Abuin, and A. Ciferri, *Macromolecules 27*: 6834 (1994).
109. See *e.g. Polymers as Rheology Modifiers* (D. N. Schulz, and J. E. Glass, eds.) ACS Symp. Ser. 462, Washington D.C.,1991, and references therein.
110. B. Magny, I. Iliopoulos, R. Zana, and R. Audebert, *Langmuir 10*: 3180 (1994).
111. D.B. Siano, J. Bock, P. Myer, and P.L. Valint Jr., in *Polymers in Aqueous Media* (J. E. Glass, ed.), American Chemical Society, Advances in Chemistry Series N° 223, Washington D.C. 1989, pp. 425-435.
112. A.J. Dualeh, and C.A. Steiner, *Macromolecules 23*: 251 (1990).
113. A.J. Dualeh, and C.A. Steiner, *Macromolecules 24*:112 (1991).
114. C.E. Flynn, and J.W. Goodwin, in ref 109, pp. 190-206.
115. F. Guillemet, and L. Piculell, *J. Phys. Chem. 99*: 9201 (1995).
116. K.N. Bakeev, E.A. Ponomarenko, T.V. Shishkanova, D.A. Tirrell, A.B. Zezin, and V.A. Kabanov, *Macromolecules 28*: 2886 (1995).
117. O.E. Philippova, D. Bourdet, R. Audebert, and A.R. Khokhlov, *Macromolecules 29*: 2822 (1996).
118. F.M. Winnik, S. Regismond, and E.D. Goddard, *Colloids Surf. A 106*: 243 (1996).
119. A. Sarrazin-Cartelas, I. Iliopoulos, R. Audebert, and U. Olsson, *Langmuir 10*: 421 (1994).

120. Y. Wang, and M.A. Winnik, *Langmuir 6*: 1437 (1190).
121. A. Yekta, J. Duhamel, H. Adiwidjaja, P. Brochard, and M.A. Winnik, *Langmuir 9*: 882 (1993).
122. E. Alami, M. Rawiso, F. Isel, G. Beinert, W. Binana-Limbélé, and J. François, in *Hydrophilic Polymers. Performance with Environmental Acceptablity* (J. E. Glass, ed.) Advances in Chemistry Series N°248, American Chemical Society, Washington D. C., 1996, pp. 343-362.
123. K. Persson, G. Wang, and G. Olofsson, *J. Chem. Soc. Faraday Trans. 90*: 3555 (1994).
124. R. Zana, *Curr. Opin. Colloid Interface Sci. 1*: 566 (1996).
125. A. Yekta, J. Duhamel, P. Brochard, H. Adiwidjaja, and M.A. Winnik, *Macromolecules 26*: 1829 (1993).
126. A. Yekta, B. Xu, J. Duhamel, H. Adiwidjaja, and M. A. Winnik, *Macromolecules 28*: 956 (1995).
127. K. Persson, and B.L. Bales, *J. Chem. Soc.Faraday Trans. 91*: 2863 (1995).
128. B. Richey, A. Kirk, E.K. Eisenhart, S. Fitzwater, and J. Hook, *J. Coat Technol. 63*: 31 (1991).
129. E. Alami, M. Almgren, W. Brown, and J. François, *Macromolecules 29*: 2229 (1996).
130. A. Yekta, T. Nivaggioli, S. Kanagalingam, B. Xu, Z. Masoumi, and M.A. Winnik, in *Hydrophilic Polymers. Performance with Environmental Acceptablity* (J. E. Glass, ed.) Advances in Chemistry Series N°248, American Chemical Society, Washington D. C., 1996, pp. 363-376.
131. B. Xu, A. Yekta, L. Li, Z. Masoumi, and M.A. Winnik, *Colloids Surf. A. 112*: 239 (1996).
132. Y.-Z. Hu, C.-L. Zhao, and M.A. Winnik, *Langmuir 6*: 880 (1990).
133. K. Zhang, B. Xu, M.A. Winnik, and P. Macdonald, *J. Phys. Chem. 100*: 9834 (1996).
134. W. Binana-Limbélé, F. Clouet, and J. François, *Colloid Polym. Sci. 271*: 748 (1993).
135. E. Alami, M. Almgren, and W. Brown, *Macromolecules 29*: 5026 (1996).
136. M. Antonietti, J. Conrad, and A. Thünemann, *Macromolecules 27*: 6007 (1994).
137. M. Antonietti, S. Forster, M. Zisenis, and J. Conrad, *Macromolecules 28*: 2270 (1995).
138. M. Antonietti and M. Maskos, *Macromolecules 29*: 4199 (1996).
139. M. Antonietti, A. Kaul, and A. Thünemann, *Langmuir 11*: 2633 (1995).
140. M. Antonietti, A. Wentzel, and A. Thünemann, *Langmuir 12*: 2111 (1996).
141. K. N. Bakeev, S. A. Chugunov, I. Teraoka, W.J. MacKnight, A. . Zezin, and V.A. Kabanov, *Macromolecules 27*: 3926 (1994).
142. A.V. Kabanov, V.G. Sergeev, M.S. Foster, V.A. Kasaikin, A.V. Levashov, and V.A. Kabanov, *Macromolecules 28*: 3657 (1995).

11
Solubilization of Dyes by Polymer-Surfactant Complexes

KATUMITU HAYAKAWA Department of Chemistry and BioScience, Kagoshima University, Kagoshima, Japan

I.   INTRODUCTION

Solubilization by surfactant micelles has been extensively studied. Both the physico-chemical aspects of micellar solubilization and applications of surfactant solubilization in a wide variety of fields have been thoroughly reviewed in the volumes 1, 23 and 55 of the "Surfactant Science" series [1,2]. In the presence of polymers, solubilization of small molecules by surfactants has been observed at concentrations below the surfactant critical micelle concentration (CMC) [3-5]. Potential applications are expected in the fields of cosmetics (drug solubilization by polymer/surfactant complexes), water processing (pollutants solubilization by polymer/surfactant complexes), fabric dyeing, etc, but only a limited number of physico-chemical studies on solubilization by polymer-surfactant systems have been published.   Early studies observed the solubilization of the water-insoluble dye, Yellow OB, by nonionic polymers such as polyvinylpyrrolidone and polyethylene glycol in the presence of anionic surfactants such as sodium dodecyl sulfate (SDS) and sodium octylbenzene sulfonate [3,4], and by the anionic polymer polyacrylic acid in the presence of nonionic surfactant [6,7]. In the case of ionic polymers with surfactants of opposite charge, solubilization was observed at a concentration far below the surfactant CMC and before the polyion-surfactant complexes were precipitated [8]. The solubilization at concentrations below the surfactant CMC observed in most polymer-surfactant systems was ascribed to the formation of hydrophobic domains in polymer/surfactant complexes formed in aqueous solution.

Solubilization by polymer-surfactant mixed systems has been mentioned briefly in a number of review articles on polymer-surfactant systems [9-14], but as yet no review has been devoted solely to this topic. The purpose of this chapter is to explain the physico-chemical aspects of solubilization of water-insoluble dyes by polymer-surfactant systems, using results from the author's laboratory together with those described in a selected number of recent studies from other groups.   A full review of solubilization in polymer-surfactant systems would await more complete data on a wider range of polymers, surfactants, and solubilizates. The solubilization of water-insoluble dyes is of particular interest in view of applications in drug solubilizates in medical and agriculture applications, colloid-enhanced ultrafiltration for water processing [15-18] and fabric dyeing. In addition, dye systems allow for the use of spectroscopic techniques to study both the solubilization equilibrium and the environment of the solubilized molecules or ions, yielding information not available with other solubilizates. Abbreviations for the names of polymers and surfactants mentioned are listed at the end of this chapter.

II. COOPERATIVE BINDING OF SURFACTANTS BY POLYMERS AND THE FORMATION OF HYDROPHOBIC DOMAINS IN AQUEOUS MEDIA

Interactions between polymers and surfactants in aqueous solution have been the subject of intense research effort, and a number of reviews have been published in short succession [9-12, 14,19]. Much of this material is collected and described in other chapters of this volume. Aqueous solutions of ionic polymers exhibit unusual properties such as low counter ion activity coefficients even at very low concentration, due to the counter ion condensation phenomenon. Surfactant solutions also possess unique properties such as micelle formation, counter ion binding and solubilization of water-insoluble materials. It is therefore not surprising that combined solutions of polymers and surfactants exhibit diverse and complex properties. The interaction between ionic polymers and oppositely charged surfactants is dominated by electrostatic forces, causing the association to start at a very low surfactant concentration, known as the critical aggregation concentration (CAC), usually a few orders of magnitude lower than the CMC of the free surfactant. Complexes of ionic polymers and surfactants of opposite charge exhibit a number of phenomena indicating the special nature of these aggregates, including the dissociation of solubilized dye aggregates [20-26], solubilization of water-insoluble materials [4,8,27-30], enhanced energy transfer between solubilized donors and acceptors [31], and induction of an ordered conformation of polypeptides. These properties are principally induced by the highly cooperative interaction between ionic polymers and surfactants of opposite charge in dilute solution.

Interactions of nonionic polymers with ionic surfactants, and of ionic polymers with nonionic surfactants are observed at relatively high concentrations of polymers and/or surfactants due to the weak interactions. The properties of such systems at low concentrations are reviewed by Saito in chapter 9 of this volume. At higher concentrations, these systems exhibit interesting phase behavior and rheological properties, as described in chapter 3 of this volume.

The binding isotherms of ionic surfactants by polyions of opposite charge are characterized by a very sharp rise at a critical equilibrium concentration of free surfactant. This sudden increase in the amount of bound surfactant is caused by the strong hydrophobic interactions among bound surfactant chains. Satake and Yang [32] first interpreted this highly cooperative surfactant binding process, using a linear lattice model of the polyion sites with interactions between the bound surfactants in the Bethe approximation. This model is completely equivalent to the treatment derived by Schwarz for dye stacking on linear polyions [33-35]. This treatment has been successfully applied to many polyion/surfactant systems. Reviews of such binding data and their interpretation have been presented by Goddard [9] and Hayakawa and Kwak [19], with current reviews included in chapters 4 and 5 of this book. In this cooperative binding model, two parameters are introduced for the construction of binding isotherms: the intrinsic binding constant K which measures the Gibbs energy for a surfactant binding to an isolated site on the polyion, and a cooperativity parameter u which measures the additional Gibbs energy of surfactant binding to a site adjacent to a site already occupied by a surfactant ion. The product Ku corresponds to the binding constant for a site adjacent to a site already occupied by a surfactant ion. The degree of surfactant binding was derived as:

$$\beta = \frac{1}{2}\left[1 - \frac{1-s}{\sqrt{(1-s)^2 + 4s/u}}\right] \tag{1}$$

In Eq. (1) we have:

$$s = K u [S] \tag{2}$$

where [S] denotes the equilibrium surfactant concentration. This treatment also predicts the average size of surfactant aggregates in the polyion domain:

$$m = \sqrt{\frac{u}{s}\left(\frac{\beta}{1-\beta}\right)} + \left(\frac{\beta}{1-\beta}\right) \tag{3}$$

This equation indicates that the aggregation number m of the surfactant aggregates in the polyion domain increases as the cooperative parameter u increases.

The size of the surfactant aggregates in the polyion domain may influence a number of properties of polymer-surfactant systems, such as their ability to solubilize of water-insoluble materials, the energy transfer efficiency of solubilized donor-acceptor pairs, and the spectroscopic properties of solubilized dyes in polyion/surfactant complexes. Aggregation of the bound surfactant ions can be demonstrated by means of dynamic photophysical techniques using hydrophobic probe molecules [27,36-49]. It can also be proven indirectly from binding isotherms, solubilization of water-insoluble dyes [28], spectroscopy of the solubilizates [25], and induction of an ordered conformation of ionic polypeptides [50,51].

III. SOLUBILIZATION OF WATER-INSOLUBLE DYES

A. Solubilization Equilibrium

The formation of hydrophobic domains by polymer-surfactant complexes in aqueous solution may be expected to solubilize water-insoluble materials. Solubilization by polymer-surfactant systems has been observed in various polymer-surfactant systems [3,4,8-12,15,27,29,52-54]. Although solubilization in micellar systems has been extensively studied, leading to a detailed understanding of the molecular mechanism and thermodynamic parameters for this process, only a limited number of papers have been published concerning the solubilization mechanism in polymer-surfactant systems.

Thalberg and Lindman measured the solubilization of Orange OT in the system TTAB-Hy [55]. They observed two solubilization zones (Figure 1). In the first zone, Hy-TTAB complexes solubilize Orange OT and at a same time phase separation occurs due to the increasing hydrophobicity of the aggregate as surfactant binding increases. When the concentration of the surfactant in the

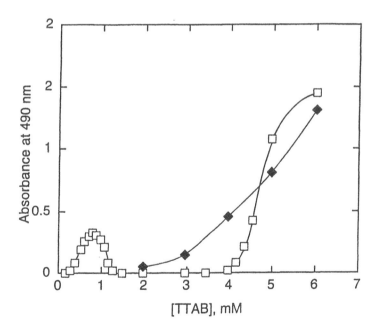

FIG. 1 Solubilization of orange OT by TTAB-Hy complexes (open symbols) and TTAB micelles (closed symbols). From ref. 55 with permission.

supernatant solution exceeds the CMC, a second solubilization zone is observed for the polymer-surfactant lower phase, where surfactant micelles solubilize the dye.

Almgren *et al.* reported extensive studies of polyelectrolyte-surfactant mixed systems by time-resolved fluorescence quenching techniques and estimated the average aggregation numbers of the surfactant on the polyions [38,56-59]. The measured values of the surfactant aggregation number for surfactant aggregates on the polyion were high, and comparable to the aggregation number of the surfactant micelles in the absence of polyions, as listed in Table 1. However, these authors observed very low solubilizing power of PSS/DTAB and PSS/CTAB complexes for dimethylbenzophenone (DMBP), compared to free surfactant micelles [38].

Hayakawa *et al.* reported a study of the solubilization of an oil-soluble dye, o-(2-amino-1-naphthylazo)toluene (OY), and 1-pyrenecarbaldehyde (PyA) by complexes of PVS, DxS and PSS with DTA and TTA. They proposed a concerted cooperative binding model for this process [28,30]. These studies demonstrated that the solubilization capacity is strongly related to the cooperativity of the binding of surfactant to polymer. This cooperativity, in turn, is related to the aggregate size in the mixed system as shown in Eq. (3). These results are summarized in Figure 2 [30].

The low cooperativity of surfactant binding in the PSS/surfactant system would predict a low solubilization ability and hence high Gibbs energy for the solubilization of both OY and PyA. A low solubility of DMBP in PSS/DTAB and PSS/CTAB mixed systems was also reported by Almgren *et al.* [38]. These

TABLE 1 Aggregation Numbers of Surfactants Bound to Polymers, from Almgren *et al.* (Collected from refs. 38, 56-59).

System	Quencher	Temperature, °C	Aggregation Number
PVS - DTAB	DPC	50	120
DxS - DTAB	DPC	50	156
	DPC	25	107
	DMBP	26	113
	DMBP	40	135
PAA - DTAB	DPC	25	59
PSS - DTAB	DMBP	25	38
PVS - TTAB	CPC	50	184
DTAB micelle			
0.044 m	DPC	25	62
26 wt %	DPC	25	82
TTAB micelle			
0.020 m	CPC	25	75

authors estimated the surfactant aggregation number in PSS/DTAB mixed systems as 38 at the half-bound point of surfactant. The DxS/surfactant and PVS/surfactant systems exhibit high cooperativity in the binding isotherms [58, 60-62] and show more effective solubilization for both OY and PyA, compared to the PSS/surfactant system. The aggregation numbers determined by Almgren *et al.* were 110 for DxS/DTAB and 120 for PVS/DTAB, both values taken at the half-bound point of surfactant, *i.e.* at half neutralization of the polyion charge [58]. These results suggest that the solubilization capacity is related to the aggregate size of the surfactant in the polymer-surfactant complex.

The DxS/surfactant systems show more effective solubilization for OY than the corresponding micellar systems in the absence of polymer, while for PyA the solubilization efficiency is comparable to that of micellar solubilization. This difference in solubilization behavior between OY and PyA in polymer/surfactant and micellar systems may be related to the different locations of the OY and PyA dyes in the micelles compared to polymer-surfactant complexes.

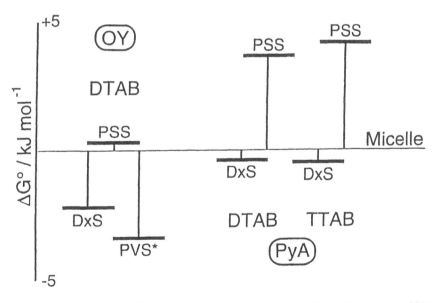

FIG. 2 Comparison of Gibbs energies of solubilization for OY and PyA in DxS, PVS, and PSS complexes with DTAB (OY, PyA) and TTAB (PyA) , relative to solubilization in surfactant micelles. Negative $\Delta G°$ values indicate that solubilization in the polymer-surfactant complex is more favorable than solubilization in the micelle. From ref. 30, with permission.

B. Model for the Solubilization Mechanism

In our studies on the solubilization of OY in solutions of anionic polyelectrolytes and cationic surfactants we found that in the presence of OY the CAC observed in the binding isotherm of DTAB shifts to lower free surfactant concentrations [28].   The binding isotherms for the DTAB/PVS systemin the absence and presence of OY are compared in Figure 3. This figure also includes the solubilization curves of OY by PVS/DTAB complexes. In the presence of water-insoluble OY, the surfactant binding isotherms shift to lower equilibrium concentrations of the surfactant.   This observation indicates that the surfactant/polymer complex induces the solubilization of OY and at the same time the solubilized OY enhances the surfactant binding through the hydrophobic interaction.   Based on this observation a concerted cooperative binding model was proposed which introduces a new parameter in the model for surfactant binding presented in Section II of this chapter, i.e. the statistical weight for a site occupied by dye, d. Since no dye is bound to the polyion in the absence of bound surfactant, d represents the following binding equilibria:

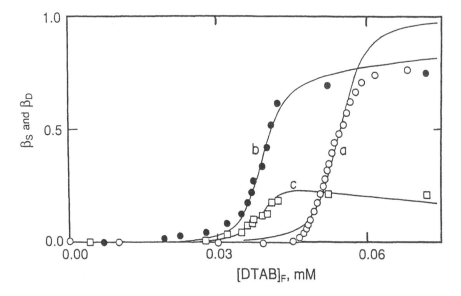

FIG. 3 Binding of DTAB by PVS (a,O; b,●) and of solubilization of OY in DTAB/PVS (c,□). Solid curves calculated from Eqs. (1), (6), and (7). From ref. 28, with permission.

$$-SO- \; + \; D \; \longrightarrow \; -SD- \qquad (4)$$

$$-DO- \; + \; D \; \longrightarrow \; -DD- \qquad (5)$$

Applying an eigenvalue method for the statistical weight matrix of a binding site on the polyion to calculate the partition function of polymer, the following expression of the degree of binding of surfactant β_S and dye β_D are derived [28]:

$$\beta_S = \frac{s}{2(s+d-s/u)} \left\{ 1 - 1/u + \frac{s+d-1+(1+d-s)/u}{\sqrt{(s+d-1)^2+4s/u}} \right\} \qquad (6)$$

$$\beta_D = \frac{s}{2(s+d-s/u)} \left\{ 1 - 1/u + \frac{s+d-1+2s/d}{\sqrt{(s+d-1)^2+4s/u}} \right\} \qquad (7)$$

The solid lines in Figure 3 show the best fit curves obtained by the use of values for Ku, u and d listed in Table 2.

The present approach assumes site binding of the dye molecule and the surfactant ion, in a linear Ising approximation with only nearest-neighbor interactions. Although such a model may seem inconsistent with the concept of "solubilization" of the dye by polyion-surfactant complexes, the success in

TABLE 2   Cooperative Binding Parameters for Solubilization of OY by PVS-CnTAB Complexes

[NaCl] / mM	Ku (± 1%)	u (± 10%)	d (±10%)
	PVS - DeTAB		
0	3800	1200	0.52
5	2320	250	0.37
10	1680	200	0.20
20	1180	200	0.13
	PVS - DTAB		
5	26000	150	0.35
10	18300	400	0.29
20	14500	700	0.25
	DxS - DTAB		
5	52	30	0
20	15	50	0.2
40	9	350	0.25
	PSS - DTAB		
20	54	9	0
40	120	10	0

describing the binding data, as shown in Figure 3, shows that the model may be applicable to the present system. Equally, the approach taken here should be useful in describing the competitive binding of hydrophobic counterions and surfactant ions to a polyion.

Detailed structure of the aggregates was revealed from NMR studies by Kwak et al. for the PMA-BVE/DTAB mixed system solubilizing benzene [63]. They observed an effect of solubilized benzene on the N-methyl, α, β and γ-methylene proton chemical shifts of the DTA surfactant ion, but no effect on methylene protons far from the headgroup and the ω-methyl protons. These authors concluded that benzene was solubilized in the vicinity of the cationic surfactant head group, which in turn is located close to the polymer ionic group.

IV. SPECTROSCOPY OF SOLUBILIZED DYES

Many dye molecules form aggregates in aqueous solution, with their molecular shape leading to a "stacking" arrangement. Such dyes tend to show cooperative binding to polyions of opposite charge. Schwarz successfully derived a theoretical equation for this cooperative binding of stacking dyes on polyions, which is substantially identical to treatment later developed by Satake and Yang for surfactant binding to polymers, as mentioned in Section II [33-35,64]. The two cooperative binding models (i.e. for surfactants in the polymer domain, and for dyes in the polymer domain) are schematically compared in Figure 4. The interaction of a cationic dye with an anionic polymer begins at dye concentration lower than the surfactant cac, indicative of the stronger intrinsic binding of the dye cation. The cooperativity of dye binding is, however, less than that found for surfactant binding.

The stacking tendency of the dye ions in the polymer domain can be demonstrated for instance from the fluorescence spectrum of pyrene covalently bound to polymer, which exhibits excimer emission in the absence of surfactant. Winnik et al. synthesized pyrene-labeled water-soluble polymers, hydroxypropylcellulose (HPC) and poly(N-isopropylacrylamide) (PNIPAM), and observed relatively intense excimer emission in their fluorescence spectra in the absence of surfactant [20,47,48,65-68]. A detailed review of these studies is presented in Chapter 7 of this volume, for the present purpose it is useful to compare the results obtained by this group to the case of dye binding discussed here. Winnik et al. showed that addition of surfactants to the pyrene-labeled HPC solution induced reduction of excimer emission and enhancement of monomer emission around the CAC as demonstrated in Figures 5 and 6 [48]. Similar results were observed for PNIPAM polymers. This phenomenon was

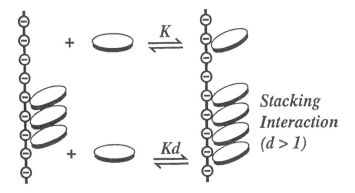

A: Polyion/Surfactant Cooperative Interaction

B: Polyion/Dye Stacking Interaction

FIG. 4 Schematic representation depicting the cooperative binding of cationic surfactant and cationic dye by polyanion.

interpreted as pyrene aggregation without surfactant and the solubilization into monomeric form in surfactant aggregates in the polymer domain [66].

It is indeed of interest to compare these results for polymers covalently labeled with pyrene to the case of polyions binding oppositely charged dye ions. The dyes rhodamine 6G (R6G), proflavine (PF) and acridine orange (AO), present a range of molecular structures with different stacking properties.

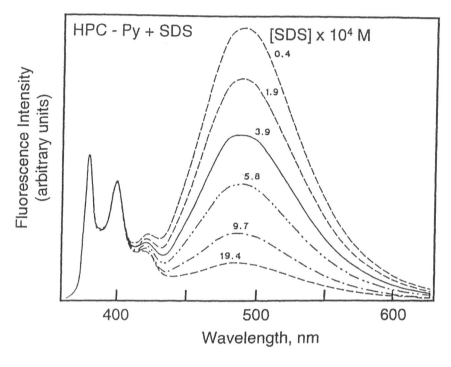

FIG. 5 Fluorescence spectra of HPC-Py at various concentrations of SDS. From ref. 48 with permission.

Absorption spectra of these dyes solubilized in polyion/surfactant complexes were measured as a function of surfactant concentration [24-26]. Figure 7 shows the spectra of R6G in the presence of PMA/DTAB, PSS/DTAB and Pectate/DTAB complexes as a function of DTAB concentration. The addition of PMA to an aqueous solution of R6G induces a reduction in the absorbance and the appearance of a new band at a shorter wavelength due to the aggregated form of the dye. The new band corresponding to the aggregate disappears upon addition of DTAB. The addition of DTAB also induces a recovery in intensity and a shift of the monomer band to a longer wavelength, indicating a more hydrophobic environment of R6G. Similar behavior was observed in the systems of DxS/DTAB, PMA-E/DTAB and PMA-S/DTAB.

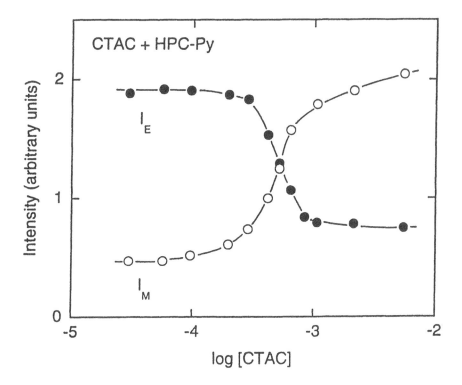

FIG. 6 Dependence of the intensity of excimer (I_E) and monomer (I_M) emission of HPC-Py on the concentration of added SDS. From ref. 48 with permission.

This behavior can be explained as follows. In surfactant-free polyelectrolyte solution the cationic dye binds cooperatively to the polyanion, leading to stacking of the dye in the polyion domain (*cf.* Figure 4). With the addition of surfactant, and the formation of surfactant aggregates above the surfactant cac, the stacked dye molecules dissolve in monomeric form in the hydrophobic polyion/surfactant complexes.

Quite different behavior is observed in pectate and PSS solutions. In PSS the absorption spectrum of R6G in the absence of DTAB does not show a blue-shift but a red-shift. DTAB addition induces only a minor change in the wavelength of maximum absorbance and a decrease in absorbance. This spectral change suggests a specific interaction of R6G with PSS and a minor effect of DTAB binding to PSS. On the other hand, in pectate a relatively small decrease in absorbance for monomeric R6G at 527 nm is observed, indicating a weak interaction between R6G and pectate. DTAB addition induces the recovery of the spectrum of monomeric R6G in an aqueous environment. This behavior can be attributed to a competition for binding sites on the pectate polyion between R6G and DTAB, without stacking of R6G bound to the pectate polyion, and DTAB exhibiting stronger binding to pectate [24].

We have also investigated the spectral change of PF bound to these polyions as a function of DTAB concentration as shown in Figure 8 [24]. Different from R6G, PF exhibits similar behavior in PSS and pectate solutions. Stacking of PF was also observed in DxS, PMA-S, PMA-E, PAA, and PMA solutions. The dissolution ability of the polyion/DTAB complexes for PF varies with polyion structure. The PMA, PAA and PMA-E - DTAB complexes show weak dissolution ability for PF, as indicated by a weak red-shift and an increase in absorbance at the band peak with increasing surfactant concentration. These differences may be ascribed to the different molecular structure and stacking ability of PF (compared to R6G), and to the differences in cooperativity of DTAB binding to the various polyions.

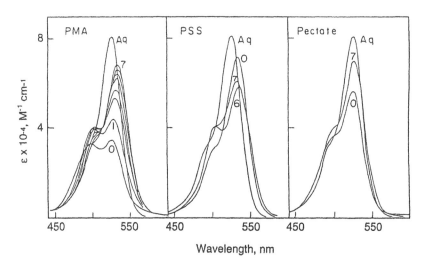

FIG. 7 Absorption spectra of R6G in solutions of polyelectrolytes PMA, PSS, and pectate, with added DTAB. Polyion concentration 5×10^{-4} m, [R6G] = 1.1×10^{-5} m, Aq: [polymer] = 0; [DTAB] = 0, 0,5, 1.0, 1.5, 2.0, 2.5, 3.0, 4.0×10^{-4} m (curves 0 to 7 respectively, not all curves shown for clarity). Reprinted from ref. 26, *Colloids Surf. 50*, 309 (1990), with permission from Elsevier Science-NL, Sara Burgerhartstrat 25, 1055 KV Amsterdam, the Netherlands.

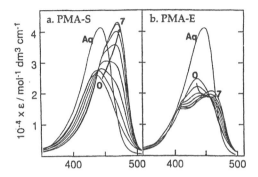

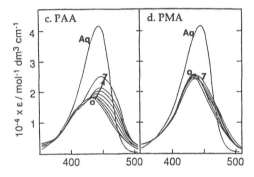

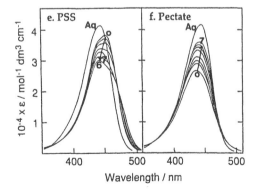

FIG. 8 Absorption spectra of PF in polyion-DTAB mixed solutions. Aq: aqueous only, no polymer. Curves 0-7: [polymer] = 0.5 mmol/kg, [DTAB] = 0 (curve 0), 0.05, 0.1, 0.15, 0.2, 0.25, 0.3, 0.4 (curve 7) mmol/kg. From ref. 24 with permission.

V. ENHANCED ENERGY TRANSFER BETWEEN SOLUBILIZED PROBES

Fluorescence of stacked dye aggregates in the polymer domain is completely quenched. As shown in Section IV, dye aggregates are solubilized in monomeric form by polymer/surfactant complexes. As a result, solubilization of the dye in polymer surfactant aggregates induces a strong enhancement of the emission intensity of the dye. When both donor and acceptor of a donor-acceptor dye pair are solubilized in the small hydrophobic domains of the polymer/surfactant complexes, the energy transfer between the dye pairs should be enhanced. Such enhanced, dye induced energy transfer was indeed observed for emission probes dissolved in lipid vesicles and in premicelles [69-74].

Our group has reported similar enhancements of energy transfer from excited PF to R6G in the presence of a DxS/DTAB complex [31]. Figure 9A shows the intense PF emission at 512 nm (spectra a, b) and weak R6G fluorescence at 558 nm (spectrum c) in aqueous solution, and the increasing intensity of both PF and R6G inDxS/DTAB mixed solutions, while Figure 9B shows the PF single-dye spectra in DxS/DTAB mixed solutions. In all systems containing DxS the DxS ionic concentration is 2 mmol/L, [PF] = 1.6×10^{-6} M, [R6G] = 1×10^{-6} M.

In DxS solution in the absence of surfactant (spectra 0 in 9A and 9B) PF emission is strongly quenched due to the non-fluorescent stacking pairs. Addition of DTAB first results in a further small reduction of PF emission (spectrum 0→1), followed by recovery in intensity of the PF emission (spectra 1-11 in 9B). In mixed dye solutions, the recovery of the PF intensity is lower (spectra 1-12 in 9A) and accompanied by strong enhancement of R6G emission.

We have observed similar energy transfer between pyrene and PF both solubilized in DxS/DTAB, PVS/DTAB and PSS/DTAB complexes [75]. Effective enhancement of PF fluorescence was observed upon irradiation of pyrene in DxS/DTAB and PVS/DTAB complex systems, but only a weak enhancement of PF emission was observed in the PSS/DTAB complex. Application of a kinetic model to the inter-cluster energy transfer process in the polyion domain allows for a crude estimate of aggregate sizes which is nevertheless sufficient to discern the effect of polyion hydrophobicity on aggregate size [75]. This method results in an estimate for the surfactant aggregation number on polyion of around 200 for DxS/DTAB and PVS/DTAB complexes but only around 20 for the PSS/DTAB complex. This finding would be supported by the lower cooperativity in the binding isotherms of DTAB with PSS compared to binding to DxS and PVS [60].

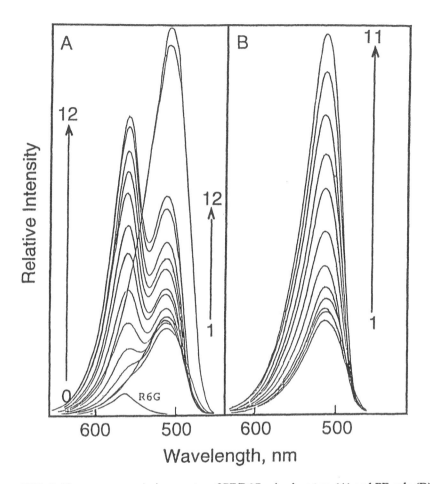

FIG. 9 Fluorescence emission spectra of PF/R6G mixed system (A) and PF only (B) in the presence of DxS/DTAB complexes. See text for explanation. From ref. 31 with permission.

VI. CONCLUSIONS

This chapter focused on solubilization of dyes in polyion-surfactant complexes, considering the solubilization of water insoluble dyes, spectroscopy of solubilizates, and energy transfer between solubilized donor-acceptor pairs. Solubilization by polymer/surfactant complexes primarily occurs in the hydrophobic domains formed in aqueous solutions. In this respect the process in polymer-surfactant systems is similar to micellar solubilization. However, in polymer-surfactant systems solubilization capacity and the effect of environment on the solubilizates including the effects on solubilizate spectroscopy varies according to the structures of surfactant, polymer and solubilizate. The studies discussed in this chapter lead to the conclusion that the cooperativity of the surfactant binding process strongly influences the dye-solubilization process, mainly because of differences in the size of the surfactant aggregates. The complex behavior shown in both absorbance and emission spectra of the dyes as function of surfactant concentration can be explained by considering the effect of stacking of the dye in the polymer domain in the absence of surfactant, followed in some cases by dissociation of these stacking pairs in the polymer-surfactant aggregate. This latter process depends on the nature of the dye and the size of the surfactant aggregate. The complex behavior shown in the spectroscopic properties of solubilizates indicates that the further research is required for a better understanding of the solubilization functions of polyion/surfactant complexes. Concentrated polymer/surfactant systems exhibit a variety of interesting rheological properties and phase behavior. Solubilization in concentrated systems may well require different explanations compared to the dilute systems studied here. At the same time, spectroscopic studies of dyes in such systems should be of considerable interest.

ABBREVIATIONS

Surfactants

DeTAB	Decyltrimethylammonium Bromide
DTAB	Dodecyltrimethylammonium Bromide
DTAC	Dodecyltrimethylammonium Chloride
TTAB	Tetradecyltrimethylammonium Bromide
CTAB	Hexadecyltrimethylammonium Bromide
CTAC	Hexadecyltrimethylammonium Chloride
DPC	Dodecylpyridinium Chloride
CPC	Hexadecylpyridinium Chloride
SDS	Sodium Dodecyl Sulfate

Polymers

PAA	Poly(acrylic acid)
PMA	Poly(methacrylic acid)

PSS	Poly(styrenesulfonic acid)
PVS	Poly(vinylsulfuric acid)
PMA-E	Poly(maleic acid-co-ethylene)
PMA-St	Poly(maleic acid-co-styrene)
PMA-BVE	Poly(maleic acid-co-butylvinylether)
PNIPAM	Poly(N-isopropylacrylamide)
Hy	Hyaluronan
DxS	Dextransulfate
HPC	Hydroxypropylcellulose

REFERENCES

1. T. Nakagawa, in *Nonionic Surfactants Vol. 1* (M.J. Schick, ed.), Marcel Dekker, New York, 1966, pp. 558-603. A. Mackay, in *Nonionic Surfactants: Physical Chemistry* (M.J. Schick, ed.), Marcel Dekker, New York, 1987, pp. 297-368.

2. *Solubilization in Surfactant Aggregates,* (S.D. Christian and J.F. Scamehorn, eds.), Marcel Dekker, New York, 1995.

3. H. Arai, M. Murata, and K. Shinoda, *J. Colloid Interface Sci. 37*: 223 (1971).

4. F. Tokiwa and K. Tsujii, *Bull. Chem. Soc. Jpn. 46*: 2684 (1973).

5. E.D. Goddard and R.B. Hannan, in *Micellization, Solubilization, and Microemulsions* (K. L. Mittal, ed.), Plenum, New York, 1977, pp. 835-845.

6. S. Saito and T. Taniguchi, *J. Colloid Interface Sci. 44*: 114 (1973).

7. S. Saito, *Coll. Polym. Sci. 257*: 266 (1979).

8. P.S. Leung, E.D. Goddard, C. Han, and C.J. Glinka, *Colloids Surf. 13*: 47 (1985).

9. E.D. Goddard, *Colloids Surf. 19*: 301 (1986).

10. E.D. Goddard, *Colloids Surf. 19*: 255 (1986).

11. I.D. Robb, in *Anionic Surfactants: Physical Chemistry of Surfactant Action* (E.H. Lucassen-Reynders, ed.), Marcel Dekker, New York, 1981, pp. 109.

12. S. Saito, in *Nonionic surfactants: Physical Chemistry,* (M. J. Schick, ed.), Marcel Dekker, New York and Basel, 1987, p. 881.

13. E.D. Goddard, in *Interactions of surfactants with polymers and proteins* (E.D. Goddard and K.P. Ananthapadmanabhan, eds.), CRC Press, Boca Raton, Fla.,1993, pp. 171-201.

14. E.D. Goddard, in *Interactions of surfactants with polymers and proteins* (E.D. Goddard and K.P. Ananthapadmanabhan, eds.), CRC Press, Boca Raton, Fla., 1993, pp. 123-169.

15. B.-H. Lee, S.D. Christian, E.E. Tucker, and J.F. Scamehorn, *Langmuir 7*: 1332 (1991).

16. M. Tunçay, S.D. Christian, E.E. Tucker, R.W. Taylor, and J.F. Scamehorn, *Langmuir 10*: 4688 (1994).

17. M. Tunçay, S.D. Christian, E.E. Tucker, R.W. Taylor, and J.F. Scamehorn, *Langmuir 10*: 4693 (1994).

18. H. Uchiyama, S.D. Christian, E.E. Tucker, and J.F. Scamehorn, *J. Colloid Interface Sci. 163*: 493 (1994).

19. K. Hayakawa and J.C.T. Kwak, in *Cationic Surfactants: Physical chemistry* (D.N. Rubingh and P.M. Holland, eds.), Marcel Dekker, New York, 1991, pp. 189-248.

20. F.M. Winnik, *Langmuir 6*: 522 (1990).

21. E.A. Bekturov, S.E. Kudaibergenov, V.A. Frolova, R.E. Khamzamulina, R.C. Schulz, and J. Zöller, *Macromol. Chem., Rapid Commun. 12*: 37 (1991).

22. J. Kido, Y. Imamura, N. Kuramoto, and K. Nagai, *J. Colloid Interface Sci. 150*: 338 (1992).

23. S. Saito, *J. Colloid Interface Sci. 158*: 77 (1993).

24. K. Hayakawa, I. Satake, and J.C.T. Kwak, *Colloid Polym. Sci. 272*: 876 (1994).

25. K. Hayakawa, J. Ohta, T. Maeda, I. Satake, and J.C.T. Kwak, *Langmuir 3*: 377 (1987).

26. K. Hayakawa, I. Satake, J.C.T. Kwak, and Z. Gao, *Colloids Surf. 50*: 309 (1990).

27. K.P. Ananthapadmanabhan, P.S. Leung, and E.D. Goddard, *Colloids Surf. 13*: 63 (1985).

28. K. Hayakawa, T. Fukutome, and I. Satake, *Langmuir 6*: 1495 (1990).

29. E.A. Sudbeck, P.L. Dubin, M.E. Curran, and J. Skelton, *J. Colloid Interface Sci. 142*: 512 (1991).

30. K. Hayakawa, S. Shinohara, S. Sasawaki, I. Satake, and J.C.T. Kwak, *Bull. Chem. Soc. Jpn. 68*: 2179 (1995).

31. K. Hayakawa, T. Ohyama, T. Maeda, I. Satake, and J.C.T.Kwak, *Langmuir 4*: 481 (1988).

32. I. Satake and J.T. Yang, *Biopolymers 15*: 2263 (1976).

33. G. Schwarz, *Eur. J. Biochem. 12*: 442 (1970).

34. G. Schwarz, S. Klose, and W. Balthasar, *Eur. J. Biochem. 12*: 454 (1970).

35. G. Schwarz and W. Balthasar, *Eur. J. Biochem. 12: 461* (1970).

36. J.C. Scaiano, E.B. Abuin, and L.C. Stewart, *J. Am. Chem. Soc. 104*: 5673 (1982).

37. D.-Y. Chu and J.K. Thomas, *Macromolecules 17*: 2142 (1984).

38. M. Almgren, P. Hansson, E. Mukhtar, and J. van Stam, *Langmuir 8*: 2405 (1992).

39. W. Binana-Limbele and R. Zana, *Macromolecules 20*: 1331 (1987).

40. L.-S. Choi and O.-K. Kim, *Langmuir 10*: 57 (1994).

41. C. Maltesh and P. Somasundaran, *J. Colloid Interface Sci. 157*: 14 (1993).

42. K. Shirahama, M. Tohdo, and M. Murahashi, *J. Colloid Interface Sci. 86*: 282 (1982).

43. N.J. Turro, B.H. Baretz, and P.-L. Kuo, *Macromolecules 17*: 1321 (1984).

44. N.J. Turro and P.-L. Kuo, *Langmuir 3*: 773 (1987).

45. N.J. Turro, X.-G. Lei, K.P. Ananthapadmanabhan, and M. Aronson, *Langmuir 11*: 2525 (1995).

46. N.J. Turro and T. Okubo, *J. Am. Chem. Soc. 104*: 2985 (1982).

47. F.M. Winnik, H. Ringsdorf, and J. Venzmer, *Langmuir 7*: 905 (1991).

48. F.M. Winnik, M.A. Winnik, and S. Tazuke, *J. Phys. Chem. 91*: 594 (1987).
49. D.-Y. Chu and J.K. Thomas, *J. Am. Chem. Soc. 108*: 6270 (1986).
50. I. Satake and K. Hayakawa, *Chem. Lett.* : 1051 (1990).
51. K. Hayakawa, M. Fujita, S. Yokoi, and I. Satake, *J. Bioact. Compat. Polym. 6*: 36 (1991).
52. P. Bahadur, N.V. Sastry, and Y.K. Rao, *Colloids Surf. 29*: 343 (1988).
53. S. Saito, *J. Am. Oil Chem. Soc. 66*: 987 (1989).
54. Z. Gao, J.C.T. Kwak, R. Labonté, D.G. Marangoni, and R.E. Wasylishen, *Colloids Surf. 45:* 269 (1990).
55. K. Thalberg and B. Lindman, *J. Phys. Chem. 93*: 1478 (1989).
56. P. Hansson and M. Almgren, *Langmuir 10*: 2115 (1994).
57. P. Hansson and M. Almgren, *J. Phys. Chem. 99*: 16684 (1995).
58. P. Hansson and M. Almgren, *J. Phys. Chem. 99*: 16694 (1995).
59. P. Hansson and M. Almgren, *J. Phys. Chem. 100*: 9038 (1996).
60. K. Hayakawa and J.C.T. Kwak, *J. Phys. Chem. 86*: 3866 (1982).
61. K. Hayakawa and J.C.T. Kwak, *J. Phys. Chem. 87*: 506 (1983).
62. K. Shirahama and M. Tashiro, *Bull. Chem. Soc. Jpn. 57*: 377 (1984).
63. Z. Gao, J.C.T. Kwak, and R.E. Wasylishen, *Macromolecules 22*: 2544 (1989).
64. V. Vitagliano, in *Aggregation Processes in Solution* (E. Wyn-Jones and J. Gormally, eds.), Elsevier, Amsterdam, 1983, p. 271.
65. Y.-Z. Hu, C.-L. Zhao, M.A. Winnik, and P.R. Sundararajan, *Langmuir 6*: 880 (1990).
66. F. M. Winnik, H. Ringsdorf, and J. Venzmer, *Langmuir 7*: 912 (1991).
67. F.M. Winnik, in *Interactions of Surfactants with Polymers and Proteins* (E.D. Goddard and K.P. Ananthapadmanabhan, eds.), CRC Press, Boca Raton, 1993, pp. 367-394.
68. F. M. Winnik, S.T.A. Regismond, and E.D. Goddard, *Colloids Surf., A 106*: 243 (1996).
69. K. Kano, H. Kawazumi, and T. Ogawa, *J. Phys. Chem. 85*: 2998 (1981).
70. K. Kano, Y. Ueno, and S. Hashimoto, *J. Phys. Chem. 89*: 3161 (1985).
71. K. Kano, *Hyomen 26*: 243 (1988).
72. Y. Kusumoto, *Bunkoukenkyuu 37*: 96 (1988).
73. H. Sato, M. Kawasaki, and K. Kasatani, *J. Phys. Chem. 87*: 3759 (1983).
74. T. Sato, M. Kurahashi, and Y. Yonezawa, *Langmuir 9*: 3395 (1993).
75. K. Hayakawa, T. Nakano, I. Satake, and J.C.T. Kwak, *Langmuir 12*: 269 (1996).

Index